OXFORD MATHEMATICAL MONOGRAPHS

Series Editors

J. M. BALL   E. M. FRIEDLANDER   I. G. MACDONALD
L. NIRENBERG   R. PENROSE   J. T. STUART

OXFORD MATHEMATICAL MONOGRAPHS

A. Belleni-Moranti: *Applied semigroups and evolution equations*

M. Rosenblum and J. Rovnyak: *Hardy classes and operator theory*

J. W. P. Hirschfeld: *Finite projective spaces of three dimensions*

A. Pressley and G. Segal: *Loop groups*

J. C. Lennox and S. E. Stonehewer: *Subnormal subgroups of groups*

D. E. Edmonds and W. D. Evans: *Spectral theory and differential operators*

Wang Jianhua: *The theory of games*

S. Omatu and J. H. Seinfeld: *Distributed parameter systems: theory and applications*

J. Hilgert, K. H. Hofmann, and J. D. Lawson: *Lie groups, convex cones, and semigroups*

S. Dineen: *The Schwarz lemma*

B. Dwork: *Generalized hypergeometric functions*

S. K. Donaldson and P. B. Kronheimer: *The geometry of four-manifolds*

T. Petrie and J. Randall: *Connections, definite forms, and four-manifolds*

R. Henstock: *The general theory of integration*

D. W. Robinson: *Elliptic operators and Lie groups*

A. G. Werschulz: *The computational complexity of differential and integral equations*

P. N. Hoffman and J. F. Humphreys: *Projective representations of the symmetric groups*

I. Györi and G. Ladas: *The oscillation theory of delay differential equations*

J. Heinonen, T. Kilpelainen, and O. Martio: *Non-linear potential theory*

B. Amberg, S. Franciosi, and F. de Giovanni: *Products of groups*

M. E. Gurtin: *Thermomechanics of evolving phase boundaries in the plane*

I. Ionescu and M. Sofonea: *Functional and numerical methods in viscoplasticity*

N. Woodhouse: *Geometric quantization* 2nd edition

U. Grenander: *General pattern theory*

J. Faraut and A. Koranyi: *Analysis on symmetric cones*

I. G. Macdonald: *Symmetric functions and Hall polynomials* 2nd edition

B. L. R. Shawyer and B. B. Watson: *Borel's methods of summability*

D. McDuff and D. Salamon: *Introduction to symplectic topology*

M. Holschneider: *Wavelets: an analysis tool*

Jacques Thévenaz: *G-algebras and modular representation theory*

P. D. D'Eath: *Black holes: gravitational interactions*

Black Holes
Gravitational Interactions

P. D. D'EATH

Department of Applied Mathematics and Theoretical Physics
University of Cambridge

CLARENDON PRESS · OXFORD
1996

Oxford University Press, Walton Street, Oxford OX2 6DP

Oxford New York
Athens Auckland Bangkok Bombay
Calcutta Cape Town Dar es Salaam Delhi
Florence Hong Kong Istanbul Karachi
Kuala Lumpur Madras Madrid Melbourne
Mexico City Nairobi Paris Singapore
Taipei Tokyo Toronto
and associated companies in
Berlin Ibadan

Oxford is a trade mark of Oxford University Press

Published in the United States by
Oxford University Press Inc., New York

A catalogue record for this book is available from the British Library

Library of Congress Cataloging in Publication Data
(Data applied for)

ISBN 0 19 851479 4

Typeset by the author using TeX

Printed in Great Britain by
Bookcraft (Bath) Ltd
Midsomer Norton, Avon

For Stephen and Roger

and for Kathy

PREFACE

The study of black holes has blossomed from the early 1960s to the present. A great deal has been learnt during this period, from the global structure of equilibrium black holes to linear perturbation theory and black-hole thermodynamics to new black-hole models incorporating matter through scalar fields. Introductory material on black holes is described in Chapter 2. This is intended to act as a basis for understanding the treatment of black holes in the rest of the book. However, one should note that in most work on the subject the black hole is normally regarded as an isolated object. One might wish to describe instead a black hole interacting with an external strong-field system, such as a background universe (Chapter 3) or another black hole (Chapter 4). These cases can be treated by the method of matched asymptotic expansions. In the case of the background universe, one has one (internal) perturbation scheme, which describes the black hole over a length-scale of order the mass of the black hole, perturbed because of the curvature of the background. Conversely, the black hole produces only a small perturbation of the background geometry, giving an external perturbation scheme. One then matches these two schemes in a region of common validity. This leads to the equations of motion and spin propagation for the black hole. A similar approach is used in Chapter 4, leading to post-Newtonian and higher-order corrections to the equations of motion and spin propagation for the two black holes.

One can treat black-hole interactions more generally, studying scattering at possibly relativistic speeds. Most dramatic is the case of a collision or close encounter at nearly the speed of light (Chapter 5), which can again be treated by a suitable perturbation method. Remarkably, one can deduce information about strong-field gravitational radiation. For a head-on collision, there is some detailed structure in the gravitational radiation emitted near the symmetry axis, and one finds a nearly constant radiation pattern just outside the axis. This would give 25% efficiency for the conversion of initial mass–energy into radiation, if the radiation were isotropic with this profile. This is to be compared with an upper bound of 29% efficiency, following a calculation by Penrose based on the Cosmic Censorship Hypothesis. Results for other impact parameters are also given. In Chapters 6, 7 and 8, an analogous calculation is done for a head-on collision at exactly the speed of light. The radiation is given by a series in $\sin^2 \theta$ about the axis. The first term is that already found in Chapter 5; here, using second-order perturbation theory, the second term is found numerically. There are indications that the radiation described here is not the only radiation in the space–time, but that it is only a 'first burst', to be followed much later by a 'second burst'. Nevertheless, one can still take the standard expression for energy emitted in gravitational waves with the first two terms in the angular series above, to find an improved estimate of 16.4% for the efficiency. These processes should be among the most powerful in the universe.

In Chapter 9, as well as giving a summary of the book, the work of 't Hooft on ultra-high energy quantum scattering is briefly described. This uses the ideas of Chapters 5–8, and demonstrates the profound relation between the quantum and the classical: at Planckian energies, quantum processes are determined with the help of the null gravitational shock wave of a particle moving at the speed of light, as in Chapter 5. At even higher energies, quantum processes are described by the classical space–times studied in Chapters 5–8.

Cambridge   P.D.D.
December 1995

ACKNOWLEDGEMENTS

My deepest gratitude is to Roger Penrose and to Stephen Hawking, whose ideas about speed-of-light collisions and cosmic censorship were inspirational, and are now the basis for the work of Chapters 5–8. Roger's seminar in Cambridge in 1974 was particularly suggestive of what turned out to be invaluable directions for further reflection and research. To both of these colleagues, I also owe a warm friendship of nearly two decades, and much encouragement and support of a professional nature as well, throughout that long period.

The work in Chapters 6–8 was carried out jointly with my graduate student, Philip Payne, whose contribution cannot be fully enough acknowledged here. His fine doctoral dissertation at Cambridge constituted an important further extension of this research, begun in the 1970's on colliding black holes by Geoff Curtis (a student of Roger Penrose) at the Institute at Oxford. Another Cambridge graduate student, Geoff Chapman, continued studying the collision problem, following Penrose's and Curtis' lead.

The early main chapters of the book were written under the supervision of Stephen Hawking, who listened patiently and advised incomparably. Dennis Sciama made helpful suggestions for improvement of the work for publication, while Jürgen Ehlers was also interested in its development, both in the earliest stages and much later, when I visited his institute at Garching. (Bernd Schmidt also offered advice about how to clarify and improve the work.) Thibault Damour has done more than anyone to develop and to extend further our understanding of matched asymptotic expansions, by refining the treatment of the interactions among compact bodies.

Kip Thorne provided simply wonderful hospitality for a year at Cal Tech in the mid-1970's, and a continuing professional interest, and warm friendship. Jim Hartle has also been unfailingly generous with his welcome at Santa Barbara, whether for sabbatical terms or shorter visits, and unstinting with his time and energy. Gerard 't Hooft, more recently, took our knowledge of black- hole collisions into the quantum realm, and has opened up a new and fascinating field of work.

Gabriele Veneziano, at CERN, used string theory to extend 't Hooft's discoveries, along with Roberto Pettorino. To both, I owe once again hospitality and much generosity with time and ideas. Grisha Vilkovisky, whether in Moscow, Cambridge, Naples, or wherever he is, is always ready to listen, argue, and clarify. Finally, Stanley Deser's enthusiasm, both at Amsterdam in 1992 and at Cambridge in 1993, did more than he can be aware of to bring this book into being.

To Cambridge University, I am indebted for the freedom which terms of sabbatical leave gave me to initiate, develop, and complete this work. The Department of Applied Mathematics and Theoretical Physics was also helpful in the early stages. Stuart Rankin, our expert on computing, amongst other things, cannot be praised adequately for his kindness, good will, and determination to get this manuscript into publishable form. Larry Smarr, Norbert Straumann, Igor Volovich, Valery Frolov, Victor Berezin, Alexei Starobinsky, Brandon Carter, Bernard Whiting, and others have contributed directly or indirectly to the continuing effort to understand the subject area of this book.

CONTENTS

1

INTRODUCTION

Black holes are regions of space–time where the gravitational field is so strong that not even light can escape. As will be seen in Chapter 2, it appears to be inevitable that black holes will form as a result of gravitational matter collapse. This is expected to occur for a star of sufficiently large mass, of the order of a few solar masses, once it has burnt its nuclear fuel. Black holes can also be formed by gravitational collapse of the inner regions of a galaxy, or by collisions. This leads one to look first for equilibrium (time-independent) black-hole states. The stationary (time-independent) black holes, say in a theory containing only gravity and Maxwell electromagnetism, are Schwarzschild (with mass), Reissner–Nordström (with mass and charge), Kerr (with mass and angular momentum), and Kerr–Newman (with mass, angular momentum, and charge). It is a remarkable result that, in such a non-linear theory as general relativity, the stationary Einstein–Maxwell black holes should be exactly known, and be given in terms of only three parameters (Hawking and Ellis 1973; Robinson 1975; Wald 1984). The structure of these space–times is investigated with the help of Penrose conformal diagrams (Chapter 2). It is now known that the stationary (time-independent) black holes are stable against small perturbations (Whiting 1989). Hence it seems reasonable to expect that a matter collapse will lead at late times to a stationary black-hole space–time together with radiation which propagates out toward infinity. Much can then be learnt from studying the standard stationary black-hole models, together with linearized perturbations propagating in them (Chapter 2).

An important law of black-hole dynamics can be found by considering variations among nearby members of the rotating Kerr family. This will be used later. It is found that one can extract energy from a rotating black hole, with (say) the help of suitably arranged processes involving test particles. The extracted energy is the rotational energy of the black hole. More generally, in any process with particles falling into the black hole, one finds that the surface area of the black hole should increase with time. This is the second of four laws of black holes, which are described in Section 2.5. These laws imply an analogy with thermodynamics, which was fulfilled when Hawking (1975) showed that black holes radiate thermally at a certain temperature, so finding the factor relating the surface area to the entropy.

Turning to space–time geometries which are close but not equal to the Kerr metric, it is again remarkable that the geodesic equations for test particles in motion around a Kerr black hole are separable. Further, the perturbation equations for fields of different spin in the Kerr geometry are also separable (Teukolsky 1973). As mentioned above, one expects that, some time after the formation of a black hole, the hole will be close to one of the stationary models above, together with some radiation which either propagates out to infinity or into the black hole. This gives a good approximate description of a dynamical black hole in isolation. But there are also more powerful perturbation methods which allow the description of a much wider variety of phenomena involving black holes. One would like, for example, to understand how a black hole interacts with a surrounding 'background' universe, of which the black hole is only a small part (Chapter 3). In any perturbation problem in general relativity, one has to work with a one- (or many-) parameter family of space–times. For the case of a black hole in a background, one expects that, as one comes sufficiently close to the black hole, the space–time curvature due to the black hole is much greater than the space–time curvature due to the background universe. This says that the black-hole length-scale of order M, the mass in geometrical units, is much less than the length-scale over which the background varies. Hence we should study a one-parameter family of space–times, with a black hole being superimposed non-linearly on a background universe, where the black hole 'mass' M is taken to be the small parameter, with $M \to 0$. As one moves sufficiently far away from the black hole, one finds that its far field only produces a small perturbation of the background universe. Conversely, the background universe induces a small perturbation of the black hole, principally of a quadrupole nature. One has here the interaction between two strong-field systems, and it can be dealt with by the method of matched asymptotic expansions (Nayfeh 1973). The full metric can be viewed either on the background length-scale (giving the external perturbation series) or on a length-scale with radius r comparable with the size, roughly M, over which the internal structure varies (internal perturbation series). One then matches these together in an intermediate region such as $r \sim M^{\frac{1}{2}}$. This yields the internal and external series, order by order. As an example, one can show that, as $M \to 0$, a rotating Kerr black hole moves approximately on a geodesic, and its spin is approximately parallel-transported along the geodesic.

Another area in which one would have an interaction of two strong-field objects in general relativity concerns the motion of two black holes of comparable mass. (In the case that one black hole has much greater mass than the other, the small black hole can be considered as moving in the background metric of the large one, so that the work of Chapter 3 applies.) The simplest case, when the masses are comparable, is that in which the relative motion is roughly Newtonian, so that the black holes are quite well separated. Again, one employs matched asymptotic expansions (Chapter 4). There are now three main perturbation regions: the internal perturbation scheme for each black hole and the external scheme around both of them. If one wishes to calculate the gravitational radiation produced by this system, then one also has to take account of the wave-zone region near infinity (Burke 1971) by matching with the external scheme. The expressions for gravitational radiation at low orders are well known (Misner *et al.* 1973; Epstein and Wagoner 1975; Wagoner and Will 1976). Taking then the geometry well inside the wave region, one must match the internal geometries with the external geometry. The matching to the radiation zone will induce radiative potentials at high order in the external scheme, which lead to radiation reaction forces on the black holes. As in Chapter 3, one matches in two intermediate regions. One can derive an approximate form of the metric, both in the external and internal regions. For example, this will give the deformation of one black hole due to the other. It will also give the relative motion of the two black holes—the information for this is contained in the metric. One finds agreement with the Newtonian equations of motion and with the post-Newtonian correction to these equations of motion (Einstein *et al.* 1938). Further corrections which can be derived include spin–orbit and spin–spin terms in the equations of motion. Correspondingly, one can also derive equations of spin propagation. This demonstrates the power of matched asymptotic expansions: here we find the space–time geometry in an asymptotic series and then can deduce the equations of motion and spin propagation.

Note that the subject of the dynamics of two black holes is of astrophysical relevance. Suppose that two (or more) black holes form in a galactic nucleus. By tidal friction, the black holes will gradually sink to the centre of the galactic nucleus and find each other, forming a binary. The spin axis of a black hole in such a system can be observed from the jet which emerges along the axis, as a result of the accretion of matter by the black hole. In some cases this jet is observed to have an S-shape, and this could be explained by the relative motion and precession of one black hole in the field of the other (Begelman *et al.* 1980).

Now that the benefits of the matching method in the nearly Newtonian case have been seen, one would like to try to use similar methods in higher-speed encounters of two black holes, which will lead to scattering or collisions (Chapter 5). First take the case of a fixed impact parameter with fixed incoming velocities, with masses M_1, M_2 tending to zero. Thus one studies a fairly distant encounter, at possibly relativistic speeds. Kovács and Thorne (1978) have studied this problem analytically; they did not explicitly use matching, but matching provides boundary conditions on the gravitational field near the black holes. Kovács and Thorne calculated the leading correction to the linearized gravitational field, which consists of the sum of two far-field Schwarzschild geometries, suitably boosted. In particular, they found the part of the gravitational field at order $M_1 M_2$; this gives the main contribution to gravitational radiation. One has to be careful at speeds approaching the speed of light (D'Eath 1978); the condition for the calculation of Kovács and Thorne to be valid turns out to be

$$(M_1 + M_2) \ll b(1 - v^2), \tag{1.1}$$

where b is the impact parameter and v is a typical incoming speed. Kovács and Thorne (1978) found beamed bremsstrahlung as the incoming γ-factors $\to \infty$, with a particular pattern. Here $\gamma(v) = (1 - v^2)^{-1/2}$.

In this example one only uses matched asymptotic expansions in matching out the leading Schwarzschild far fields to find the source of the linearized perturbations h_{ab}. But there is one remaining limit accessible to matched asymptotic expansions, when one considers gravitational interactions between black holes: black-hole encounters or collisions in the limit that the incoming γ-factors tend to infinity (Chapter 5). We shall see that gravitational radiation near the forward and backward directions can be computed analytically with the help of matched asymptotic expansions, taking the small parameter to be γ^{-1}. Remarkably, this radiation gives a power output per unit solid angle of the order $c^5/G \simeq 3.6 \times 10^{59}$ erg/sec, characteristic of strong-field non-linear gravitational radiation. One expects these high-speed collisions or encounters to be among the most violent local processes in the universe. Further, a new avenue of research has more recently opened up, following the work of 't Hooft (1987). At high energies in quantum scattering processes, the interaction is approximated by the classical gravitational collision at the speed of light. Thus there are deep connections between the strictly classical methods described in Chapters 2–8 and the extreme high-energy quantum-theory limit to be described in Section 9.2.

Chapter 5 is concerned with the limit of large γ. For convenience, we scale the masses inversely to γ, with $M\gamma = \mu = $ const., where M is a typical mass. One can study various different régimes of impact parameter. The incoming metrics are given by an asymptotic series, and at high speed resemble 'pancakes' concentrated near a null shock surface. One can study the evolution of the shocks: before the collision, they are planar, being surrounded by regions of flat space–time. After the collision, as one finds by matching, the shocks are curved, and one can study the evolution of the detailed shock structure with thickness of order γ^{-1}. There is also a caustic region where parts of the curved shocks cross over each other. This gives a region of very strong curvature of $O(\gamma^{\frac{3}{2}})$. After passing through the caustic, where the detailed shock structure is modified, the curved shock moves on towards infinity giving gravitational radiation.

The main technique for computing gravitational radiation is to study wave generation in the far-field curved shocks. These are given a head start, at the collision, over the near-field shocks by an amount $8\mu \log P$, where P is the transverse distance from the centre of the other shock. For large P, the shock has been deflected by a small angle. One can compute the radiative part of the field at small angles $\theta = \psi\gamma^{-1}$ from the z-axis, taking the z-axis to give the direction of the initial black-hole motion. One finds that there is structure in the radiation on the corresponding angular scale of $O(\gamma^{-1})$. In the case of a head-on collision, one finds that the radiation is not beamed into a narrow cone of angular width a few multiples of γ^{-1}. Rather, as $\psi \to \infty$, the radiation pattern tends to a limiting shape, which is the radiation pattern near the axis of a collision at exactly the speed of light. The power in the radiation would give 25% efficiency of conversion of initial mass–energy into gravitational wave energy, if the radiation were isotropic. For a collision with non-zero impact parameter of order μ, one has the same radiation pattern near the axis, but a different pattern at angles further away. By increasing the impact parameter to be of the order of $M\gamma^2$, one finds a different régime, where there is a non-axisymmetric radiation pattern, again with angular structure on scales of $O(\gamma^{-1})$. The radiation is now beamed, with amplitude $\to 0$ as $\psi \to \infty$, and it still has strong-field power within the beam. If one increases the impact parameter still further, to be $\gg M\gamma^2$, one can recover the results of Kovács and Thorne (1978).

One can also use these methods for collisions at exactly the speed of light (Chapters 6,7, and 8). Penrose (1974) first studied this problem by examining the initial data on the pair of plane-fronted incoming shocks. He found apparent horizons (locally defined indications that one is inside a black hole) (Hawking and Ellis 1973; Wald 1984) for a range of (impact parameter/incoming energy). Assuming the cosmic censorship hypothesis (Wald 1984), Penrose deduced upper bounds on the fraction of energy which could be emitted in gravitational waves, if the eventual space–time were a stationary black hole together with gravitational radiation. The largest upper bound on the energy efficiency for converting incoming mass–energy into outgoing radiation is 29% for the head-on collision. This can be compared with the estimate above of 25%.

Here we study only the axisymmetric head-on collision at the speed of light. One proceeds by applying a large boost in the z-direction, so that one has a weak shock scattering off a strong shock (Curtis 1978a, b; Chapman 1979). When boosted back to the centre-of-mass frame, this will give a good description of the radiation near the axis. The axisymmetric gravitational radiation is described by the *news function* $c_0(\hat{\tau}, \hat{\theta})$ of retarded time $\hat{\tau}$ and angle $\hat{\theta}$ (Bondi *et al.* 1962). The rate of emission of mass–energy in gravitational waves is given by

$$d(\text{mass})/d\hat{\tau} = -\frac{1}{2} \int_0^\pi d\theta \sin\theta [c_0(\hat{\tau}, \hat{\theta})]^2. \tag{1.2}$$

One expects, and can prove under certain conditions, that the news function admits a convergent series expansion

$$c_0(\hat{\tau}, \hat{\theta}) = \sum_{n=0}^\infty a_{2n}(\hat{\tau}) \sin^{2n}\hat{\theta}, \tag{1.3}$$

where $\hat{\tau}$ is a retarded time coordinate. First-order perturbation theory gives an expression for $a_0(\hat{\tau})$ in agreement with that found previously by studying the finite-γ collisions. Second-order perturbation theory gives $a_2(\hat{\tau})$ as a complicated integral expression. A new mass-loss formula is derived, which shows that if the end result of the collision is a single Schwarzschild black hole at rest, plus gravitational radiation which is (in a certain precise sense) accurately described by the above series for $c_0(\hat{\tau}, \hat{\theta})$, then the final mass can be determined from knowledge only of $a_0(\hat{\tau})$ and $a_2(\hat{\tau})$. This leads to an interesting test of the cosmic censorship hypothesis (Wald 1984), which loosely asserts that the complete gravitational collapse of a body always results in a black hole rather than a naked singularity.

The numerical calculation of $a_2(\hat{\tau})$ is made practicable by analytical simplifications introduced in Chapter 7. At each order of perturbation theory there is a conformal symmetry. Hence one can reduce the axisymmetric perturbation problem to one in two independent variables at each order in perturbation theory.

One obtains a complicated but still manageable expression for $a_2(\hat{\tau})$.

Chapter 8 summarizes results on the speed-of-light black-hole collision, giving the numerical form of $a_2(\hat{\tau})$. The new mass-loss formula of Chapter 6 implies that the 'final mass' exceeds 2μ, the total incoming mass–energy. Hence the assumptions of the new mass-loss formula must not all hold. The most likely explanation is that there is a 'second burst' of radiation present in the space–time, centred for small angles $\hat{\theta}$ on retarded times roughly $|\, 8\mu \log \hat{\theta}\, |$ later than the 'first burst' described above. The 'second burst' is expected to be generated in the centre of the space–time, whereas the 'first burst' is generated in the far-field curved shocks; the time delay may be calculated from the delay $8\mu \log P$ induced in the curved shock as a result of going through the other shock. One can nevertheless obtain a more realistic crude estimate of the energy emitted in gravitational waves by taking the Bondi expression in eqn (1.2), and including only the first two terms a_0 and a_2 in c_0. This gives an estimated efficiency of 16.4% for gravitational wave generation.

Chapter 9 summarizes the work described in the book. It also describes, as mentioned already, a new avenue of research ('t Hooft 1987) in quantum field theory, based on the underlying ideas appearing in Chapters 5–8. The scattering process of two pointlike particles at centre-of-mass energies of the order of the Planck energy $(\hbar c^5/G)^{1/2} \simeq 10^{19}\mathrm{GeV}$, is calculable, since the main effect is the logarithmic time delay discussed above. In particle-physics language, one finds that graviton exchange dominates all other interaction processes. At energies much higher than the Planck energy, the semi-classical approximation is given by the classical space–times studied in Chapters 5–8: i.e. black-hole production sets in as a quantum process, accompanied by the coherent emission of real gravitons. Further references are given in Chapter 9.

2

BLACK HOLES

2.1 Introduction

One of the major applications of general relativity is to the study of black holes formed by the gravitational collapse of a massive body (Misner *et al.* 1973) or in the early universe by the collapse or collision at high speeds of regions containing large amounts of energy (Barrow and Carr 1978). Black holes are an intrinsically non-linear phenomenon, depending on the strong-field behaviour of the gravitational field. In a black hole, the gravitational field is so strong that even light cannot escape. Black holes should be contrasted with another generic feature of the vacuum gravitational field—gravitational waves. There one solves approximately linear perturbation equations, perhaps about a non-trivial background space–time, to describe the waves. Indeed, at least in the non-cosmological case, one might conjecture that a generic vacuum gravitational field could be built up by 'superposing' in a non-linear way the fields of various black holes and gravitational radiation. This superposition is described in Chapters 3 and 4. One can also extend this to the case in which there is matter in the space–time.

It is a remarkable feature of general relativity that the space–time geometry produced by a collapse or a collision apparently settles down to a final equilibrium state, parametrized by the mass M, charge Q and angular momentum Ma (Misner *et al.* 1973). The uniqueness results for the final equilibrium states (Hawking and Ellis 1973; Robinson 1975), combined with the stability property of the rotating Kerr black hole (Whiting 1989), guarantee the astrophysical importance of black holes, provided that general relativity is an accurate theory of nature (Will 1993). Hence, fortunately, one can learn a great deal from studying the stationary black-hole space–times—the massive Schwarzschild black hole, the charged Reissner–Nordström black hole, and the rotating Kerr black hole (Hawking and Ellis 1973; Misner *et al.* 1973) (there is also the charged and rotating Kerr–Newman black hole (Newman *et al.* 1965)).

In the remainder of this section, order-of-magnitude calculations will be studied, leading to the conclusion that there is an upper limit to the mass of a spherical star at electron or neutron densities, which has exhausted its nuclear fuel. This suggests that stars more massive than this limit might eventually collapse and form a black hole. In Section 2.2, an example of spherical collapse is studied, showing how the space–time geometry outside the collapsing matter is dragged down to a Schwarzschild singularity, but that the singularities both inside and outside the matter are concealed behind an *event horizon*, which separates the region of the space–time invisible from infinity from the region visible from infinity. This provides an example of *cosmic censorship*, which will recur in subsequent discussion, particularly the treatment of the collision of two black holes at an approach velocity close to the speed of light. In Section 2.3, the notion of a Penrose conformal diagram is described and applied to the case of flat Minkowski space–time and to Schwarzschild space–time. The charged Reissner–Nordström space–time is treated in Section 2.4. In Section 2.5, we turn to the rotating Kerr metric, and discuss its maximal extensions in terms of its conformal diagram. The rotating geometry near the black hole is treated—the event horizon is located, and there is a region outside the horizon known as the *ergoregion*, such that all observers in the ergoregion moving on timelike paths must co-rotate with the black hole. This section also considers the extraction of rotational energy, and the notion of reversible and irreversible transformations of a black hole. The result (Hawking 1971, 1972a; Penrose 1971;), that the area of a dynamical black hole can only increase with time or stay constant, is used to bound the amount of energy emitted in gravitational waves in a collision of black holes. A calculation by Penrose (1974) of this type gives an upper bound on the gravitational wave energy emitted in a black-hole collision at the speed of light, as studied here in Chapters 5–8. The Kerr black holes to be studied later, in Chapters 3 and 4, can be made dynamical in either of two ways: by considering geodesics (free-particle paths) or by considering linear wave equations of various types, corresponding for example to spin 0 (scalar), spin 1 (electromagnetic), or spin 2 (gravitational perturbations), or to fermionic modes. The geodesic equations (Section 2.6) are treated via the Hamilton–Jacobi equation. The perturbative wave equations are described in the formalism of Teukolsky (1973) in Section 2.7. Remarkably, the perturbations are described by a decoupled separable equation. Further (Chrzanowski 1975; Wald 1978), the gravitational metric perturbations or electromagnetic potentials can be recovered from the decoupled quantity above by differential operations. This will be essential in treating fully the motion of two black holes in Chapter 4.

Now let us consider some general order-of-magnitude estimates, due to Carter (1972)—see (Hawking and Ellis 1973)—which suggest the inevitability of gravitational collapse to a black hole, for a spherically symmetric object with mass somewhat greater than the Chandrasekhar mass of 1.4 times the solar mass (Chandrasekhar 1939). Take the case of a 'cold' non-rotating star which has exhausted all its nuclear fuel, and is in its ground state. At low pressures, the star will be made up purely of iron, since this gives the state which is most favourable energetically. At higher pressures, with density $\rho > 10^5$ g/cm^3, one has fermion degeneracy, first of the electrons, and then of neutrons. One can estimate the pressure straightforwardly. Suppose that there are n fermions per cm^3, each with mass m. Then degeneracy implies that the typical momentum of a fermion $\sim \hbar n^{1/3}$. Hence a typical fermion velocity $\sim \hbar n^{1/3}/m$ for the non-relativistic case, where m is the fermion mass. In the relativistic case, one has velocity ~ 1, in geometrical units with $c = G = 1$. The pressure is of the order of (momentum) $\times$ (velocity) $\times$ (number density). Hence pressure $\sim \hbar^2 n^{5/3} m^{-1}$ in the non-relativistic case with $\hbar n^{1/3} < m$. In the relativistic case, with $\hbar n^{1/3} > m$, the pressure $\sim \hbar n^{4/3}$. Four régimes are covered by the above cases. Where the density ρ is measured in g/cm^3, one has for $10^5 < \rho < 10^7$ the case in which non-relativistic electrons provide the main pressure. For $10^7 < \rho < 4 \times 10^{11}$, one has relativistic electrons. Stars supported by electron degeneracy are known as white dwarfs. For $4 \times 10^{11} < \rho < 10^{13}$, one has non-relativistic neutrons, which are formed at high energies by inverse beta decay, in which electrons and protons combine to form a neutron and a neutrino. For densities $\rho > 10^{13}$, the main source of pressure is relativistic neutron degeneracy and strong interactions. Stars supported by neutron degeneracy pressure are known as neutron stars.

We can now estimate the mass of each of these types of star. The mass density gives in order of magnitude

$$R \sim M^{1/3} n^{-1/3} m_n^{-1/3}, \tag{2.1.1}$$

where R is the radius and M is the mass of the star. The pressure balance may be approximated by a Newtonian expression, giving the average pressure P as

$$P = M^2/R^4 \simeq M^{2/3} n^{4/3} m_n^{4/3}. \tag{2.1.2}$$

The four cases outlined above can now be treated separately:

(a) Non-relativistic electrons have $n < m_e^3 \hbar^{-3}$. Hence

$$P = \hbar^2 n^{5/3} m_e^{-1} = M^{2/3} n^{4/3} m_n^{4/3}. \tag{2.1.3}$$

This implies

$$n = M^2 m_n^4 m_e^3 \hbar^{-6}. \tag{2.1.4}$$

Hence

$$M \lesssim M_L \simeq \hbar^{3/2} m_n^{-2}$$
$$\simeq 1.5 \,\text{solar masses.} \tag{2.1.5}$$

(b) For relativistic electrons, with $n > m_e^3 \hbar^{-3}$, pressure balance gives

$$P = \hbar n^{4/3} = M^{2/3} n^{4/3} m_n^{4/3}. \tag{2.1.6}$$

Hence

$$M = M_L, \tag{2.1.7}$$

with n undetermined.

(c) Non-relativistic neutrons have $n < m_n^3 \hbar^{-3}$. As in (a), one obtains

$$n = M^2 m_n^7 \hbar^{-6}. \tag{2.1.8}$$

Hence again $M < M_L$.

(d) Finally, for relativistic neutrons, with $n > m_n^3 \hbar^{-3}$, one has

$$M = M_L. \tag{2.1.9}$$

When this calculation is re-done using general relativity, there is little qualitative difference (Hartle 1978). There is a maximum mass for a star at the endpoint of its evolution, which is no greater than a few solar masses. (The original calculation that there is a maximum mass is due to Chandrasekhar (1939).) Stars heavier than this limit must either shed mass or collapse catastrophically. In the following Section 2.2 we shall see an example of a matter collapse which leads to the formation of a black hole.

2.2 Spherical collapse

Consider the simple idealization of spherical collapse of a massive body in general relativity, due to Oppenheimer and Snyder (1939). One takes dust of uniform density, starting from rest at a moment of time-reflection symmetry. The energy–momentum tensor T_{ab} is given by

$$T_{ab} = \rho u_a u_b, \tag{2.2.1}$$

where ρ is the matter density and u^a is the four-velocity of the matter, normalized by $u_a u^a = -1$. Baryon conservation is described by

$$(\rho u^a)_{;a} = 0. \tag{2.2.2}$$

Energy–momentum conservation

$$T^{ab}_{;b} = 0 \tag{2.2.3}$$

implies

$$u^b u^a_{;b} = 0, \tag{2.2.4}$$

which shows that each dust particle follows a geodesic. The solution inside the matter ($0 \leq \chi \leq \chi_s$, where χ is a radial coordinate inside the matter) is taken to be a Friedmann–Robertson–Walker $k = +1$ cosmological model (Hawking and Ellis 1973; Misner *et al.* 1973). Here

$$ds^2 = -d\tau^2 + R^2(\tau)[d\chi^2 + \sin^2 \chi (d\theta^2 + \sin^2 \theta d\phi^2)] \tag{2.2.5}$$

with

$$\rho = \frac{3R_0}{8\pi R^3(\tau)}, \qquad u^a = \delta_\tau^{a}, \tag{2.2.6}$$

where R_0 is a constant, and $R(\tau)$ is given parametrically (see Fig. 2.1) by

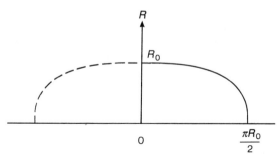

FIG. 2.1. The radius R of the Friedmann solution, as a function of proper time τ.

$$R(\eta) = \frac{1}{2}R_0(1 + \cos\eta),$$

$$\tau(\eta) = \frac{1}{2}R_0(\eta + \sin\eta).$$

(2.2.7)

The radius $R(\tau)$ obeys the Friedmann equation

$$\left(\frac{dR}{d\tau}\right)^2 = -1 + \frac{8\pi}{3}R^2\rho$$

$$= -1 + \frac{R_0}{R}.$$

(2.2.8)

Note that the matter density ρ and hence the scalar curvature $R_a{}^a$, which is proportional to ρ by the Einstein field equations

$$R_{ab} - \frac{1}{2}g_{ab}R_c{}^c = 8\pi T_{ab},$$

(2.2.9)

reach infinite values after a finite proper time $\frac{1}{2}\pi R_0$ measured along geodesic particle worldlines. Thus there is a barrier to extending the space–time. The timelike geodesics followed by the dust particles are incomplete, since they are prevented from reaching the region where the curvature is infinite. Geodesic incompleteness is usually taken as the signal for the existence of a space–time singularity (Penrose 1968, 1972a; Hawking and Ellis 1973).

It is well known (Birkhoff's theorem—see Hawking and Ellis (1973)) that the only spherically symmetric solution of the vacuum Einstein field equations is the Schwarzschild solution, even when one allows the space–time in principle to be dynamical. Hence, in our example, the geometry expressed in Schwarzschild coordinates outside the matter is

$$ds^2 = -\left(1 - \frac{2M}{r}\right)dt^2 + \frac{dr^2}{\left(1 - \frac{2M}{r}\right)} + r^2(d\theta^2 + \sin^2\theta d\phi^2).$$

(2.2.10)

One also needs to understand the matching across the boundary at $\chi = \chi_s$ between the matter and vacuum regions. One can compare with the Newtonian analogue in which the density ρ is discontinuous across the boundary, while the Newtonian potential ϕ and its spatial gradient $\nabla\phi$ are continuous. The second-order Einstein field equations (2.2.9) have a similar property, that the metric g_{ab} and its first derivatives $g_{ab,c}$ are continuous across the boundary, when expressed in terms of suitable coordinates (Misner et al. 1973). Clearly, one identifies r with $R(\tau)\sin\chi_s$ on the boundary. One can also find the mass M in terms of the matter parameters. One uses the property that a dust particle on the boundary follows a geodesic with respect both to the interior matter geometry, and to the exterior Schwarzschild geometry, because of the continuity of g_{ab} and $g_{ab,c}$ above. This leads (Misner et al. 1973) to

$$M = \frac{1}{2}R_0 \sin^3 \chi_s. \qquad (2.2.11)$$

Note that, since we identify r on the boundary with $R(\tau)\sin\chi_s$, the interior geometry is well-behaved as the matter boundary passes through the Schwarzschild radius with $r = R(\tau)\sin\chi_s = 2M$. One can loosely regard the interior geometry as dragging the exterior geometry down through the Schwarzschild radius, and one must then extend the external Schwarzschild geometry through $r = 2M$. One way to do this is to define ingoing Eddington–Finkelstein coordinates for the Schwarzschild geometry, (v, r, θ, ϕ), where $v = $ const. with θ, ϕ constant labels an ingoing radial null geodesic:

$$v = t + r + 2M\log(r - 2M). \qquad (2.2.12)$$

The Schwarzschild metric then takes the form

$$ds^2 = -\left(1 - \frac{2M}{r}\right)dv^2 + 2dvdr + r^2(d\theta^2 + \sin^2\theta\, d\phi^2). \qquad (2.2.13)$$

At this point it is helpful to introduce some notation concerning three-dimensional hypersurfaces embedded in a four-dimensional space–time. Suppose that the hypersurface is given by a condition of the form $\{x = \text{const.}\}$, where x is some coordinate. If the intrinsic three-dimensional metric is positive-definite, with canonical form $\text{diag}(+,+,+)$, we say that the surface is spacelike. If the canonical form is $\text{diag}(-,+,+)$, we say that the surface is timelike. If the canonical form is $\text{diag}(0,+,+)$, the surface is said to be null. In the null case, there is a null tangent vector ℓ^a to the hypersurface, unique up to a multiplicative factor (which may vary over the surface). The integral curves of ℓ^a in the null hypersurface are the *null generators* of the surface. One can prove (Wald 1984) that the null generators are null geodesics in the four-dimensional space–time. In the case (2.2.13) above of the Schwarzschild geometry in (v, r, θ, ϕ) coordinates, one can see that the surfaces $\{v = \text{const.}\}$ are null.

We see from eqn (2.2.13) that the Schwarzschild geometry is regular across

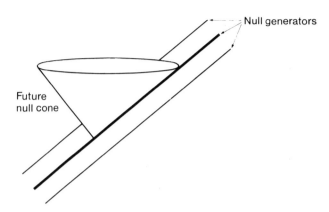

Future null cone

Null generators

F_IG. 2.2. Future-directed particles can only cross the null surface $\{r = 2M\}$ moving inwards.

the surface $\{r = 2M\}$. The external Schwarzschild geometry does not reach a singularity until $r = 0$. One can see this by evaluating the curvature invariant

$$R_{abcd}R^{abcd} = \frac{48M^2}{r^6}. \tag{2.2.14}$$

This shows that one could not extend the space–time past the region $r = 0$ in any smooth way, and hence that there is a singularity there. This can also be verified by studying timelike and null geodesics, and finding that those which approach the region $\{r = 0\}$ cannot be continued through it. Because of the divergence of the scalar polynomial invariant in eqn (2.2.14) as $r \to 0$, one says that the Schwarzschild geometry has a *curvature singularity* at $r = 0$.

Next, note that the Schwarzschild singularity is not visible from infinity, since $\{r = 2M\}$ is a null surface. Future-directed null or timelike curves can only cross $\{r = 2M\}$ when moving inwards to smaller r-values, as can be seen in Fig. 2.2. A diagram of the space–time structure is given in Fig. 2.3. Here, in the external vacuum region with coordinates v, r, each point represents a two-sphere. The straight lines $\{v = \text{const.}\}$, which give the paths of radially infalling photons, are drawn at 45 degrees. The boundary between the matter region and the vacuum region is a timelike geodesic. The singularity starts on the axis of symmetry in the matter region, and then 'joins on' to the Schwarzschild singularity at $r = 0$. Projected null cones, where the full null cones in four dimensions are projected as in Fig. 2.2, are given in the vacuum region. One notes that non-spacelike curves having $ds^2 \leq 0$ obey

$$-\left(1 - \frac{2M}{r}\right) dv^2 + 2dvdr \leq 0. \tag{2.2.15}$$

Hence the future null cone obeys

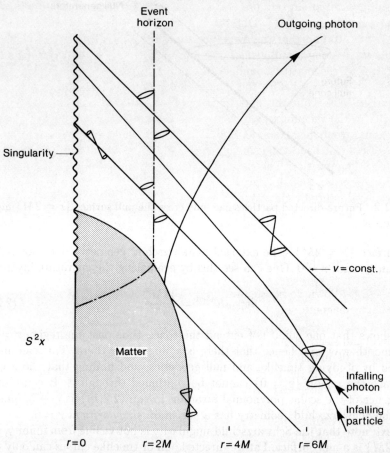

FIG. 2.3. The collapse in the matter region and the corresponding collapse of the
vacuum region outside to a singularity. The Schwarzschild vacuum region is viewed
in ingoing Eddington–Finkelstein coordinates (v, r). Note that the light cones turn
more and more inwards as the event horizon $\{r = 2M\}$ is approached, and are
tangential to the horizon.

$$dv \geq \frac{2dr}{\left(1 - \frac{2M}{r}\right)}, \qquad dv \geq 0. \qquad (2.2.16)$$

Points in the external region with $r > 2M$ are visible from infinity, e.g. on
outgoing radial null geodesics, with

$$u = t - r - 2M \log(r - 2M) = \text{const.} \qquad (2.2.17)$$

Correspondingly, we say that, in the vacuum region, $\{r \leq 2M\}$ is inside a *black hole.* The boundary of the black hole, which separates those events in space–time which are visible from infinity and invisible from infinity, is known as the *future event horizon.* In our present example, the future event horizon in the vacuum region is $\{r = 2M\}$. One can also carry on this investigation into the matter region, to find that the event horizon there begins on the axis and then moves outward along a null geodesic to join up with the vacuum event horizon at $r = 2M$ at the matter boundary. The event horizon in the matter region is formed on the axis before the matter collapses to a singularity which also starts on the axis. Thus the singularity inside the matter, in common with the singularity in the vacuum region, is invisible from infinity.

Following from this example, one is led to ask whether generic matter collapses possess (a) space–time singularities, and (b) an event horizon which surrounds the singularities. The powerful Hawking–Penrose singularity theorems imply (a) (Penrose 1968, 1972a; Hawking and Ellis 1973), based on a study of space–time structure by global methods. But there is only limited or circumstantial evidence for (b), which is known as the Cosmic Censorship Hypothesis; for discussion see Wald (1984). There are many different mathematical formulations of cosmic censorship, but much of the known information comes from examples. If (b) were to fail, then the theory of general relativity would have lost predictive power. We shall take the 'conservative viewpoint' here, namely that generic collapses settle down to stable stationary black holes plus radiation (gravitational, electromagnetic, etc.) which propagates outward near infinity. The only stationary black holes in the Einstein–Maxwell theory (working again, say, with gravitation and electrodynamics) are characterized by mass, angular momentum, and charge. The massive black hole is the Schwarzschild geometry, the massive rotating black hole is the Kerr geometry, and the rotating charged black hole is the Kerr–Newman geometry (Wald 1984). The stability against linearized perturbations of the Kerr metric (which includes the case of the Schwarzschild metric) was shown by Whiting (1989). This suggests that gravitational collapse of rotating matter which leads to an oscillating Kerr solution will eventually settle down to a nearly exact Kerr solution together with radiation near infinity. The radiation near infinity will consist partly of the quasi-normal modes of the Kerr black hole, which correspond to complex frequencies giving exponential decay at late times (Detweiler 1979). In Chapters 6–8, an example of a high-energy head-on collision of two black holes will lead to an interesting test of cosmic censorship.

2.3 Conformal diagrams and causal properties

2.3.1 *Asymptotically simple space–times*

In the treatment in the previous Section 2.2, the notion of infinity was used rather loosely, in deciding which events in space–time were visible from infinity and which were not. A more precise definition of infinity is needed. To motivate this, consider the simplest case of flat Minkowski space–time (Penrose 1968; Hawking and Ellis 1973), written in polar coordinates as

$$ds^2 = -dt^2 + dr^2 + r^2(d\theta^2 + \sin^2 \theta \, d\phi^2). \qquad (2.3.1.1)$$

Now adapt the coordinates to the radially outgoing and ingoing directions, taking

$$u = t - r, \quad v = t + r. \tag{2.3.1.2}$$

The metric then becomes

$$ds^2 = -dudv + \frac{1}{4}(u - v)^2(d\theta^2 + \sin^2\theta\, d\phi^2). \tag{2.3.1.3}$$

The final transformation brings infinity in to a finite region in the final coordinates. For this, one defines new null variables p, q by

$$u = \tan q, \quad v = \tan p. \tag{2.3.1.4}$$

After this transformation, the metric takes the form

$$ds^2 = \sec^2 p \sec^2 q[-dpdq + \frac{1}{4}\sin^2(p - q)(d\theta^2 + \sin^2\theta\, d\phi^2)]. \tag{2.3.1.5}$$

The domain for p and q is

$$-\frac{\pi}{2} < p < \frac{\pi}{2}, \quad -\frac{\pi}{2} < q < \frac{\pi}{2}, \quad p \geq q. \tag{2.3.1.6}$$

Thus the original Minkowski metric g_{ab} can be written as a conformal factor Ω^{-2} multiplying a conformal metric $\tilde{g}_{ab}$:

$$g_{ab} = \Omega^{-2}\tilde{g}_{ab}, \tag{2.3.1.7}$$

where

$$\Omega = \cos p \cos q,$$
$$d\tilde{s}^2 = -dpdq + \frac{1}{4}\sin^2(p - q)(d\theta^2 + \sin^2\theta\, d\phi^2). \tag{2.3.1.8}$$

Since the region near infinity has been brought in to a finite region, one can draw a *Penrose conformal diagram* of a two-space $\{\theta = \text{const.}, \phi = \text{const.}\}$—see Fig. 2.4. The intrinsic metric

$$ds^2 = -\sec^2 p \sec^2 q\, dp\, dq \tag{2.3.1.9}$$

is conformally flat, so that one can draw null lines at 45 degrees to the vertical. There are three boundaries to each figure. The boundary on the left is simply the spatial origin $p = q$, as one would expect in polar coordinates. The upper right boundary is the endpoint of future-directed null geodesics, which is known as $\mathcal{I}^+$, since all null geodesics have $q \to$ const. as they move out to large radius, with $p \to \pi/2$. Similarly, the lower right boundary, known as $\mathcal{I}^-$, represents the endpoints of past-directed null geodesics, which have $q \to -\pi/2$ with $p \to$ const. The conformal nature of the construction implies that null geodesics are

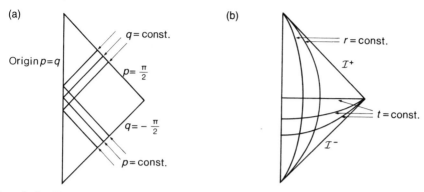

FIG. 2.4. The Penrose conformal diagram of flat Minkowski space–time. In (a), paths
of null rays are sketched. In (b), lines of constant t and r are shown.

identical in the Minkowski metric g_{ab} and in the conformal metric $\tilde{g}_{ab}$ (Wald
1984). If desired, one could view any null geodesic of Minkowski space–time
(projected) in the conformal diagram; the 'points at infinity' of the null geodesic
will be visible on the right-hand boundary. To summarize, if we write $(\mathcal{M}, g_{ab})$ for
Minkowski space–time, then the conformal metric $\tilde{g}_{ab}$ can be defined on a larger
manifold $\tilde{\mathcal{M}}$. $\mathcal{M}$ has boundary $\partial\mathcal{M}$, known as null infinity, consisting of the two
regions $p = \pi/2, -\pi/2 < q < \pi/2$, denoted by $\mathcal{I}^+$, called future null infinity,
and $q = -\pi/2, -\pi/2 < p < \pi/2$, denoted by $\mathcal{I}^-$, called past null infinity. The
conformal metrics $g_{ab}, \tilde{g}_{ab}$ have the same light cones and the same null geodesics,
and hence the same causal structure.

This motivates the following definition (Penrose 1968; Hawking and Ellis
1973): a space–time $(\mathcal{M}, g_{ab})$ is *asymptotically simple* if there exists another
space–time $(\tilde{\mathcal{M}}, \tilde{g}_{ab})$ and a continuous imbedding $\theta : \mathcal{M} \to \tilde{\mathcal{M}}$ where $\theta(\mathcal{M})$ is
an open submanifold of $\tilde{\mathcal{M}}$ with smooth boundary $\partial\mathcal{M}$ in $\tilde{\mathcal{M}}$, such that:

(i) There is a smooth function Ω on $\tilde{\mathcal{M}}$ with $\Omega > 0$ on $\theta(\mathcal{M})$ and $\tilde{g}_{ab} = \Omega^2 g_{ab}$.

(ii) On $\partial\mathcal{M}$, $\Omega = 0$ and $\Omega_{,a} \neq 0$.

(iii) Every null geodesic in $\mathcal{M}$ has two endpoints on $\partial\mathcal{M}$.

In the above example, one can easily verify that Minkowski space–time is asymptotically simple. The following remarks concerning the general asymptotically
simple case should also be noted:

(a) If the vacuum Einstein equations $R_{ab} = 0$ hold near $\partial\mathcal{M}$, then Ω
is asymptotically a constant times r^{-1}, where r is an asymptotic radial
coordinate (Bondi *et al.* 1962).

(b) If again $R_{ab} = 0$ near $\partial\mathcal{M}$, it can be proved that $\partial\mathcal{M}$ is a null hypersurface in $\tilde{\mathcal{M}}$.

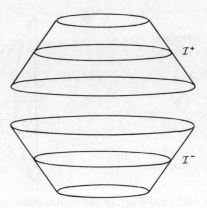

FIG. 2.5. A schematic representation of $\mathcal{I}^+$ and $\mathcal{I}^-$, which are null surfaces, each with topology $\mathbb{R} \times S^2$.

(c) Then $\partial \mathcal{M}$ has two disconnected components, $\mathcal{I}^+$ and $\mathcal{I}^-$, each with topology $\mathbb{R}^1 \times S^2$.

This definition is intended to capture, in particular, the behaviour of many space–times which have gravitational fields of moderate strength in their interior, and which contain gravitational radiation. An existence theorem for weak gravitational fields of this type has been proved by Christodoulou and Klainerman (1993), by taking an initial-value approach. This theorem shows that the guess as to the asymptotic form of generic vacuum fields, given by Bondi *et al.* (1962) and Sachs (1962) and by the definition of asymptotic simplicity above, is valid for weak but non-linear gravitational fields. A sketch of the form of $\mathcal{I}^+$ and $\mathcal{I}^-$ is given in Fig. 2.5.

It is clear that the condition (iii) above in the definition of asymptotic simplicity may fail, as in the case of space–times containing black holes, which is precisely the case of most interest to us here. In this case one modifies the definition of asymptotic simplicity, saying that a space–time $(\mathcal{M}, g_{ab})$ is *weakly asymptotically simple* if there is an asymptotically simple space–time $(\mathcal{M}', g'_{ab})$ and a neighbourhood $\mathcal{U}'$ of $\partial \mathcal{M}'$ in $\bar{\mathcal{M}}'$ such that $\mathcal{U}' \cap \theta(\mathcal{M}')$ is isometric to an open set $\mathcal{U}$ of $\mathcal{M}$. Essentially, weakly asymptotically simple space–times are those which have a region near null infinity which is identical to that in an asymptotically simple space–time. This is depicted in Fig. 2.6. In such a space–time, one can draw a Penrose diagram of any two-dimensional timelike surface which approaches null infinity. The null directions can be taken to be at 45 degrees to the vertical; note that any two-dimensional metric is conformally flat (Wald 1984). This construction will be applied to help understand the causal structure of the Schwarzschild, Reissner–Nordström, and Kerr black-hole geometries, in what follows.

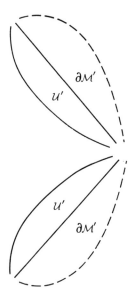

FIG. 2.6. The region near $\mathcal{I}^+$ and $\mathcal{I}^-$ in a weakly asymptotically simple space–time.

2.3.2 Schwarzschild geometry

We now look for ways in which to extend the Schwarzschild geometry, by analogy with the use of Eddington–Finkelstein coordinates in eqns (2.2.12) and (2.2.13). We shall find the complete analytic extension of the Schwarzschild geometry in this way. This will help us to understand the causal structure of the Schwarzschild geometry.

As before, the Schwarzschild metric in Schwarzschild coordinates is

$$ds^2 = -\left(1 - \frac{2M}{r}\right) dt^2 + \frac{dr^2}{\left(1 - \frac{2M}{r}\right)} + r^2(d\theta^2 + \sin^2\theta \, d\phi^2). \qquad (2.3.2.1)$$

Here, for the sake of definiteness, we shall for the moment consider only the case $M > 0$. As in eqns (2.2.12) and (2.2.13), one may take

$$v = t + r + 2M \log(r - 2M), \qquad (2.3.2.2)$$

where v is constant on radially ingoing null geodesics, giving the ingoing extension

$$ds^2 = -\left(1 - \frac{2M}{r}\right) dv^2 + 2dvdr + r^2(d\theta^2 + \sin^2\theta \, d\phi^2). \qquad (2.3.2.3)$$

Now the Schwarzschild geometry is time-symmetric (invariant under $t \to -t$), and so may also be rewritten using the null coordinate

$$u = t - r - 2M \log(r - 2M), \tag{2.3.2.4}$$

which is constant on radially outgoing null geodesics. This gives the outgoing extension

$$ds^2 = -\left(1 - \frac{2M}{r}\right) du^2 - 2du\,dr + r^2(d\theta^2 + \sin^2\theta\, d\phi^2). \tag{2.3.2.5}$$

One can then attempt to make both ingoing and outgoing extensions together. This leads to the double-null form

$$ds^2 = -\left(1 - \frac{2M}{r}\right) du\,dv + r^2(d\theta^2 + \sin^2\theta\, d\phi^2). \tag{2.3.2.6}$$

Here r is given implicitly as a function of u and v through

$$\frac{1}{2}(v - u) = r + 2M \log(r - 2M). \tag{2.3.2.7}$$

The logarithm implies that $r(u, v)$ is badly behaved at $r = 2M$.

The appropriate transformation is given by redefining the null coordinates (Kruskal 1960):

$$\begin{aligned} u' &= -\exp(-u/4M), \\ v' &= \exp(v/4M). \end{aligned} \tag{2.3.2.8a}$$

The metric then takes the form

$$ds^2 = -\left(1 - \frac{2M}{r}\right) \frac{dv}{dv'} \frac{du}{du'} du'\,dv' + r^2(d\theta^2 + \sin^2\theta\, d\phi^2). \tag{2.3.2.8b}$$

Now define

$$r' = \frac{1}{2}(v' - u'), \qquad t' = \frac{1}{2}(v' + u'). \tag{2.3.2.9}$$

The metric then becomes

$$ds^2 = F^2(t', r')(-dt'^2 + dr'^2) + r^2(t', r')(d\theta^2 + \sin^2\theta\, d\phi^2), \tag{2.3.2.10}$$

where $r(t', r')$ is given implicitly by

$$(t')^2 - (r')^2 = -(r - 2M)\exp(r/2M). \tag{2.3.2.11}$$

This expression is now invertible for $0 < r < \infty$. The curvature singularity at $r = 0$, described in Section 2.2, implies that one should only consider values of r with $r > 0$. In eqn (2.3.2.10), one also has

$$F^2(t', r') = \frac{16M^2}{r} \exp\left(\frac{-r}{2M}\right), \tag{2.3.2.12}$$

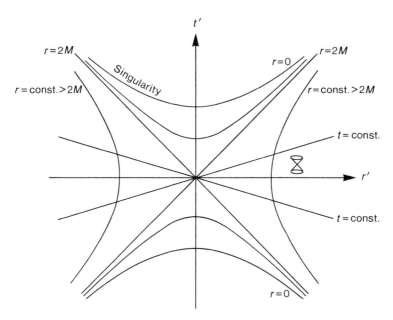

FIG. 2.7. The Kruskal diagram of the Schwarzschild geometry. Light cones are at 45 degrees. The future event horizon can again be identified as a surface with $r = 2M$. There are two distinct singularities $r = 0$, one in the past and one in the future.

$$\frac{t'}{r'} = \tanh\left(\frac{t}{4M}\right).\qquad(2.3.2.13)$$

The domain of validity of the form (2.3.2.10) of the Schwarzschild geometry is

$$\{t',r' : r(t',r') > 0\} = \{t',r' : (t')^2 - (r')^2 < 2M\}.\qquad(2.3.2.14)$$

One arrives at the *Kruskal diagram* of the maximally extended Schwarzschild geometry, depicted in Fig. 2.7. Light cones are at 45 degrees to the vertical. Note that there are two null surfaces with $r = 2M$, which meet at a point in the centre of the figure, representing a two-sphere. There are also two singular 'surfaces' with $r = 0$. One cannot regard a singular region as part of the space–time manifold, but it should nevertheless be evident from Fig. 2.7 that the two singularities are in some sense 'spacelike'.

One can also construct a Penrose conformal diagram for the Schwarzschild geometry, by changing variables from the metric in Kruskal coordinates. One defines (Penrose 1968; Hawking and Ellis 1973)

$$u'' = \tan^{-1}\left(\frac{u'}{\sqrt{(2M)}}\right), \qquad v'' = \tan^{-1}\left(\frac{v'}{\sqrt{(2M)}}\right).\qquad(2.3.2.15)$$

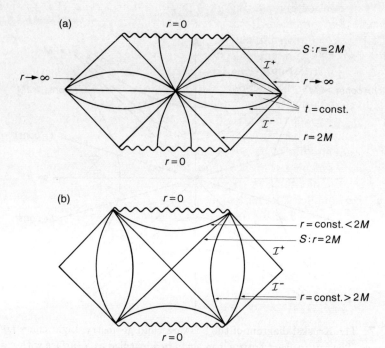

FIG. 2.8. The Penrose diagram of the maximally extended Schwarzschild geometry. In (a), lines of constant t are shown. In (b), lines of constant r are shown. The future event horizon is the surface marked S.

This gives the maximally extended vacuum Schwarzschild solution, with null infinity brought in to finite coordinates. The domain is then

$$-\frac{\pi}{2} < u'' + v'' < \frac{\pi}{2}; \qquad -\frac{\pi}{2} < u'' < \frac{\pi}{2}; \qquad -\frac{\pi}{2} < v'' < \frac{\pi}{2}.$$

The resulting Penrose conformal diagram is given in Fig.2.8, which shows in addition lines of constant t and r respectively. Here, as in the case above of Minkowski space, each point in the diagram represents a two-sphere S^2, with (θ, ϕ) being the angular coordinates over the two-sphere. One can also verify from this construction that Schwarzschild space–time is weakly asymptotically simple, having two sets of asymptotic regions, either $\mathcal{I}^-, \mathcal{I}^+$ as on the right-hand side of Fig. 2.8(a), or $\mathcal{I}^-, \mathcal{I}^+$ on the left-hand side of Fig. 2.8(a). From Fig.2.8, one can read off immediately that Schwarzschild space–time contains a black hole, namely the region above the null surface S with $r = 2M$. This region cannot send signals to $\mathcal{I}^+$. As remarked earlier, the singularities $r = 0$ are 'spacelike'. The final singularity is hidden from view from $\mathcal{I}^+$ by the future event horizon S. But the initial singularity at $r = 0$ is visible from $\mathcal{I}^+$, so that the extended vacuum Schwarzschild geometry has a naked singularity. This difficulty is avoided in certain spherically symmetric matter collapses (see Fig. 2.12). The extension given in Fig.2.8 is maximal, in the sense that any geodesic in it cannot be continued. It can be verified that any geodesic either hits the singularity at $r = 0$ with finite affine parameter, or else goes to infinity with affine parameter tending to infinity. Geodesics in the Kerr metric and *a fortiori* in the Schwarzschild metric will be treated in Section 2.6 (see also Misner *et al.* (1973) for both cases). The domain for the maximally extended Schwarzschild geometry can be divided naturally into four regions I, II, III, IV. Figure 2.9 shows how different groupings of these regions are obtained from the different coordinate systems which have been used.

One would also like to have some understanding of the geometry of spacelike hypersurfaces which span Fig.2.8. There are two asymptotic regions joined by a bridge. For example, consider the spacelike hypersurface $\{t' = 0\}$, with topology $\mathbb{R}^1 \times S^2$. For simplicity, let us visualize the equator $\theta = \pi/2$, by embedding it in flat $\mathbb{R}^3$. From eqn (2.3.2.10), the intrinsic geometry is

$$ds^2 = \frac{16M^2}{r} \exp\left(\frac{-r}{2M}\right) dr'^2 + r^2 d\phi^2, \tag{2.3.2.16}$$

with $-\infty < r' < +\infty$, which is embedded as in Fig. 2.10.

We also note an alternative characterization of the Schwarzschild event horizon in terms of its symmetry properties. We define a Killing vector field k^a to be the generator of an infinitesimal isometry of a space–time $(\mathcal{M}, g_{ab})$. Equivalently

$$\mathcal{L}_k g_{ab} = 0,$$

i.e. the Lie derivative of g_{ab} with respect to k^a vanishes. This condition can be written out as

$$k_{a;b} + k_{b;a} = 0.$$

Thus a space–time has a Killing vector field if its metric is unchanged under the map

$$x^a \rightarrow x^a + \epsilon k^a,$$

to first order in ϵ. In the case of the Schwarzschild metric, one has (for example) the time-translations $(t, r, \theta, \phi) \rightarrow (t+\epsilon, r, \theta, \phi)$. The corresponding Killing vector field k^a is represented by $k^a \equiv \partial/\partial t$ in region I, $k^a \equiv \partial/\partial v$ in region II, and $k^a \equiv \partial/\partial u$ in region III. The direction of k^a is depicted in Fig. 2.11. This points along the direction of lines of constant r. Note that the time-translation Killing vector field k^a tips over and becomes spacelike in the region with $0 < r < 2M$. In the left-hand region of Fig. 2.11, k^a is past-directed. One says that either surface S with $r = 2M$ is a *Killing horizon* since (i) $k^a k_a = 0$ on the surface S, (ii) S is a null surface, (iii) k^a is tangent to the null generators of S. This gives a local characterization of the Schwarzschild event horizon. One can similarly treat other static metrics (metrics where there is a time-translation Killing vector field which is orthogonal to a one-parameter family of hypersurfaces). In fact, for static metrics, condition (i) above implies conditions (ii) and (iii) (Wald 1984).

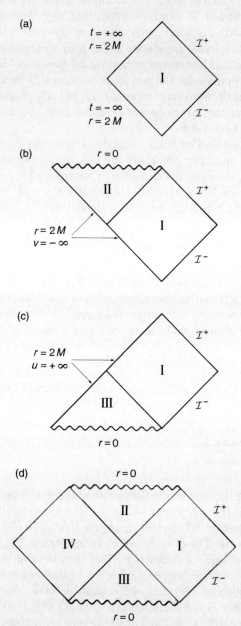

FIG. 2.9. Four diagrams showing the different regions in the Schwarzschild geometry covered by different coordinates. (a) Schwarzschild coordinates; (b) (v, r) Eddington-Finkelstein coordinates; (c) (u, r) Eddington-Finkelstein coordinates; (d) Kruskal coordinates.

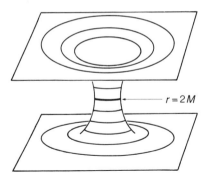

FIG. 2.10. The intrinsic spatial geometry of a section $t =$ const., $\theta = \pi/2$ through the maximally extended Schwarzschild geometry, visualized by embedding in flat Euclidean three-space.

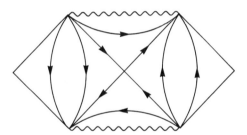

FIG. 2.11. The 'time-translation' Killing vector field k^a in the Schwarzschild geometry is represented. It points along lines of constant r.

The Penrose diagram of spherical collapse, with matter infalling from infinity with non-zero velocity, is shown in Fig. 2.12. This shows that only the regions I and II in the vacuum Schwarzschild geometry normally have physical relevance. One can also see that in this case cosmic censorship holds, in that the singularity in both the matter and vacuum regions is invisible from $\mathcal{I}^+$, being concealed by the future event horizon. The Penrose diagram of the Oppenheimer–Snyder collapse studied in Section 2.2 is shown in Fig. 2.13. The positions of the points at which the past and future horizons meet the axis $\chi = 0$ depend on the boundary coordinate χ_s. In this case the singularity in the past is visible from $\mathcal{I}^+$. This occurs because the space–time, being based in the matter region on a $k = +1$ Friedmann universe, is reflection-symmetric about a moment of time symmetry. If the Cosmic Censorship Hypothesis is to hold, then generic collapses should resemble that in Fig. 2.12. This can be verified at least for slightly non-spherical models of gravitational collapse (Moncrief *et al.* 1979).

Note finally that the collapse through the surface $r = 2M$ appears to take an

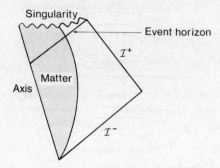

FIG. 2.12. The Penrose diagram of spherical collapse, with matter infalling from infinity.

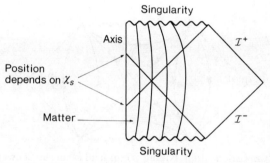

FIG. 2.13. The Penrose diagram of Oppenheimer–Snyder collapse. The space–time is symmetrical about a surface of time-reflection symmetry.

infinite time, as viewed from $\mathcal{I}^+$. By contrast, the collapse through this surface takes only a finite amount of proper time as measured by a freely falling observer on the matter boundary. Hence, viewed from infinity, the boundary of the matter appears to hover just outside $\{r = 2M\}$, and the light from the matter becomes infinitely red-shifted at very late times as measured at null infinity.

2.4 The Reissner–Nordström metrics

Although our main aim in studying examples of black holes in this chapter is to understand the rotating Kerr geometry, as this appears in Chapters 3 and 4 on the motion of rotating black holes, much can be learnt as a preliminary by studying the Reissner–Nordström geometries, which are static with mass and charge. As with the Schwarzschild geometry, the static condition is associated with the metric being stationary (time-independent) and the time-translation

Killing vector field being orthogonal to a family of spacelike hypersurfaces. The Reissner–Nordström metric is

$$ds^2 = -\left(1 - \frac{2M}{r} + \frac{Q^2}{r^2}\right) dt^2 + \left(1 - \frac{2M}{r} + \frac{Q^2}{r^2}\right)^{-1} dr^2$$
$$+ r^2 (d\theta^2 + \sin^2\theta \, d\phi^2). \tag{2.4.1}$$

Here M and Q are constants which will be identified below. The Reissner–Nordström geometry with suitable Maxwell field obeys the coupled Einstein–Maxwell field equations. The electromagnetic field tensor or two-form

$$F_{ab} = F_{[ab]} \tag{2.4.2}$$

obeys the Maxwell field equations in the absence of currents:

$$F_a{}^b{}_{;b} = 0, \tag{2.4.3}$$
$$F_{[ab;c]} = F_{[ab,c]} = 0. \tag{2.4.4}$$

Eqn (2.4.4) implies that, at least in a local region of space–time, F_{ab} can be written in the form

$$F_{ab} = 2A_{[b;a]} = 2A_{[b,a]}. \tag{2.4.5}$$

The Einstein field equations are

$$R_{ab} - \frac{1}{2} g_{ab} R = 8\pi T_{ab}, \tag{2.4.6}$$

with

$$T_{ab} = \frac{1}{4\pi}\left(F_{ac}F_b{}^c - \frac{1}{4} g_{ab} F_{cd} F^{cd}\right). \tag{2.4.7}$$

In the Reissner–Nordström case, the only non-zero potential and field components are

$$A_t = -\frac{Q}{r}, \tag{2.4.8}$$

$$F_{tr} = -F_{rt} = -\frac{Q}{r^2}. \tag{2.4.9}$$

From the $1/r$ part of the potential A_t, one reads off that Q is the charge of the space–time. Similarly, from the $1/r$ part of the metric component g_{tt} one finds that M is the mass of the space–time. This is because the geodesic equation describing the Newtonian motion of non-relativistic bodies in the far field depends only on the far-field g_{tt}, which can be identified with $-1 - 2\phi$, where ϕ is the Newtonian potential (Misner et al. 1973).

As with the Schwarzschild case, one can consider extensions of the Reissner–Nordström geometry. This will help us later in studying extensions of the Kerr

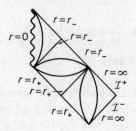

Fɪɢ. 2.14. A Penrose conformal diagram showing the region in the Reissner–Nordström space–time covered by ingoing coordinates $(v, r*)$.

metric. First consider the case $M >| Q |$. There are coordinate singularities in
eqn (2.4.1) at

$$r = r_{\pm} = M \pm (M^2 - Q^2)^{\frac{1}{2}}. \tag{2.4.10}$$

To make an ingoing extension, define

$$
\begin{aligned}
r^* &= \int dr / \left(1 - \frac{2M}{r} + \frac{Q^2}{r^2}\right) \\
&= r + \frac{r_+^{\,2}}{(r_+ - r_-)} \log(r - r_+) - \frac{r_-^{\,2}}{(r_+ - r_-)} \log(r - r_-),
\end{aligned}
\tag{2.4.11}
$$

for $r > r_+$. Proceeding by analogy with the Schwarzschild case, one defines
$v = t + r^*$. Then the metric takes the form

$$ds^2 = 2dvdr - \frac{(r - r_-)(r - r_+)}{r^2}dv^2 + r^2(d\theta^2 + \sin^2\theta\, d\phi^2). \tag{2.4.12}$$

This provides an ingoing extension across the null surfaces $\{r = r_+\}, \{r = r_-\}$.
As with the Schwarzschild geometry, there is a curvature singularity at $r = 0$. The
Penrose diagram of the part of the space–time covered by the coordinates (v, r)
is shown in Fig. 2.14. Because of the time-symmetry of the Reissner–Nordström
geometry, one can also make the outgoing extension, by defining $u = t - r^*$. This
has the Penrose diagram given in Fig. 2.15. Combining the null coordinates u
and v, one has the double null form

$$ds^2 = -\frac{(r - r_+)(r - r_-)}{r^2}dudv + r^2(d\theta^2 + \sin^2\theta\, d\phi^2). \tag{2.4.13}$$

One can find Kruskal-type transformations $u \to U^{\pm}(u), v \to V^{\pm}(v)$ (Hawking
and Ellis 1973). These lead to the conformal diagrams of Figs 2.16 and 2.17.
When these methods of extension are combined, one arrives at the Penrose
conformal diagram of Fig. 2.18.

One can check that this extension is maximal, in that all geodesics are inex-
tendible; any geodesic either runs off to infinity with infinite affine parameter,

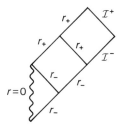

FIG. 2.15. The region in the Reissner–Nordström geometry covered by outgoing co-ordinates $(u, r*)$.

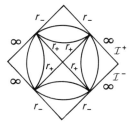

FIG. 2.16. One basic building block of the Reissner–Nordström space–time.

or else hits the singularity at $r = 0$ at a finite affine parameter. Figure 2.18 shows that the maximally extended Reissner–Nordström geometry consists of an infinite chain of asymptotic regions connected by 'wormholes' between the singularities. It also shows that the space–time contains a black hole, with event horizon at $r = r_+$; further, the event horizon is also a Killing horizon in the sense of Section 2.3. In contrast to the case of the Schwarzschild geometry, the singularities $r = 0$ here are timelike. This behaviour is connected with the existence

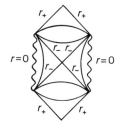

FIG. 2.17. The other basic building block of the Reissner–Nordström space–time.

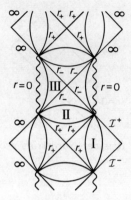

FIG. 2.18. The maximally extended Reissner–Nordström space–time.

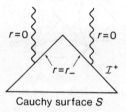

Cauchy surface S

FIG. 2.19. A surface S in the Reissner–Nordström geometry is shown, which is a Cauchy surface only for the region of the space–time up to the surface $r = r_-$. We say that there is a Cauchy horizon in the space–time.

of a Cauchy horizon at $r = r_-$; see Fig. 2.19. One cannot predict what happens beyond the surfaces $r = r_-$ in the figure, for a matter field with prescribed initial data on the Cauchy surface S. Because an infinite amount of proper distance on S corresponds to a small finite amount near the Cauchy horizon $r = r_-$, one expects that perturbations on S will be infinitely blue-shifted on $r = r_-$. Hence non-spherical collapse should produce infinite curvature before $r = r_-$ is reached.

One can visualize the intrinsic geometry of the hypersurfaces of constant t, such as the surface S above, as in the Schwarzschild case. Taking the equator $\theta = \pi/2$, the intrinsic metric is as in Fig. 2.10, except for the throat being somewhat longer and narrower. As Q^2 increases towards M^2, the throat becomes arbitrarily long.

Now consider the case $Q^2 = M^2$, known as the extreme Reissner–Nordström geometry. Here the roots r_+, r_- coincide:

$$r_+ = r_- = M. \tag{2.4.14}$$

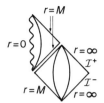

FIG. 2.20. The portion of the extreme $Q^2 = M^2$ Reissner–Nordström geometry covered by ingoing coordinates (v, r).

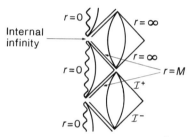

FIG. 2.21. The maximal extension of the extreme $Q^2 = M^2$ Reissner–Nordström geometry.

By analogy with earlier cases, define

$$r* = \int dr / \left(1 - \frac{M}{r}\right)^2$$

$$= r + 2M \log(r - M) - \frac{M^2}{(r - M)}. \tag{2.4.15}$$

With

$$v = t + r^*, \tag{2.4.16}$$

one has the ingoing extension corresponding to the metric

$$ds^2 = 2dvdr - \frac{(r - M)^2}{r^2} dv^2 + r^2 (d\theta^2 + \sin^2\theta\, d\phi^2). \tag{2.4.17}$$

The Penrose diagram of the ingoing extension is given in Fig. 2.20, and the maximal extension is shown in Fig. 2.21.

For $Q^2 = M^2$, the Reissner–Nordström geometry has a black hole, with event horizon at $r = M$. As can be seen from Figs 2.20 and 2.21, the Killing horizon

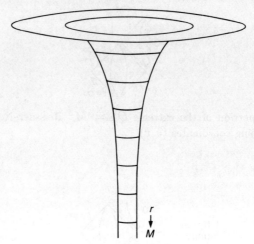

FIG. 2.22. A diagram of the equator $\theta = \pi/2$ in a section of constant t for the extreme Reissner–Nordström geometry, shown embedded in flat $\mathbb{R}^3$. As $r \to M$, an infinite tube develops.

at $r = M$ is degenerate, in the sense that the corresponding time-translation Killing vector field k^a obeys

$$(k^b k_b)_{;a} = 0 \tag{2.4.18}$$

there. Because of the Killing equation $k_{a;b} + k_{b;a} = 0$, one deduces

$$k^b k_{a;b} = 0 \tag{2.4.19}$$

at the horizon. Thus the horizon null generators with tangent $k^a = dx^a/dv$ obey the geodesic equation with affine parameter v. Hence the horizon null generators have infinite affine length to the past, since $v \to -\infty$ as one moves to the past along a generator. This part of the extreme Reissner–Nordström geometry constitutes an 'internal infinity'. To see this in another way, consider a spacelike surface of constant t. On the equator $\theta = \pi/2$, this has the metric

$$ds^2 = \frac{dr^2}{\left(1 - \frac{r}{M}\right)^2} + r^2 d\phi^2. \tag{2.4.20}$$

The spatial geometry becomes that of a tube of constant radius as $r \to M$, as shown in Fig. 2.22.

Finally, in the case $Q^2 > M^2$, the coordinates (t, r, θ, ϕ) cover the region $0 < r < \infty$, which gives the entire Penrose diagram shown in Fig. 2.23. There is no black hole in this case, but there is a naked singularity. However, one expects that in the case of a charged matter collapse with $Q^2 > M^2$, the charge will

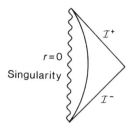

FIG. 2.23. In the Reissner–Nordström geometry with $Q^2 > M^2$, there is a naked singularity and no black hole.

prevent the formation of a singularity in the matter region, visible from infinity (Hawking and Ellis 1973).

2.5 The Kerr metric

2.5.1 *Maximal extensions*

The Kerr metrics are rotating generalizations of the Schwarzschild geometry, and are the unique black-hole endstates of uncharged stellar collapse (Hawking and Ellis 1973; Robinson 1975; Wald 1984). Further, they are stable against small perturbations (Whiting 1989). These properties guarantee the astrophysical importance of the Kerr geometry. If charge is included in the system, as well as mass and angular momentum, then one has the Kerr–Newman solution (Misner *et al.* 1973; Wald 1984), which generalizes both the Kerr and the Reissner–Nordström solutions.

The Kerr metric is

$$ds^2 = \rho^2 \left(\frac{dr^2}{\Delta} + d\theta^2 \right) + (r^2 + a^2)\sin^2\theta \, d\phi^2$$
$$- dt^2 + \frac{2Mr}{\rho^2}(a\sin^2\theta \, d\phi - dt)^2, \tag{2.5.1.1}$$

where

$$\rho^2(r,\theta) = r^2 + a^2\cos^2\theta,$$
$$\Delta(r) = r^2 - 2Mr + a^2, \tag{2.5.1.2}$$

in Boyer–Lindquist coordinates (t, r, θ, ϕ) (Boyer and Lindquist 1967). This is a solution of the vacuum Einstein equations (Kerr 1963). One can interpret the two parameters M, a in the solution by examining the metric in the asymptotic region in which $r \to \infty$. First, the r^{-1} part of g_{tt} is $2M/r$. As in Section 2.4, this implies that M is the mass of the Kerr geometry, as measured at spatial infinity (Misner *et al.* 1973). Second, one can deduce that the angular momentum in the space–time is $\mathbf{S} = aM\mathbf{e_z}$, where $\mathbf{e_z}$ is a unit vector at infinity, pointing in the z-direction. This follows from studying the time–space metric cross-term $g_{t\phi}$.

Changing to 'Cartesian' coordinates $x = r \sin \theta \cos \phi, y = r \sin \theta \sin \phi, z = r \cos \theta$, the cross-terms are

$$\frac{4aM}{r^3}(ydx - xdy)dt + O(r^{-3}) = -4\epsilon_{jkl}S^k \frac{x^l}{r^3}dx^j dt + O(r^{-3}). \qquad (2.5.1.3)$$

For weak gravitational fields, it can be proved that S^k so defined equals the angular momentum of the matter in the space–time (Misner *et al.* 1973). For strong gravitational fields, one defines the angular momentum of an isolated system, such as is described by the Kerr geometry, to be S^k as in eqn (2.5.1.3), from the $O(r^{-2})$ part of g_{tj}.

The mass and spin of a weakly asymptotically simple space–time, such as the Kerr geometry (see below) or geometries fairly close to it, do not correspond to propagating gravitational perturbation modes. Only the quadrupole and higher $\ell \geq 2$ modes propagate. One is led to a scenario for rotating gravitational collapse in which the geometry at late times is nearly Kerr plus $\ell \geq 2$ modes which radiate both out to infinity and down the black hole—see Section 2.7 for a description of gravitational and other perturbations, as found by Teukolsky (1973).

Now consider the symmetries of the Kerr geometry. As with the Schwarzschild and Reissner–Nordström geometries, there is a time-translation Killing vector field $k^a \equiv \partial/\partial t$. This obeys $k^a k_a < 0$ at large r. For these reasons we say that the Kerr metric is stationary. But, unlike the examples of the Schwarzschild and Reissner–Nordström geometries, the Kerr geometry is not static. As suggested above, we adopt the definition that a space–time is static if there is a timelike Killing vector field k^a which is orthogonal to a family of hypersurfaces. Equivalently, the space–time is static if there is a timelike Killing vector field k^a which can be written in the form $k_a = fh_{,a}$, where f and h are functions; here the contours $h = $ const. give the family of hypersurfaces. Equivalently again, the space–time is static if there is a timelike Killing vector field k^a which obeys

$$k_{[a;b}k_{c]} = 0. \qquad (2.5.1.4)$$

The latter condition is equivalent to the others by virtue of Frobenius' theorem (Wald 1984). One can verify that the expression on the left-hand side of eqn (2.5.1.4) is non-zero for the time-translation Killing vector field k^a of the Kerr metric. Hence the Kerr geometry with $a \neq 0$ is stationary but not static.

The other Killing vector field is the rotational Killing vector field $m^a \equiv \partial/\partial\phi$. At large r, one has $m^a m_a \geq 0$; note that $m^a = 0$ on the symmetry axis. The two vector fields k^a and m^a commute:

$$m^b k^a{}_{;b} - k^b m^a{}_{;b} = (\mathcal{L}_m k)^a = -(\mathcal{L}_k m)^a = 0, \qquad (2.5.1.5)$$

where $\mathcal{L}$ denotes the Lie derivative (Wald 1984), since

$$k^a \equiv \partial/\partial t, \qquad m^a \equiv \partial/\partial\phi. \qquad (2.5.1.6)$$

Correspondingly, the integral curves to which k^a and m^a are tangent lie in flat two-dimensional surfaces with $r = $ const.$, \theta = $ const. There are also two discrete isometries, one interchanging $(t, \phi) \rightarrow (-t, -\phi)$, and one reflecting in the equatorial plane $\theta \rightarrow \pi - \theta$.

As in the case of the Reissner–Nordström geometry, one can find the maximal extension of the Kerr geometry. First, suppose $a^2 < M^2$. There are coordinate singularities at $\Delta(r) = r^2 - 2Mr + a^2 = 0$. These correspond to the zeros

$$r = r_\pm = M \pm \sqrt{(M^2 - a^2)}. \tag{2.5.1.7}$$

One defines the ingoing Kerr coordinates (r, θ, ϕ_+, u_+) by

$$
\begin{aligned}
du_+ &= dt + \frac{(r^2 + a^2)}{\Delta} dr, \\
d\phi_+ &= d\phi + \frac{a}{\Delta} dr.
\end{aligned}
\tag{2.5.1.8}
$$

The resulting metric is

$$
\begin{aligned}
ds^2 &= \rho^2 d\theta^2 - 2a \sin^2\theta \, drd\phi_+ + 2drdu_+ \\
&\quad + X d\phi_+{}^2 + 2W d\phi_+ du_+ - V du_+{}^2,
\end{aligned}
\tag{2.5.1.9}
$$

where

$$X = g_{\phi\phi}, \qquad W = g_{t\phi}, \qquad V = -g_{tt}. \tag{2.5.1.10}$$

Note here that $(\partial/\partial r)\,|_{\theta,\phi_+,u_+}$ is a null vector, and in fact a line with (θ, ϕ_+, u_+) constant gives an ingoing null geodesic. However, the coordinate u_+ is not null: the surfaces with $u_+ = $ constant are timelike, since the three-metric $^{(3)}g_{ab}$ obeys $\det(^{(3)}g_{ab}) = -\rho^2 a^2 \sin^4\theta < 0$.

The ingoing metric (2.5.1.9) is well-behaved for $-\infty < r < \infty, -\infty < u_+ < \infty, 0 \leq \theta \leq \pi, 0 \leq \phi_+ < 2\pi$, except for points on the axis $\theta = 0, \pi$—where a coordinate transformation can be found such that the metric is regular there (Carter 1973)—and $\rho^2 = r^2 + a^2 \cos^2\theta = 0$, which gives what can be shown to be a curvature singularity at $r = 0, \theta = \pi/2$(Carter 1973). Further investigation (Carter 1973; Hawking and Ellis 1973) shows that the curvature singularity is in fact a ring.

It is helpful, in considering the ingoing extension of the Kerr metric, to consider the surfaces with $r = $ const. For these surfaces,

$$\det(^{(3)}g_{ab}) = -\rho^2(VX + W^2) = -\sin^2\theta(r^2 - 2Mr + a^2)\rho^2. \tag{2.5.1.11}$$

These surfaces are timelike for $r > r_+$, spacelike for $r_- < r < r_+$, and timelike for $r < r_-$. This is analogous to the behaviour of the Reissner–Nordström metrics in Section 2.4. When $r = r_\pm$, the intrinsic three-dimensional geometry is

$$ds^2 = (r^2 + a^2\cos^2\theta)d\theta^2 + \frac{\sin^2\theta}{(r^2 + a^2\cos^2\theta)}(2Mrd\phi_+ - adu_+)^2, \tag{2.5.1.12}$$

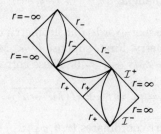

FIG. 2.24. The region of the axis $\theta = 0$ in the Kerr solution with $a < M$ covered by ingoing coordinates (u_+, r).

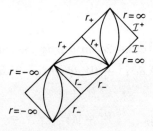

FIG. 2.25. The region of the Kerr geometry with $a < M$ covered by outgoing coordinates (u_-, r).

and hence the surfaces with $r = r_\pm$ are null. The Penrose conformal diagram of the axis $\theta = 0$, with r and u_+ varying, is given in Fig. 2.24. Similarly, one can work with outgoing Kerr coordinates (r, θ, ϕ_-, u_-), defined by

$$du_- = dt - (r^2 + a^2)\Delta^{-1}dr,$$
$$d\phi_- = d\phi - a\Delta^{-1}dr. \tag{2.5.1.13}$$

This leads to the Penrose diagram in Fig. 2.25. Then, by analogy with the Reissner–Nordström case in Section 2.4, one can patch together regions of type I $(r_+ < r < +\infty)$, type II $(r_- < r < r_+)$, and type III $(-\infty < r < r_-)$ to obtain the conformal diagram of the symmetry axis in Fig. 2.26. The structure at other polar angles is similar. One can show, by studying the Penrose diagram and corresponding properties at other polar angles, that the Kerr geometry describes a black hole with event horizon at $r = r_+$ (Carter 1973). This will be discussed further in the following subsection 2.5.2.

In a realistic rotating matter collapse, as mentioned earlier, the geometry in the vacuum region outside the matter is expected to be only approximately equal to the Kerr metric. The geometry will only tend to Kerr at late times after the wavelike perturbations have propagated away to $\mathcal{I}^+$ and into the black

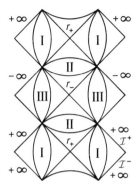

FIG. 2.26. A conformal diagram of the symmetry axis $\theta = 0$ in the maximally extended Kerr space–time with $a^2 < M^2$.

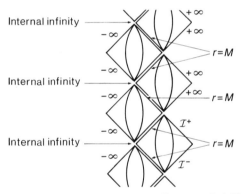

FIG. 2.27. A conformal diagram of the maximally extended Kerr geometry for the extreme case $a^2 = M^2$. There is a degenerate horizon, with corresponding internal infinity.

hole. Further, as in the Reissner–Nordström case, there is a Cauchy horizon at $r = r_-$ which is unstable against the formation of a nearly infinite amount of curvature, due to the infinite blue-shifting of perturbations on an initial Cauchy surface. Hence one expects that the region in the Kerr geometry with $r < r_-$ is not relevant in a realistic matter collapse; in particular, the region around the Kerr singularity at $r = 0$ is expected to be unphysical.

As in the Reissner–Nordström case, there is an analogous extreme Kerr solution, with $a^2 = M^2$. One can again use Kerr coordinates to extend the exterior geometry, giving the Penrose diagram of Fig. 2.27. There is a black hole with event horizon at $r = M$. The horizon has infinite affine length in both directions,

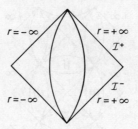

FIG. 2.28. The Kerr space–time for $a^2 > M^2$. There is no black hole, and the singularity at $\rho^2 = 0$ is visible from $\mathcal{I}^+$.

corresponding to an 'internal infinity', such as that sketched in Fig. 2.22.

For the case $a^2 > M^2$, the Penrose conformal diagram is depicted in Fig. 2.28. There is no black hole, and the singularity at $\rho^2 = 0$ is visible from $\mathcal{I}^+$—i.e. there is a naked singularity. If the cosmic censorship conjecture holds, then centrifugal forces are expected to prevent the formation of a Kerr geometry with $a^2 > M^2$. Centrifugal forces might force the rotating body to shed its outer regions, so that it then obeys $a^2 < M^2$, and can collapse to a Kerr black hole. This problem can only be treated numerically, and it will be very interesting to see the results of numerical work on rotating collapse (or indeed collision) over the next few years.

2.5.2 *Ergoregion and event horizon*

In this subsection, we restrict attention to the case $a^2 < M^2$. Consider

$$V = -k^a k_a = -g_{tt} = 1 - \frac{2Mr}{(r^2 + a^2 \cos^2\theta)}. \qquad (2.5.2.1)$$

Note that V becomes negative outside the event horizon, for

$$r_+ \leq r < M + \sqrt{(M^2 - a^2\cos^2\theta)}. \qquad (2.5.2.2)$$

This region is called the *ergoregion* or *ergosphere*. The outer boundary

$$r = M + \sqrt{(M^2 - a^2\cos^2\theta)} \qquad (2.5.2.3)$$

is known as the (outer) *ergosurface*. A schematic representation of the ergosurface in relation to the event horizon is given in Fig. 2.29. The two surfaces touch at the poles. Inside the ergosurface, particles cannot exist with r, θ, ϕ held constant. To view this in more detail, draw the equatorial plane with projected null cones. The projection is via the Killing vector field k^a, as indicated in Fig. 2.30. That is, one picks a spacelike hypersurface, and maps it to another, infinitesimally close, by moving a vector separation $\delta t\, k^a$; then one considers the light cone of a typical point P on the initial hypersurface, and projects this back from the final to the initial hypersurface. The result is sketched in Fig. 2.31. One finds

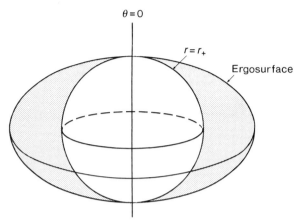

FIG. 2.29. The ergosurface and the event horizon of the Kerr geometry with $a^2 < M^2$.

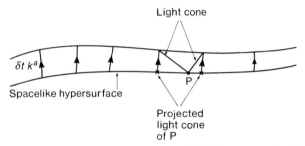

FIG. 2.30. Given a spacelike hypersurface, one finds the projected null cones by dragging along the Killing vector field k^a.

the following:

- (a) Outside the ergosurface, all spatial directions are permitted, but the null cones are partially 'dragged around' by the black hole's rotation.

- (b) At the ergosurface, the projected null cone includes the source point. One has $d\phi/dt > 0$ for timelike orbits, but dr/dt may be of either sign (in fact, the ergosurface is a timelike three-surface).

- (c) Between the event horizon and the ergosurface, one cannot have (r, θ, ϕ) constant on a non-spacelike orbit. For further understanding, consider the invariants

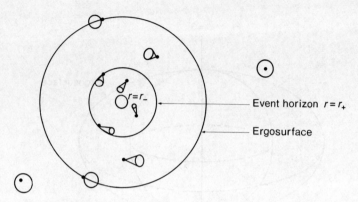

FIG. 2.31. The projected null cones in the equatorial plane of the Kerr geometry. Inside the ergosurface, the null cones are swept around by the rotation of the black hole. Inside the future event horizon $r = r_+$, the cones tip inwards.

$$X = m^a m_a = g_{\phi\phi} = (r^2 + a^2)\sin^2\theta + \frac{2Ma^2 r \sin^4\theta}{(r^2 + a^2\cos^2\theta)},$$

$$W = k^a m_a = g_{t\phi} = \frac{-2Mar\sin^2\theta}{(r^2 + a^2\cos^2\theta)}. \tag{2.5.2.4}$$

From the metric (2.5.1.1) in Boyer–Lindquist coordinates, one finds that

$$X d\phi^2 + 2W d\phi dt - V dt^2 \le 0, \tag{2.5.2.5}$$

for a non-spacelike curve. But $X > 0, W < 0, V < 0$ in the ergoregion. Hence $d\phi/dt > 0$, so that a particle must co-rotate with the black hole in the ergoregion. As in (b) above, dr/dt can have either sign, so that, for example, a particle can still escape to infinity. The determinant

$$VX + W^2 = \sin^2\theta(r^2 - 2Mr + a^2) \tag{2.5.2.6}$$

is positive in the ergoregion, showing that the two-dimensional surface with parameters t and ϕ is timelike.

- (d) At the surface with $r = r_+$, the null cone just touches the surface. On future-directed non-spacelike paths, one has $dr \le 0$, with $dr = 0$ only for a null curve along the horizon. It can be seen from Fig. 2.31 that $r = r_+$ gives the future event horizon of the Kerr space–time.
- (e) Inside the horizon, with $r_+ > r > r_-$, r must decrease along future directed non-spacelike paths, since the hypersurfaces with $r = $ const. are spacelike there.

Now consider the *intrinsic geometry* of the event horizon. This is given by the degenerate three-metric

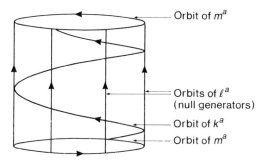

FIG. 2.32. The orbits of the Killing vectors k^a and m^a on an equatorial section of the event horizon. The null generators of the horizon are directed along the Killing vector field ℓ^a.

$$ds^2 = (r_+{}^2 + a^2\cos^2\theta) + \frac{\sin^2\theta}{(r_+{}^2 + a^2\cos^2\theta)}(2Mr_+d\phi_+ - adu_+)^2. \qquad (2.5.2.7)$$

The null geodesic generators of the horizon have tangents

$$\frac{\partial}{\partial u_+} + \frac{a}{2Mr_+}\frac{\partial}{\partial \phi_+} \equiv k^a + \Omega m^a = \ell^a, \qquad (2.5.2.8)$$

say, where

$$\Omega = \frac{a}{2Mr_+} \qquad (2.5.2.9)$$

is the *angular velocity* of the black hole. Here Ω is constant over the horizon; this is an example of the *rigidity* of the rotation of the event horizon of a stationary (non-static) space–time, which is a general property of rotating black holes (Carter 1973; Hawking and Ellis 1973). Note that the event horizon is a Killing horizon, because Ω is constant. Note also that both k^a and m^a are tangent to the horizon, but are spacelike there. This can be visualized as in Fig. 2.32 by drawing an equatorial section of the horizon with $\theta = \pi/2$. One can see that the horizon, with null generators ℓ^a, rotates with respect to infinity, as given by the spacelike generators k^a.

One can also examine the intrinsic spatial geometry of the horizon. Consider the two-dimensional section with $u_+ = $ const. This has the intrinsic positive-definite geometry

$$ds^2 = (r_+{}^2 + a^2\cos^2\theta)d\theta^2 + \frac{4M^2(r_+)^2\sin^2\theta}{(r_+{}^2 + a^2\cos^2\theta)}d\phi_+{}^2. \qquad (2.5.2.10)$$

The embedding of this surface in flat $\mathbb{R}^3$, as in Fig. 2.33, shows that there is an equatorial bulge, as one might have expected. The area of the instantaneous two-geometry (2.5.2.10) is

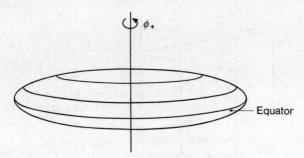

FIG. 2.33. An embedding diagram in flat Euclidean three-space of the intrinsic geometry of a Kerr black hole. Note the flattening at the poles.

$$A = \int \int d\theta d\phi_+ \{\det[g_{ab}(\theta, \phi_+)]\}^{\frac{1}{2}}$$
$$= 8\pi M r_+ \tag{2.5.2.11}$$
$$= 8\pi M[M + (M^2 - a^2)^{\frac{1}{2}}].$$

The quantity $\mathcal{A}$ is called the *area of the event horizon*; note that is the same for any section through the horizon, of the form $\{\theta, \phi_+, u_+ = f(\theta, \phi)\}$, namely

$$\int_0^\pi \int_0^{2\pi} d\theta d\phi_+ (2Mr_+ \sin\theta - a\frac{\partial f}{\partial\phi_+}\sin\theta) = 8\pi M r_+. \tag{2.5.2.12}$$

The area and its variations under small perturbations of the black hole (such as infalling test particles) will be studied in the following subsection 2.5.3.

2.5.3 *Extraction of rotational energy; reversible and irreversible transformations*

Consider now a process known as the Penrose process for extracting energy from rotating black holes. Note that in a stationary space–time, with time-translation Killing vector field k^a, the quantity $E = -k_a p^a$ is constant along the geodesic path of a test particle with four-momentum p^a. This can be verified directly from the geodesic equation and Killing vector equation:

$$p^b(k_a p^a)_{;b} = p^b p^a k_{a;b} + k_a p^b p^a{}_{;b}$$
$$= 0, \tag{2.5.3.1}$$

where the first term vanishes by the Killing equation $k_{(a;b)} = 0$ and the second term vanishes by the geodesic equation. For a particle at large distances from the black hole, where the geometry is nearly flat, E is essentially the relativistic energy (mass) γ of the particle, where $\gamma = \gamma(v) = (1 - v^2)^{-1/2}$, with v being the speed of the particle near infinity. Since E is conserved in the particle's

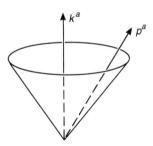

FIG. 2.34. Outside the ergosurface, the energy $E = -k_a p^a > 0$ for future-directed momentum p^a.

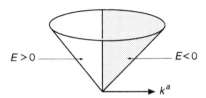

FIG. 2.35. In the ergoregion, $E = -k_a p^a$ can have either sign.

motion, one also defines E to be the particle's energy in general. Now k^a is timelike outside the ergosurface, so that $E > 0$ there for a future-directed timelike momentum p^a (see Fig. 2.34). But in the ergoregion k^a is spacelike, and $E = -k_a p^a$ can have either sign (see Fig. 2.35). The condition $E < 0$ means that the particle's binding energy exceeds its rest energy. One can now consider a process which utilizes such a particle.

As in Fig. 2.36, take the process in which a particle A is sent on a timelike

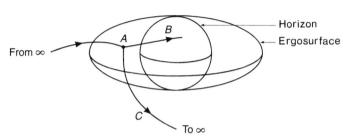

FIG. 2.36. This process will extract energy from the black hole, provided that $E_B < 0$.

geodesic from infinity, so that it has positive energy E_A, such that it enters the ergoregion. There it explodes into two particles, B and C, where B then falls inside the horizon, while C moves back out to infinity. As described above, it can be arranged that $E_B < 0$. By local energy–momentum conservation, one has

$$(p_A^a)_{\text{just before explosion}} = (p_B^a + p_C^a)_{\text{just after}}. \qquad (2.5.3.2)$$

Hence

$$\begin{aligned}
E_A &= E_B + E_C, \\
L_{zA} &= L_{zB} + L_{zC},
\end{aligned} \qquad (2.5.3.3)$$

where $L_z = p_a m^a$, with m^a being the rotational Killing vector; here L_z is the angular momentum of the test particle. The angular momentum result will be needed shortly. One can also use global energy conservation (Misner *et al.* 1973), noting that energy is additive for widely separated non-interacting systems, to write

$$M_{\text{hole,initial}} + E_A = M_{\text{hole,final}} + E_C \qquad (2.5.3.4)$$

and hence deduce

$$\delta M_{hole} = E_B. \qquad (2.5.3.5)$$

There is also a global conservation law for linear momentum. There will further be gravitational waves generated during the process, some of which will fall down the hole, and others which will be emitted to infinity; these are of quadratic and higher order in the small masses, and so do not contribute at the lowest, linear order being studied here. Similarly, global angular momentum conservation gives

$$\delta L_{zhole} = L_{zB}. \qquad (2.5.3.6)$$

As remarked above, one can arrange that $E_B < 0$. Then

$$\delta M_{\text{hole}} < 0 \qquad (2.5.3.7)$$

and

$$E_C > E_A. \qquad (2.5.3.8)$$

In such a process, observers near infinity can gain energy from the rotating black hole.

One can now find efficiency bounds on a process of the above type. That is, one can find constraints on the change of the black hole's energy and angular momentum, when a particle falls through the horizon on a geodesic. Let ℓ^a be the Killing vector

$$\ell^a = k^a + \Omega m^a \qquad (2.5.3.9)$$

tangent to the null generators of the event horizon, as in Fig. 2.37(a). One sees

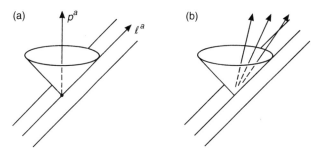

FIG. 2.37. In (a), a massive particle falling into the Kerr black hole leads to an irreversible transformation with $\delta\mathcal{A} > 0$, where $\mathcal{A}$ is the area of the black hole. In (b), a sequence of irreversible transformations can only approximate reversibility.

from Fig. 2.37(a) that

$$p_a\ell^a \leq 0, \tag{2.5.3.10}$$

with equality only if p^a is directed along ℓ^a. From the definitions $E = -p_a k^a$ and $L_z = p_a m^a$ and from eqn (2.5.3.9), one finds

$$E \geq \Omega L_z \tag{2.5.3.11}$$

for the infalling particle. Equality holds only for a massless particle directed along a horizon null generator, in which case the particle is not actually falling in. Since $\Omega = a/2Mr_+$, one has for an infinitesimal transformation of a Kerr black hole by means of test particles:

$$\delta M \geq \frac{\mathbf{a}.\delta\mathbf{S}}{2Mr_+}, \tag{2.5.3.12}$$

where

$$\mathbf{S} = M\mathbf{a}. \tag{2.5.3.13}$$

In the Penrose process, $\delta M < 0$ implies that $\mathbf{a}.\delta\mathbf{S} < 0$. Thus one can only gain energy by slowing down the black hole's rotation; the extracted energy comes from the rotational energy of the black hole. The inequality (2.5.3.12) can most effectively be expressed in terms of variations of the area of the black hole. As shown in subsection 2.5.2, the area of the horizon of a Kerr geometry is

$$\mathcal{A} = 8\pi M[M + (M^2 - a^2)^{\frac{1}{2}}]. \tag{2.5.3.14}$$

The variation of $\mathcal{A}$ is then

$$\delta\mathcal{A} = \frac{8\pi}{(M^2 - a^2)^{\frac{1}{2}}}(2Mr_+\delta M - \mathbf{a}.\delta\mathbf{S}). \tag{2.5.3.15}$$

Hence, from the inequality (2.5.3.12), one has

$$\delta \mathcal{A} \geq 0 \qquad\qquad (2.5.3.16)$$

in any process involving test particles. A transformation with $\delta \mathcal{A} = 0$ is *reversible*. A transformation with $\delta \mathcal{A} > 0$ is *irreversible*. As can be seen from Fig. 2.37b, transformations using massive particles must be irreversible; one can only approximate reversibility.

The expression (2.5.3.14) for the area can be rewritten as

$$M^2 = \frac{\mathcal{A}}{16\pi} + \frac{4\pi S^2}{\mathcal{A}}. \qquad\qquad (2.5.3.17)$$

The first term on the right-hand side is the 'irreducible contribution' to M^2; this can be held almost constant by making almost reversible transformations. The second term on the right-hand side of eqn (2.5.3.17) is the 'rotational energy contribution' to M^2; this can be reduced to zero reversibly. Hence, if we define the irreducible mass of the black hole as

$$M_{ir} = \sqrt{(\mathcal{A}/16\pi)}, \qquad\qquad (2.5.3.18)$$

we cannot reduce M below M_{ir}. We can only extract the maximum energy $M - M_{ir}$ by a sequence of reversible transformations leading to $a = 0$ (the Schwarzschild metric). Thus the Schwarzschild geometry can be regarded as a ground state of the Kerr geometries.

One can consider more general uncharged rotating stationary black holes, which might have an equilibrium distribution of rotating matter outside the horizon. It can be shown (Carter 1973, 1979) then that

$$\delta M = \frac{\kappa}{8\pi}\delta \mathcal{A} + \mathbf{\Omega} . \, \delta \mathbf{S}, \qquad\qquad (2.5.3.19)$$

the *first law of black hole mechanics* (which can be generalized to the charged case). Here κ is known as the *surface gravity*, and is constant over the horizon (the *zeroth law* of black-hole mechanics). For the Kerr case, κ can be read off from eqn (2.5.3.15). For general stationary black holes, the angular velocity Ω is also constant over the horizon ('rigidity'). Up to constant factors, κ corresponds to a temperature for the black hole, and $\mathcal{A}$ corresponds to an entropy. Further, for any dynamical black hole, $\delta \mathcal{A} \geq 0$, the *second law*, from the global theory of black holes (Penrose 1971; Hawking 1972a; Hawking and Ellis 1973). The *third law* states that one cannot attain an extremal ($\kappa = 0$) black hole such as the Kerr metric with $a = M$ by a finite number of steps, starting from $a < M$.

Quantum field theory (Hawking 1975) shows that black holes emit particles thermally, with temperature

$$T = \kappa \hbar / 2\pi k c, \qquad\qquad (2.5.3.20)$$

where k is Boltzmann's constant and κ has dimensions of acceleration. The corresponding entropy is

$$S = c^3 A / 4G\hbar. \qquad (2.5.3.21)$$

These results show that a black hole can be in equilibrium with a heat bath (Hartle and Hawking 1976), and one can determine the temperature of the black hole and its contribution to the entropy.

The area increase theorem $\delta A \geq 0$ (Hawking and Ellis 1973) makes the assumption of cosmic censorship, i.e. that all singularities of the space–time are hidden inside the black hole. One can use this theorem, assuming cosmic censorship, to set bounds on the fraction of the initial mass–energy which is emitted in gravitational waves in a collision and merger of two black holes. Suppose (for simplicity) that one starts with a pair of widely separated Schwarzschild black holes at rest, each having mass M_1, and that finally one has a single black hole at rest, of mass M_2, together with gravitational radiation propagating out towards future null infinity $\mathcal{I}^+$. Let $A_{\text{final}} = 16\pi(M_2)^2$ be the area of the event horizon of the final black hole, which is assumed to be a Schwarzschild black hole, and let $A_1 = 16\pi(M_1)^2$ be the initial area of each black hole. Then the area increase theorem gives $A_{\text{final}} > 2A_1$, i.e.

$$(M_2)^2 > 2(M_1)^2, \qquad (2.5.3.22)$$

i.e. $M_2 > \sqrt{2}M_1$. The energy M_2 must also be less than or equal to the initial energy $2M$, since the sum of M_2 and the final energy in gravitational waves must equal $2M$. The efficiency ϵ of production of gravitational waves is given by

$$\epsilon = \frac{(2M_1 - M_2)}{2M_1} = 1 - \frac{M_2}{2M_2} < 1 - \frac{1}{\sqrt{2}} \simeq 29\%. \qquad (2.5.3.23)$$

The numerical calculations of Smarr (1979) show that, for initial data given by two time-symmetric black holes (i.e. two black holes initially at rest), the true efficiency of gravitational wave production is several orders of magnitude smaller than the bound (2.5.3.23) on ϵ.

If one takes the other extreme, in which two black holes collide at or near the speed of light, then Penrose (1974) (see Chapter 6) has shown that there is the same bound of 29% on the efficiency of production of gravitational waves, assuming cosmic censorship. However, in this case the bound of 29% based on initial data gives a much more reasonable estimate of the exact efficiency. It is found in Chapter 8 that a good estimate of the radiation described by the first two angular harmonics corresponds to 16.8% efficiency for the speed-of-light head-on collision.

2.6 Geodesic equations in the Kerr metric

Geodesics in the Kerr metric are an example of perturbations of the Kerr geometry. For example, as we shall see in Chapter 3, if one has a low-mass ('small') Kerr black hole moving in a background universe, then the 'small' black hole moves approximately on a timelike geodesic of the background universe, and its spin is parallelly propagated along the geodesic. In the case of the interaction of a low-mass Kerr black hole with a large-mass Kerr black hole (Chapter 4), the low-mass black hole will move to a good approximation along a geodesic in the large-mass 'background' Kerr geometry.

Geodesic particle motion may be derived from the Hamiltonian

$$H = \frac{1}{2}g^{ab}p_a p_a, \tag{2.6.1}$$

where

$$p_a = g_{ab}(dx^b/d\lambda) \tag{2.6.2}$$

is the canonical momentum, and λ is an affine parameter along the particle path. The Hamiltonian equations of motion give

$$p^a = \frac{dx^a}{d\lambda}, \qquad \frac{dp_a}{d\lambda} = -\frac{\partial H}{\partial x^a}.$$

The second equation gives

$$\frac{dp_a}{d\lambda} + \frac{1}{2}\frac{\partial g^{bc}}{\partial x^a}p_b p_c = 0,$$

and hence

$$\frac{dp^a}{d\lambda} + \Gamma^a_{bc}p^b p^c = 0,$$

the geodesic equation.

As shown in Chapter 10 of Goldstein (1980), the general solution of a Hamiltonian system may be found by solving the associated Hamilton–Jacobi equation. This is the clearest way to proceed for geodesics in the Kerr geometry (Misner et al. 1973). One starts by replacing each appearance of the momentum p_a by the gradient $\partial S/\partial x^a$ of the Hamilton–Jacobi function S, which may be regarded as an action function depending on x^a and λ. The Hamilton–Jacobi equation is

$$-\frac{\partial S}{\partial \lambda} = H\left[x^a, \frac{\partial S}{\partial x^b}\right]$$
$$= \frac{1}{2}g^{ab}\frac{\partial S}{\partial x^a}\frac{\partial S}{\partial x^b}. \tag{2.6.3}$$

One can prove that, if $S(x^a, \lambda)$ is the action for going from some specified initial data $(x_I^a, \lambda = 0)$ to the final data (x^a, λ), then S obeys the Hamilton–Jacobi equation (2.6.3). However, it is fruitful to consider more general solutions of eqn (2.6.3).

Specializing now to the Kerr metric in Boyer–Lindquist coordinates, the Hamilton–Jacobi equation (2.6.3) reads

$$-\frac{dS}{d\lambda} = -\frac{1}{2}\frac{1}{\Delta\rho^2}\left[(r^2 + a^2)\frac{\partial S}{\partial t} + a\frac{\partial S}{\partial \phi}\right]^2$$

$$+ \frac{1}{2} \frac{1}{\rho^2 \sin^2 \theta} \left[\frac{\partial S}{\partial \phi} + a \sin^2 \theta \frac{\partial S}{\partial t} \right]^2 + \frac{1}{2} \frac{\Delta}{\rho^2} \left(\frac{\partial S}{\partial r} \right)^2 + \frac{1}{2\rho^2} \left(\frac{\partial S}{\partial \theta} \right)^2 . \quad (2.6.4)$$

One solves this by separation of variables: since there is no explicit dependence on λ, ϕ, or t in the equation, the solution, if separable, would have to take the form

$$S = \frac{1}{2}\mu^2 \lambda - Et + L_z\phi + S_r(r) + S_\theta(\theta). \quad (2.6.5)$$

Here the values of the integration constants μ^2, E, L_z follow from $\partial S/\partial \lambda = -H, \partial S/\partial t = p_t, \partial S/\partial \phi = p_\phi$. Here μ is the particle's rest mass, E its energy, and L_z its angular momentum about the symmetry axis. One finds that the Hamilton–Jacobi equation (2.6.4) is indeed separable, with

$$S_r(r) = \int \Delta^{-1}\sqrt{R}\,dr, \qquad S_\theta(\theta) = \int \sqrt{\Theta}\,d\theta, \quad (2.6.6)$$

where

$$R = P^2 - \Delta[\mu^2 r^2 + (L_z - aE)^2 + Q], \quad (2.6.7)$$
$$\Theta = Q - \cos^2 \theta[a^2(\mu^2 - E^2) + (L_z^2/\sin^2 \theta)], \quad (2.6.8)$$

where

$$P = E(r^2 + a^2) - L_z a, \quad (2.6.9)$$

and Q is an 'integration constant' or constant of the motion, given by

$$Q = p_\theta^2 + \cos^2 \theta[a^2(\mu^2 - E^2) + (L_z^2/\sin^2 \theta)]. \quad (2.6.10)$$

An integral form of the equations of motion may be derived (Misner it et al. 1973) by setting $\partial S/\partial[Q + (L_z - aE)^2], \partial S/\partial\mu^2, \partial S/\partial E$, and $\partial S/\partial L_z$ to zero. One obtains

$$\int^\theta \frac{d\theta}{\sqrt{\Theta}} = \int^r \frac{dr}{\sqrt{R}}, \quad (2.6.11)$$

$$\lambda = \int^\theta \frac{a^2 \cos^2 \theta}{\sqrt{\Theta}}\,d\theta + \int^r \frac{r^2}{\sqrt{R}}\,dr, \quad (2.6.12)$$

$$t = \int^\theta \frac{-a(aE \sin^2 \theta - L_z)}{\sqrt{\Theta}}\,d\theta + \int^r \frac{(r^2 + a^2)P}{\Delta\sqrt{R}}\,dr, \quad (2.6.13)$$

$$\phi = \int^\theta \frac{-(aE \sin^2 \theta - L_z)}{\sin^2 \theta\sqrt{\Theta}}\,d\theta + \int^r \frac{aP}{\Delta\sqrt{R}}\,dr. \quad (2.6.14)$$

After differentiating these equations, one obtains the particle's contravariant momentum components (i.e. first integrals of the motion)

$$\rho^2 \frac{d\theta}{d\lambda} = \sqrt{\Theta}, \quad (2.6.15)$$

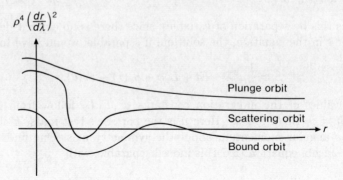

FIG. 2.38. The qualitative form of particle orbits in the Kerr metric can be read off from $\rho^4 (dr/d\lambda)^2 = R(r)$.

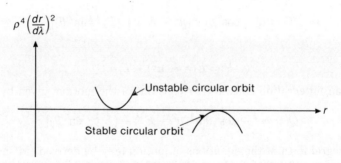

FIG. 2.39. The form of $\rho^4 (dr/d\lambda)^2 = R(r)$ for stable and unstable circular orbits to occur.

$$\rho^2 \frac{dr}{d\lambda} = \sqrt{R}, \tag{2.6.16}$$

$$\rho^2 \frac{d\phi}{d\lambda} = -[aE - (L_z/\sin^2 \theta)] + (a/\Delta)P, \tag{2.6.17}$$

$$\rho^2 \frac{dt}{d\lambda} = -a(aE \sin^2 \theta - L_z) + (r^2 + a^2)P/\Delta. \tag{2.6.18}$$

One can read off the qualitative radial motion by squaring eqn (2.6.16), to give $\rho^4 (dr/d\lambda)^2$ as a function of r; see Figs 2.38 and 2.39.

2.7 Perturbations of the Kerr metric

As a simple example of a perturbation of the Kerr geometry, consider a scalar field $\psi(x^a)$ obeying

$$\Box \psi \equiv \psi^{;a}{}_{;a} = 0. \tag{2.7.1}$$

As shown by Brill et al.(1972), one may separate variables to find a solution in the Kerr geometry, of the form

$$\psi = (r^2 + a^2)^{-\frac{1}{2}} u_{\ell m}(r) S_{m\ell}(-i\omega a, \cos\theta) e^{i(m\phi - \omega t)}. \tag{2.7.2}$$

Here m and ℓ are integers with $0 \le |m| \le \ell$, $S_{m\ell}$ is a spheroidal harmonic (Abramowicz and Stegun 1972)—see eqn (2.7.20)—and $u_{\ell m}$ obeys the radial equation

$$-\frac{d^2 u}{dr^{*2}} + Vu = 0. \tag{2.7.3}$$

Here r^* is the 'tortoise' coordinate, originally defined for the Schwarzschild metric (Regge and Wheeler 1957), here given by

$$dr^* = \Delta^{-1}(r^2 + a^2)dr. \tag{2.7.4}$$

The coordinate r^* runs from $-\infty$ to $+\infty$ as r runs from the horizon r_+ to $+\infty$. The effective potential $V(r^*)$ is given by

$$\begin{aligned}
V = &- \left[\omega - \frac{ma}{(r^2 + a^2)}\right]^2 + [(m - \omega a)^2 + Q](r^2 + a^2)^{-2}\Delta \\
&+ 2(Mr - a^2)(r^2 + a^2)^{-3}\Delta + 3a^2(r^2 + a^2)^{-4}\Delta^2.
\end{aligned} \tag{2.7.5}$$

The quantity Q in eqn (2.7.5) is a constant, analogous to the constant of the motion Q in eqns (2.6.7)–(2.6.10) for geodesics in the Kerr metric. It is given by

$$Q = \lambda_{m\ell} - m^2, \tag{2.7.6}$$

where $\lambda_{m\ell}$ is the eigenfunction of the spheroidal harmonic.

Remarkably, similar separations of variables arise for spins $s = \frac{1}{2}$ (neutrino), $s = 1$ (Maxwell) and $s = 2$ (gravitational perturbations) about a Kerr background. For definiteness, consider the $s = 2$ gravitational case (Teukolsky 1973). One writes all gravitational field quantities in terms of their tetrad components (Newman and Penrose 1962). This involves setting up a tetrad l^a, n^a, m^a and $\bar{m}^a$ of vectors throughout the space–time, where l^a and n^a are real null vectors, and m^a and $\bar{m}^a$ are complex null. The complete set of tetrad relations is

$$\begin{aligned}
l_a l^a = m_a m^a = \bar{m}_a \bar{m}^a = n_a n^a &= 0, \\
l_a n^a = -m_a \bar{m}^a &= 1, \\
l_a m^a = l_a \bar{m}^a = n_a m^a = n_a \bar{m}^a &= 0.
\end{aligned} \tag{2.7.7}$$

In the Kerr geometry, one may take the Kinnersley tetrad (1969)

$$\begin{aligned}
l^a &= [(r^2 + a^2)/\Delta, 1, 0, a/\Delta], \\
n^a &= [r^2 + a^2, -\Delta, 0, a]/2\Sigma, \\
m^a &= [ia\sin\theta, 0, 1, i/\sin\theta]/\sqrt{2}(r + ia\cos\theta).
\end{aligned} \tag{2.7.8}$$

This is adapted to the curvature of the Kerr metric, in that the only non-zero tetrad component of the Weyl tensor

$$C_{abcd} = R_{abcd} + (g_{a[d}R_{c]b} + g_{b[c}R_{d]a}) + \frac{1}{3}Rg_{a[c}g_{d]b}, \tag{2.7.9}$$

where R_{abcd} is the Riemann tensor, R_{ab} is the Ricci tensor and R is the Ricci scalar (both the latter zero for Kerr), is

$$\begin{aligned}\Psi_2 &= -\frac{1}{2}C_{abcd}(l^a n^b l^c n^d - l^a n^b m^c \bar{m}^d) \\ &= M\rho^3,\end{aligned} \tag{2.7.10}$$

for the Kinnersley tetrad, where

$$\rho = -1/(r - ia\cos\theta). \tag{2.7.11}$$

The Kinnersley tetrad is regular on the past horizon, but singular on the future horizon. There is another tetrad,

$$\begin{aligned}l^a &= [-(r^2 + a^2)/\Delta, 1, 0, -a/\Delta], \\ n^a &= [-(r^2 + a^2), -\Delta, 0, -a]/2\Sigma, \\ m^a &= [-ia\sin\theta, 0, 1, -i/\sin\theta]/\sqrt{2}(r + ia\cos\theta),\end{aligned} \tag{2.7.12}$$

which is regular on the future horizon.

The connection of the Kerr space–time is described, following Newman and Penrose (1962), by the spin coefficients. For example, ρ appearing in eqns (2.7.10) and (2.7.11) is given by

$$\rho = l_{a;b}m^a\bar{m}^b. \tag{2.7.13}$$

The remaining non-zero spin coefficients of the Kerr geometry in the Kinnersley tetrad are (Teukolsky 1973)

$$\begin{aligned}\beta &= -\bar{\rho}\cot\theta/2\sqrt{2}, \qquad \pi = ia\rho^2\sin\theta/\sqrt{2}, \qquad \alpha = \pi - \bar{\beta}, \\ \tau &= -ia\rho\bar{\rho}\sin\theta/\sqrt{2}, \qquad \mu = \rho^2\bar{\rho}\Delta/2, \\ \gamma &= \mu + \rho\bar{\rho}(r - M)/2.\end{aligned} \tag{2.7.14}$$

Now consider a gravitational perturbation of the Kerr geometry. By using the Newman–Penrose formalism, Teukolsky (1973) found decoupled equations for the perturbation quantities

$$\Psi_0 = -C_{abcd}l^a m^b l^c m^d \tag{2.7.15}$$

and

$$\rho^{-4}\Psi_4 = -\rho^{-4}C_{abcd}n^a\bar{m}^b n^c\bar{m}^d. \tag{2.7.16}$$

In the vacuum case, these are

$$
\left[\frac{(r^2 + a^2)^2}{\Delta} - a^2 \sin^2\theta\right]\frac{\partial^2\psi}{\partial t^2} + \frac{4Mar}{\Delta}\frac{\partial^2\psi}{\partial t\partial\phi}
$$
$$
+ \left[\frac{a^2}{\Delta} - \frac{1}{\sin^2\theta}\right]\frac{\partial^2\psi}{\partial\phi^2} - \Delta^{-s}\frac{\partial}{\partial r}\left(\Delta^{s+1}\frac{\partial\psi}{\partial r}\right)
$$
$$
- \frac{1}{\sin\theta}\frac{\partial}{\partial\theta}\left(\sin\theta\frac{\partial\psi}{\partial\theta}\right) - 2s\left[\frac{a(r - M)}{\Delta} + \frac{i\cos\theta}{\sin^2\theta}\right]\frac{\partial\psi}{\partial\phi}
$$
$$
- 2s\left[\frac{M(r^2 - a^2)}{\Delta} - r - ia\cos\theta\right]\frac{\partial\psi}{\partial t} + (s^2\cot^2\theta - s)\psi = 0. \tag{2.7.17}
$$

Here $s = 2$ corresponds to $\psi = \Psi_0$, and $s = -2$ to $\psi = \rho^{-4}\Psi_4$. Eqn (2.7.17) also describes perturbations of lower spin s, and the $s = 0$ case gives back the scalar case of eqns (2.7.2)–(2.7.6), while $s = \pm1$ corresponds to electromagnetic perturbations.

The decoupled perturbation equation (2.7.17) can be separated (Teukolsky 1973) by writing

$$
\psi = e^{-i\omega t}e^{im\phi}S(\theta)R(r), \tag{2.7.18}
$$

where R and S obey

$$
\Delta^{-s}\frac{d}{dr}\left(\Delta^{s+1}\frac{dR}{dr}\right)
$$
$$
+ \left(\frac{K^2 - 2is(r - M)K}{\Delta} + 4is\omega r - \lambda\right)R = 0, \tag{2.7.19}
$$

$$
\frac{1}{\sin\theta}\frac{d}{d\theta}\left(\sin\theta\frac{dS}{d\theta}\right) + \left(a^2\omega^2\cos^2\theta - \frac{m^2}{\sin^2\theta} - 2a\omega s\cos\theta\right.
$$
$$
\left. - \frac{2ms\cos\theta}{\sin^2\theta} - s^2\cot^2\theta + s + A\right)S = 0, \tag{2.7.20}
$$

where $K \equiv (r^2 + a^2)\omega - am$ and $\lambda \equiv A + a^2\omega^2 - 2am\omega$. Equation (2.7.20), subject to boundary conditions of regularity at $\theta = 0, \pi$, gives a Sturm–Liouville eigenvalue problem for the separation constant $A = {}_sA^m{}_\ell(a\omega)$. For fixed s, m and $a\omega$, one can label the eigenvalues by ℓ. The smallest eigenvalue has $\ell = \max(|m|, |s|)$. Sturm–Liouville theory shows that the eigenfunctions ${}_sS^m{}_\ell$ are complete and orthogonal on $0 \le \theta \le \pi$ for each m, s and $a\omega$. In the case $s = 0$ of a scalar field, as in eqn (2.7.2), the eigenfunctions are the spheroidal wave functions $S^m{}_\ell(-a^2\omega^2, \cos\theta)$ (Abramowicz and Stegun 1972). In the case $a\omega = 0$ of stationary perturbations (as occurs at lowest order in the study of the dynamics of black holes in Chapters 3 and 4) the eigenfunctions are the spin-weighted spherical harmonics $Y^m{}_\ell = {}_sS^m{}_\ell(\theta)e^{im\phi}$, and $A = (l - s)(l + s + 1)$ (Goldberg et al. 1967). In the general case, the eigenfunctions are referred to as 'spin-weighted spheroidal harmonics'.

As part of the boundary conditions on perturbations of the Kerr geometry, one generally imposes the condition of regularity at the future event horizon. This is because, in the full solution which describes the formation of a Kerr black hole, e.g. by matter collapse, there is no past event horizon. One takes a tetrad which is regular near the future event horizon (unlike the Kinnersley tetrad of eqn (2.7.8)), such as the tetrad described by Hartle (1974), and imposes regularity of the perturbations with respect to this tetrad (see, e.g., Section 3.3).

A further remarkable property of the Kerr metric is that the metric perturbations h_{ab} can be reconstructed by simple differential operations from Ψ_0 (or $\rho^{-4}\Psi_4$) (Chrzanowski 1975; Wald 1978; Chandrasekhar 1983). (A similar argument works for spin-1 perturbations.) The simplest argument is that due to Wald (1978). This shows that if Ψ_0 obeys the Teukolsky equation (2.7.17) with $s = 2$, then there is a complex metric perturbation h_{ab} constructed from Ψ_0 such that both $\mathrm{Re}(h_{ab})$ and $i\mathrm{Im}(h_{ab})$ satisfy the linearized Einstein equations, and lead back to (complex) multiples of Ψ_0, when we use the defining equation $\Psi_0 = C_{abcd}l^a m^b l^c m^d$. To express this, define the differential operators D, Δ, δ, and $\bar{\delta}$ by

$$D\phi = \phi_{,a}l^a, \qquad \Delta\phi = \phi_{,a}n^a,$$
$$\delta\phi = \phi_{,a}m^a, \qquad \bar{\delta}\phi = \phi_{,a}\bar{m}^a. \tag{2.7.21}$$

Then the metric perturbation h_{ab} is given in terms of Ψ_0, in a particular gauge, by

$$\begin{aligned}
h_{ab} = \rho^{-4}\{&n_a n_b(2\pi + 2\bar{\beta} + \bar{\delta})(3\pi - 4\bar{\beta} - \bar{\delta}) \\
&+ \bar{m}_a \bar{m}_b(\bar{\mu} + 4\mu - 2\gamma + \Delta)(4\gamma - \mu - \Delta) \\
&- n_{(a}\bar{m}_{b)}[(2\bar{\gamma} + 2\gamma - 4\mu - \Delta)(4\bar{\beta} - 3\pi + \bar{\delta}) \\
&+ (2\pi + 4\bar{\beta} + \bar{\tau} + \bar{\delta})(4\gamma - \mu - \Delta)]\}\Psi_0.
\end{aligned} \tag{2.7.22}$$

3

DYNAMICS OF A BLACK HOLE IN A BACKGROUND UNIVERSE

3.1 Introduction

The perturbation theory described in Chapter 2 is concerned with linear perturbations superimposed on an asymptotically flat Kerr solution, thereby ignoring the presence of the rest of the universe. However, one sometimes needs to understand the way in which a black hole fits into the surrounding space–time. For example, the rotating black hole may be interacting with matter of comparable or much greater mass, as in a galaxy or in cosmology, in which case the perturbation approach just mentioned is inadequate, and one should take into account the perturbations of the black hole produced by the matter, the perturbations of the external geometry produced by the black hole, and the motion of the black hole relative to the matter. A technique is presented here for dealing with this type of situation (D'Eath 1975a). This problem was also studied in the case of a Schwarzschild black hole by Manasse (1963) and Demiański and Grishchuk (1974). The 'external universe' may also be the geometry produced by another massive object, such as a black hole, neutron star, or other compact body. The case of two black holes in nearly Newtonian orbit around each other is described in the following Chapter 4 (D'Eath 1975b). These matching ideas are still viable for compact bodies which are not black holes (Kates 1980a; Damour and Taylor 1992).

Let us suppose that the true universe $(\mathcal{M}, g_{ab})$ can be regarded as approximately a background universe $(\mathcal{M}_0, g_{ab}^{(0)})$ on which a black hole of approximate mass M has been superimposed in some non-linear fashion. We require that M be small, in the sense that a typical length-scale M associated with the black hole is much less than a typical length-scale of the background universe, as specified by its Riemann tensor invariants, say. This condition will only fail to be satisfied in rather extreme situations, in which the shape of the black hole is strongly affected by the external gravitational field, and can be seen as part of a reasonable interpretation of our assumption that a background space–time can be separated out from $(\mathcal{M}, g_{ab})$. We assume that g_{ab} and $g_{ab}^{(0)}$ satisfy the Einstein equations with matter, except that as a preliminary simplifying assumption (which may be varied) we assume $(\mathcal{M}, g_{ab})$ is empty in a neighbourhood of the event horizon, and that $(\mathcal{M}_0, g_{ab}^{(0)})$ is empty in the region in which the superimposed field is large. Thus in the present treatment we do not attempt to allow for accretion problems, nor for rings of matter orbiting near the event horizon. However, it would of course be very interesting to have an extension of this work to allow for matter near the event horizon, as in astrophysically important examples, such as supermassive black holes in galactic nuclei (Begelman *et al.* 1980). One would also like to assume that $(\mathcal{M}, g_{ab})$ and $(\mathcal{M}_0, g_{ab}^{(0)})$ can each be given some sort of null infinity so that the notion of a black hole is meaningful. This may be too much to ask, e.g. in a cosmological context. The discussion will be concerned almost entirely with the behaviour of the gravitational field in a neighbourhood of the black hole, and we do not attempt any study of global questions. A rigorous treatment of the problem is far beyond the scope of this work, and we merely suppose that the true and background universes are sufficiently well behaved in the large that our local considerations are valid.

At points which are separated from the black hole by distances of the order of the background length-scale, the presence of the black hole will make little difference to the background metric $g_{ab}^{(0)}$. This suggests that we treat the effect of the black hole as a linear (and higher-order) perturbation on $g_{ab}^{(0)}$, at least far from the strong-field region with length-scale M, as measured, for example, by curvature invariants at distances of order a few multiples of the radius of the event horizon. Clearly, a different description is needed in the strong-field region. To motivate this description, we remark that the space–time curvature in the strong-field region is of order M^{-2}, which is much greater than the background curvature in a neighbourhood of the black hole. It is known (Robinson 1975; Whiting 1989) that the Kerr solutions are the only stationary vacuum black-hole space–times, and are stable against small perturbations. Then we are led to describe the strong-field region as a Kerr solution which is perturbed by the presence of the background. It then becomes necessary to relate these two descriptions of the gravitational field on the background length-scale and on the black-hole length-scale. One achieves this by using the technique of matched asymptotic expansions. This technique was first used in general relativity by Burke (1971) in order to relate the near-field region and the wave zone in the problem of slow-motion radiation, and is a standard method of applied mathematics for dealing with problems involving more than one length- or time-scale (Nayfeh 1973).

In order to make mathematically sensible statements about a perturbation problem of this type, we must allow the 'mass' M of the black hole to become a small parameter, varying in some interval $[0, \kappa)$. Thus, instead of making statements about a fixed space–time $(\mathcal{M}, g_{ab})$, we make asymptotic statements about the metric as $M \to 0$, which one can then attempt to interpret in $(\mathcal{M}, g_{ab})$. Following Geroch (1969), we consider a five-dimensional manifold with boundary, $\mathcal{N}$, which is built up from the space–times $(\mathcal{M}_M, g_{ab}(M))$ for $M \in [0, \kappa)$, where κ is some mass, $g_{ab}(M)$ is the metric corresponding to 'mass' M, and $g_{ab}(0) = g_{ab}^{(0)}$. More precisely, when $M = 0$ we should exclude from $\mathcal{M}_0$ a smooth timelike worldline l_0 in $(\mathcal{M}_0, g_{ab}^{(0)})$, which will later be seen to be a zeroth-order approximation to the position of the black hole. There is a natural map from contravariant tensor fields on each submanifold $\mathcal{M}_M$ to contravariant tensor fields on $\mathcal{N}$ (Hawking and Ellis 1973). In relation to this natural map, one requires that the contravariant metrics $g^{ab}(M)$ define a smooth tensor field on $\mathcal{N}$; this expresses the fact that the black hole is only a small perturbing influence on $g_{ab}^{(0)}$ far from the event horizon. We also require that if one fixes attention on a region of the space–time $(\mathcal{M}_M, g_{ab}(M))$ at a spatial separation of order M from l_0, then as $M \to 0$ the geometry becomes asymptotically that of a Kerr metric with small mass M and Kerr rotation parameter $a = \chi M$ (where $0 \leq \chi < 1$), plus small perturbations. Since this asymptotic geometry has the form $M^2 h_{ab}^{(0)}$, where $h_{ab}^{(0)}$ is the metric of a unit-mass Kerr geometry with $a = \chi$, one can express this limiting behaviour in Geroch's language by saying that, if one studies the conformal metric $M^{-2} g_{ab}(M)$, and considers the correct region of the space–time near l_0, then one obtains the Kerr metric with unit mass and spin χ as a limit.

To express this in coordinate form, let us introduce a coordinate system (τ, x, y, z, M) in an open subset of $\mathcal{N}$ such that the excised line l_0 is given locally by $x = y = z = M = 0$. Here τ is a temporal coordinate and x, y, z are spatial. Then, as $\tau, x, y, z \to$ constants, $M \to 0$, we may give $g_{ab}(M)$ an asymptotic expansion appropriate to the non-linear perturbation of $g_{ab}^{(0)}$ by the far field of a Kerr black hole; this procedure will be referred to as the external (approximation) scheme. We shall also require that as $(\tau - \tau_0)M^{-1}$, xM^{-1}, yM^{-1}, $zM^{-1} \to$ constants, $M \to 0$ for any time τ_0, the metric approximates that of a Kerr solution of mass M, where the usual Kerr parameter a satisfies $aM^{-1} \to$ constant as $M \to 0$, plus higher-order corrections; this will be referred to as the internal (approximation) scheme. Here τ_0 is a particular background time, or parameter along l_0. Thus there is one internal scheme for each background time coordinate τ_0. This is because the internal scheme is only attained by scaling all coordinates, space and time, by a factor M^{-1}. One carries out the matching by assuming that the external and internal schemes each have wider domains of validity than those just described, which overlap in such a way that both asymptotic expansions can be used in an intermediate region in which, for example, $(\tau - \tau_0)M^{-1/2}$, $xM^{-1/2}$, $yM^{-1/2}$, $zM^{-1/2} \to$ constants, $M \to 0$. This region would be at 'small' spatial separation from the origin in the external scheme, but at 'large' spatial separation from the origin in the internal scheme. One can then compare terms in the expansions and use the resulting information to impose boundary conditions on the two schemes. Thus, for example, comparison with the external scheme in the matching region will give us information on the internal scheme at large xM^{-1}, yM^{-1}, zM^{-1}. Correspondingly, comparison with the internal scheme in the matching region will give information on the external scheme at small x, y, z. A more formal description of matched asymptotic expansions on manifolds is given by Kates (1981).

We have already given the physical motivation behind the internal and external schemes, and it remains to interpret the matching region in terms of the fixed universe $(\mathcal{M}, g_{ab})$ as a domain in which a transition is effected from the black-hole régime to the background, on length-scales between M and the background length-scale. This physical point of view was adopted from the start by Manasse (1963), who studied a special case of the problem considered here, in which the background $(\mathcal{M}_0, g_{ab}^{(0)})$ is a Schwarzschild metric, and a small Schwarzschild-type black hole falls approximately along a radial geodesic l_0 of the background.

In Section 3.2 we shall discuss the asymptotic expansions and matching in some detail. This leads in Sections 3.3 and 3.4 to the conclusions that the zeroth-order worldline l_0 is a geodesic in the background universe and that, at the lowest order, the spin of the black hole is parallel propagated along l_0 in the background. We consider the distorting effects of the background on the internal structure of the black hole in Section 3.5; these cause the basic parameters of the black hole to change over a sufficiently long time-scale. Then, in Section 3.6, we consider the situation in which the small black hole is essentially non-rotating, in which case more explicit information on the internal structure can be found. Section 3.7 is devoted to the perturbations of the background caused by the far field of the black hole. In Section 3.8 we apply the same methods to the case in which a black hole moves in a background universe, under the field equations of the Brans–Dicke theory of gravitation, as an example of the applicability of matching methods to other theories of gravity. Similar ideas should work for different gravitational theories. Section 3.9 considers possible astrophysical applications, particularly to the evolution of the spin vector of a massive black hole in a galactic nucleus, to show that observable effects may correspond to the mathematical ideas studied here. The matter around a rapidly rotating massive black hole at the centre of a galactic nucleus forms an accretion disc, and some of it is not swallowed, but ejected along the poles, where it may form observable jets. Some jets are S-shaped, and it has been suggested (Begelman *et al.* 1980) that this precession may be due to a second black hole (see Chapter 4).

3.2 Asymptotic expansions and matching

As remarked in the previous section, we introduce a coordinate system

$$(\tau, x, y, z, M)$$

in an open subset of the five-dimensional manifold $\mathcal{N}$, such that a certain preferred timelike line l_0 in $(\mathcal{M}_0, g_{ab}^{(0)})$ is given by $x = y = z = M = 0$. Since the contravariant metrics $g^{ab}(M)$ are assumed to define a suitably smooth tensor field on $\mathcal{N}$, we may make the asymptotic expansion

$$g_{ab}(M)(\tau, x, y, z) \sim g_{ab}^{(0)}(\tau, x, y, z) + M g_{ab}^{(1)}(\tau, x, y, z)$$
$$+ M^2 g_{ab}^{(2)}(\tau, x, y, z) + O(M^3) \qquad (3.2.1)$$

of the one-parameter family of metrics in this coordinate system. Here $g_{ab}^{(0)}$ is the background geometry, $M g_{ab}^{(1)}$ is the first-order perturbation of the background due to the black hole, etc. This asymptotic expansion is thus describing the 'external region' in the far field of the black hole, where the geometry is dominated by the background. This expansion is valid at least as $\tau, x, y, z \to$ constants, $M \to 0$, provided that the limiting field point does not lie on l_0, and will be called the external (approximation) scheme. In the external scheme, l_0 is singled out by the fact that the perturbations $g_{ab}^{(1)}, g_{ab}^{(2)}$, etc. diverge as $(x, y, z) \to (0, 0, 0)$, and for this reason l_0 must be removed from $\mathcal{N}$. For example, if the background $g_{ab}^{(0)}$ were exactly flat, then the geometry $g_{ab}(M)$ would be exactly the Kerr solution with mass M and spin parameter $a = \chi M$ $(0 \leq \chi < 1)$. The first-order field $M g_{ab}^{(1)}$ would be the linearized Kerr solution, obtained by taking the first-order terms in $g_{ab}(M)$. The linearized field $M g_{ab}^{(1)}$ has a Schwarzschild $O(r^{-1})$ behaviour as $r^2 = (x^2 + y^2 + z^2) \to 0$, i.e. it is singular at the worldline l_0. In the typical case with a curved background, the linear perturbation $M g_{ab}^{(1)}$ still has the 'linearized Schwarzschild' behaviour above as $r \to 0$, but typically differs from the linearized Kerr metric as r moves away from 0.

We assume that each metric $g_{ab}(M)$ satisfies the Einstein equations with matter, but that as $(\tau, x, y, z) \to (\tau_0, x_0, y_0, z_0) \neq (\tau_0, 0, 0, 0)$, $M \to 0$ with (x_0, y_0, z_0) sufficiently small, then for sufficiently small M we have empty space. In particular, we make the simplifying assumption that the background is empty in a neighbourhood of l_0:

$$R_{ab}^{(0)} = 0 \tag{3.2.2}$$

near l_0, where $R_{ab}^{(0)}$ is the Ricci tensor of $g_{ab}^{(0)}$. Note that each $(\mathcal{M}_M, g_{ab}(M))$ may still contain matter for (x, y, z) sufficiently small, even though the one-parameter family of space–times satisfies the condition above. As mentioned in the previous Section 3.1, it would be very interesting to extend this work by allowing for matter on l_0 in the background geometry, where the matter falls into the black hole in the geometry $g_{ab}(M)$ via an accretion disc.

At present we do not impose any further conditions on the components $g_{ab}^{(0)}, g_{ab}^{(1)}, g_{ab}^{(2)}$, etc. However, the use here of matched asymptotic expansions in general relativity is bound to be heavily coordinate-dependent, since one needs to fix a coordinate system in which to accomplish the matching; in our case (τ, x, y, z, M) will be this common coordinate system. So if we choose a convenient coordinate system for the internal expansion, then not only genuine physical information, but also information about the coordinate system, will be communicated to the external scheme by means of the matching. Here, for example, some extra 'gauge' conditions on the components of $g_{ab}^{(0)}$ near l_0 will later be seen to be enforced by our coordinate choice for the internal expansion. As remarked by Thorne and Hartle(1985), one may be able to optimize the accuracy of the equations of motion, derived from the Einstein field equations, by suitable coordinate choice for the internal metric. As mentioned earlier, a more geometrical approach to matched asymptotic expansions in general relativity is given by Kates (1981).

To specify the behaviour of the metric near the event horizon in the internal approximation scheme, we use certain coordinates (V, R, θ, ϕ', M) which will later be related to the matching coordinates (τ, x, y, z, M). The internal expansion is then

$$ds^2 \sim M^2 h_{ab}^{(0)}(R,\theta)dx^a dx^b + M^3 h_{ab}^{(1)}(MV, R, \theta, \phi')dx^a dx^b$$
$$+ M^4 h_{ab}^{(2)}(MV, R, \theta, \phi')dx^a dx^b + O(M^5), \tag{3.2.3}$$

where $h_{ab}^{(0)}$ is a unit-mass Kerr metric, given by

$$h_{ab}^{(0)}dx^a dx^b = \Sigma d\theta^2 - 2\chi \sin^2\theta + 2dRdV + \Sigma^{-1}[(R^2+\chi^2)^2 - \Delta\chi^2\sin^2\theta]\sin^2\theta \, d\phi'^2$$
$$- 4\chi\Sigma^{-1}R\sin^2\theta \, d\phi'dV - (1 - 2R\Sigma^{-1})dV^2.$$

$$\tag{3.2.4}$$

Here χ is some constant, $0 \le \chi < 1$, with $a = \chi M$. One has to scale a at the rate $O(M)$ as $M \to 0$, in order to have a Kerr geometry $M^2 h_{ab}^{(0)}(R,\theta)dx^a dx^b$ rather than a Schwarzschild geometry in the limiting internal region as $M \to 0$. Also

$$\Sigma(R,\theta) = R^2 + \chi^2\cos^2\theta,$$
$$\Delta(R) = R^2 - 2R + \chi^2. \tag{3.2.5}$$

This expansion is assumed to be valid at least when

$$(V, R, \theta, \phi') \to (V_0, R_0, \theta_0, \phi'_0), M \to 0,$$

and $R_0 \ge \alpha R_+$, where $R_+ = 1 + (1-\chi^2)^{\frac{1}{2}}$ and $0 < \alpha < 1$. The lowest-order part $M^2 h_{ab}^{(0)}(R,\theta)$ of the internal metric is just the Kerr solution, describing a black hole of mass M and angular momentum χM^2 rotating about the axis $\theta = 0$, where MV is an ingoing Kerr null coordinate, MR is the usual radial coordinate, and θ, ϕ' are Kerr angular coordinates (Section 2.5). Here $M^3 h_{ab}^{(1)}$ describes the first-order perturbation of the black hole due to the background, $M^4 h_{ab}^{(2)}$ describes the second-order perturbation, etc. The asymptotic expansion (3.2.3) is thus describing the 'internal region' at distances of order of the horizon radius of the black hole, where the external background only generates small perturbations.

The other most significant feature of this expansion is the dependence of the internal perturbation terms $h_{ab}^{(1)}$, $h_{ab}^{(2)}$, etc. on the variables V and R. We see that the internal perturbations are allowed to vary on the same length-scale as the dominant Kerr metric, through the variable R. But the time dependence of these perturbations, through the variable MV, is in some sense 'slow motion' with respect to the Kerr metric; it is necessary to impose this time dependence in order to match to the external scheme. This is because, in the external region, the background geometry $g_{ab}^{(0)}$ varies on a τ-timescale of $O(1)$. Now MV is essentially τ, up to an additive constant, in the matching region where the internal coordinate R is large but the external coordinate r is small (see later). Hence, via the matching, the external scheme induces the slow MV time-dependence of the internal-scheme perturbations $h_{ab}^{(1)}$, $h_{ab}^{(2)}$, ... We prefer to remove a constant conformal factor M^2 from $g_{ab}(M)$ and to regard the internal scheme as describing perturbations of a unit-mass Kerr solution, of the form

$$Mh_{ab}^{(1)}(0, R, \theta, \phi') + M^2 V \frac{\partial}{\partial \omega} h_{ab}^{(1)}(\omega, R, \theta, \phi') \mid_{\omega=0}$$
$$+ M^2 h_{ab}^{(2)}(0, R, \theta, \phi') + O(M^3),$$

where we have given the perturbation terms a Taylor expansion. When viewed in this way, the internal expansion will be called the quasi-stationary scheme (QS scheme). The removal of the conformal factor need not concern us until we come to computations of surface area.

We assume that as $(V, R, \theta, \phi') \to (V_0, R_0, \theta_0, \phi'_0, M) \to 0$ with $R_0 \geq \alpha R_+$, then for sufficiently small M each $g_{ab}(M)$ is locally a vacuum metric. As stated in Section 3.1, for simplicity we do not allow the black hole to interact with matter too near to the event horizon. But we do tolerate matter in the region $R < \alpha R_+$, for example the matter involved in a stellar collapse, if the black hole is the remnant of a heavy star, say.

In order to relate the coordinates (V, R, θ, ϕ', M) to the matching coordinates (τ, x, y, z, M), we first define T and ϕ by

$$dV = dT + (R^2 + \chi^2)\Delta^{-1}dR,$$
$$d\phi' = d\phi + \chi\Delta^{-1}dR. \tag{3.2.6}$$

When we transform to coordinates (T, R, θ, ϕ, M), the lowest-order term $M^2 h_{ab}^{(0)}$ in the internal expansion becomes a Kerr solution expressed in Boyer–Lindquist coordinates (Section 2.5). Moreover, when $(T, R, \theta, \phi) \to (T_0, R_0, \theta_0, \phi_0), M \to 0$ with R_0 sufficiently large, we are still in the region of validity of the internal scheme. In fact, we have a QS scheme in these coordinates, where a unit-mass Kerr solution is given perturbations of the form

$$Mj_{ab}^{(1)}(0, R, \theta, \phi) + M^2 T \frac{\partial}{\partial \omega} j_{ab}^{(1)}(\omega, R, \theta, \phi) \mid_{\omega=0}$$
$$+ M^2 j_{ab}^{(2)}(0, R, \theta, \phi) + O(M^3).$$

This particular scheme is the most convenient for computations of internal perturbations, since we may use the work of Teukolsky (1973) on perturbations of the Kerr solution—this work is expressed in Boyer–Lindquist coordinates (see Section 2.7).

To specify the coordinates used in matching, we then require that

$$\tau = \tau_0 + MT$$

for all $R > \beta R_+$, for some constant β, where $R > \beta R_+$ implies that the point (T, R, θ, ϕ) is outside the ergosurface of the unperturbed Kerr solution; here τ_0 is some background time. For $\alpha R_+ \leq R \leq \beta R_+$, we choose $\tau = \tau(V, R, M)$ such that $\{\tau = \text{const.}\}$ labels a spacelike hypersurface of $(\mathcal{M}_M, g_{ab}(M))$ which enters inside the unperturbed event horizon $\{R = R_+\}$, and such that the surfaces $\{\tau = \text{const.}\}$ are carried into one another by the integral flow of the time-translation Killing vector of the unperturbed Kerr solution. The behaviour of τ for $R < \alpha R_+$ need not be specified. We also require that

$$x = r \sin\theta \cos\phi, \qquad y = r \sin\theta \sin\phi, \qquad z = r \cos\theta,$$

where

$$r = MR.$$

This completes the rules for the transformation between (V, R, θ, ϕ', M) and (τ, x, y, z, M). Of course, if one has a solution at low orders in perturbation theory for the internal and external metric, then one can make (say) a coordinate transformation close to the identity which will consistently give a gauge-transformed version of the perturbations.

We see that to each background time τ_0 there corresponds a whole internal scheme. Also the M dependence of the transformation $\tau = \tau_0 + MT$ $(R > \beta R_+)$ explains the quasi-stationarity of the internal perturbations. More physically, we have QS internal perturbations because gravitational waves only need a time of order M to cross the black hole, whereas the background is changing on a time-scale of order 1. Thus the black hole can adjust its gravitational field on what it feels to be a long time-scale in order to cope with the tidal field of the background.

For convenience, we chose that the rotation axis of the black hole should in some sense 'tie in' to the direction $\partial/\partial z$ at time τ_0 on the worldline l_0. But we do allow the rotation axis at another time τ_1 to 'tie in' to another direction in the background, $\lambda(\partial/\partial x) + \mu(\partial/\partial y) + \nu(\partial/\partial z)$. This can be expressed simply by rotating the dominant Kerr part of the internal metric. We do not allow the lowest-order 'mass' and 'scalar angular momentum' of the black hole to vary with τ, for reasons to be seen in Section 3.5.

The matching of the two expansions is carried out in some intermediate region, say where $M^{-1/2}(\tau - \tau_0), M^{-1/2}x, M^{-1/2}y, M^{-1/2}z \to$ constants, $M \to 0$. It will not be necessary to examine this region explicitly, for we shall assume, up to the orders in perturbation theory studied here, that all functions in the expansions are so well behaved that they can be developed in power series (involving both positive and negative powers) in the appropriate radial coordinate R or r, for large R or small r, respectively. The work of Dixon (1979), Futamase (1983), Anderson and Kegeles (1980), Kates and Kegeles (1982), and Anderson et al. (1982) shows that non-analytic $M^n \log M$ perturbation terms (and hence $R^i \log R$ terms, by matching) are to be expected at high orders in perturbation theory. These terms can be coped with, within the framework described here. But at the orders to which we shall be working, these non-analytic terms do not yet appear, and so we shall proceed by working with asymptotic power series. Thus a typical expression in the internal scheme appropriate to time τ_0 (with conformal factor M^2 removed) can be analyzed into terms of the form $M^i R^j f(\tau_0, MT, \theta, \phi)$, where i and j are integers, $i \geq 0$, and we refer the metric to coordinates $T, M^{-1}x, M^{-1}y, M^{-1}z$. Similarly, we decompose the expressions in the external scheme into terms of the form $M^k r^l g(\tau, \theta, \phi)$, where k, l are integers, $k \geq 0$. The matching then equates the functions of time and angle, while the exponents are related by

$$k = i - j, \qquad l = j.$$

The internal and external expansions can be regarded as grouping these basic terms with power-law dependence on M and a radial coordinate in two different ways; one grouping may be much more convenient than the other for the purposes of a specific computation.

It is helpful to visualize the expansion of the metric $g_{ab}(M)$ in terms of the structure (Thorne and Hartle 1985):

$$
\begin{array}{ccccccccccc}
g = & & \eta & \& & \dfrac{M}{r} & \& & \dfrac{M^2}{r^2} & \& & \dfrac{M^3}{r^3} & \& \cdots & h^{(0)} \\[2ex]
& \& & \dfrac{r}{\mathcal{R}} & \& & \dfrac{M}{\mathcal{R}} & \& & \dfrac{M^2}{r\mathcal{R}} & \& & \dfrac{M^3}{r^2\mathcal{R}} & \& \cdots & \mathcal{R}^{-1}h^{(1)} \\[2ex]
& \& & \dfrac{r^2}{\mathcal{R}^2} & \& & \dfrac{Mr}{\mathcal{R}^2} & \& & \dfrac{M^2}{\mathcal{R}^2} & \& & \dfrac{M^3}{r\mathcal{R}^2} & \& \cdots & \mathcal{R}^{-2}h^{(2)} \\[2ex]
& \& & \dfrac{r^3}{\mathcal{R}^3} & \& & \dfrac{Mr^2}{\mathcal{R}^3} & \& & \dfrac{M^2 r}{\mathcal{R}^3} & \& & \dfrac{M^3}{\mathcal{R}^3} & \& \cdots & \mathcal{R}^{-2}h^{(2)} \\[2ex]
& \& & \cdots & & & & & & & & \\[2ex]
& & g^{(0)} & & Mg^{(1)} & & M^2 g^{(2)} & & M^3 g^{(3)}. & &
\end{array}
$$

Here $\mathcal{R}$ is a typical curvature length-scale of the background geometry. One reads '&' as 'and a term of the form'. As above, this series could be modified by logarithmic dependence. Successive rows correspond to $h^{(0)}, h^{(1)}, h^{(2)}, \ldots$, the unperturbed Kerr black-hole metric $h^{(0)}_{ab}$ expanded in powers of M and its perturbations produced by the background universe. Successive columns correspond to $g^{(0)}, g^{(1)}, g^{(2)}, \ldots$, the unperturbed background metric $g^{(0)}_{ab}$ expanded in powers of $(r/\mathcal{R})$ and its perturbations produced by the black hole.

The conditions $i \geq 0, k \geq 0$ show that $k + l \geq 0$, $i - j \geq 0$. Conditions are also imposed on the external scheme by the fact that the dominant internal behaviour is of Kerr type. This appears in terms with $k + l = 0$, i.e. in the part of $g^{(k)}_{ab}(\tau, x, y, z)$ which is most singular as $(x, y, z) \to (0, 0, 0)$, for each k. For example, we find by examining the Kerr metric at large R that the most singular parts of $g^{(0)}_{ab}, g^{(1)}_{ab}, g^{(2)}_{ab}$ are given by

$$g_{ab}^{(0)} dx^a dx^b \to -d\tau^2 + dx^\alpha dx^\alpha,$$

$$g_{ab}^{(1)} dx^a dx^b \sim 2r^{-1} d\tau^2 + 2r^{-3} x^\alpha x^\beta dx^\alpha dx^\beta,$$

$$g_{ab}^{(2)} dx^a dx^b \sim -4r^{-3} \epsilon_{\alpha\beta\gamma} \chi^\beta x^\gamma dx^\alpha d\tau \qquad (3.2.7)$$

$$+ [(4r^{-4} - 2\chi^\gamma \chi^\gamma r^{-4}) x^\alpha x^\beta - \chi^\alpha \chi^\beta r^{-2} + x^\gamma \chi^\gamma$$

$$(x^\alpha \chi^\beta + \chi^\alpha x^\beta) r^{-4} + \chi^\gamma \chi^\gamma r^{-2} \delta_{\alpha\beta}] dx^\alpha dx^\beta$$

as $(x, y, z) \to (0, 0, 0)$. Here Greek indices run from 1 to 3, $x^1 = x, x^2 = y, x^3 = z$, $\epsilon_{\alpha\beta\gamma}$ is the three-dimensional alternating symbol, and we use a Cartesian summation convention for Greek indices. Also χ^α is the three-dimensional vector representing the lowest-order spin of the black hole; thus $\chi^\alpha \equiv (0, 0, \chi)$ when $\tau = \tau_0$. The condition on the components of the background metric for $(x, y, z) = (0, 0, 0)$ is just a coordinate condition which follows, via the matching, from our coordinate choice for the dominant Kerr part of the internal metric. But the conditions on $g_{ab}^{(1)}$ and $g_{ab}^{(2)}$ contain physical information about the far field of the black hole.

Similarly, the background metric determines those internal terms with $i - j = 0$; these are the terms which grow most rapidly as $R \to \infty$ at each order in the internal scheme.

The argument will proceed by studying the lowest-order internal perturbations $h_{ab}^{(1)}$, which grow at the rate $(r/\mathcal{R})$ at large r. As will be seen in the following section, one uses this boundary condition, together with the condition of regularity at the future event horizon of the black hole, to deduce that $h_{ab}^{(1)} = 0$ up to a gauge transformation. This in turn can be shown to imply that the black hole moves on a geodesic in the background metric, to leading order.

3.3   The lowest-order internal perturbations

In this section we shall show that the $O(M)$ (first-order) QS perturbations are essentially trivial. This will allow us to simplify both internal and external expansions by choice of coordinates. This will further lead in Section 3.4 to the conclusion that the black hole moves on a geodesic l_0, and that its spin is parallel transported along l_0. We proceed by examining the linearized Einstein equations for the QS scheme, and first recall the results of Teukolsky (1973) on perturbations of the Kerr metric (see also Section 2.7). In the (unit mass) Kerr metric, expressed in Boyer–Lindquist coordinates (T, R, θ, ϕ) as

$$ds^2 = -(1 - 2R\Sigma^{-1})dT^2 - 4\chi R \sin^2 \theta \Sigma^{-1} dT d\phi$$

$$+ \Sigma \Delta^{-1} dR^2 + \Sigma d\theta^2$$

$$+ \sin^2 \theta (R^2 + \chi^2 + 2\chi^2 R\Sigma^{-1} \sin^2 \theta) d\phi^2 \qquad (3.3.1)$$

where Σ, Δ are defined as before, we consider the Kinnersley null tetrad

$$l^a, n^a, m^a, \bar{m}^a,$$

where

$$l^a \equiv \left((R^2 + \chi^2)\Delta^{-1}, 1, 0, \chi\Delta^{-1} \right),$$

$$n^a \equiv (R^2 + \chi^2, -\Delta, 0, \chi)/2\Sigma, \tag{3.3.2}$$

$$m^a \equiv (i\chi \sin\theta, 0, 1, i/\sin\theta)/2^{\frac{1}{2}}(R + i\chi\cos\theta).$$

Here $l^a n_a = -1, m^a \bar{m}_a = 1$ and all other inner products are zero. We define the Newman–Penrose quantity

$$\Psi_0 = C_{abcd}l^a m^b l^c m^d, \tag{3.3.3}$$

where C_{abcd} is the Weyl tensor of any space–time perturbed about the Kerr solution. Then Ψ_0 has unperturbed value zero, and that part of Ψ_0 corresponding to the lowest-order perturbations is gauge-independent, i.e. independent of infinitesimal coordinate transformations. Also, Ψ_0 is invariant under infinitesimal Lorentz rotations of the tetrad.

Below, we shall find that Ψ_0 vanishes for the first-order $O(M)$ internal metric perturbations. Now the linear order $\Psi_0^{(1)}$, together with boundary conditions at the future event horizon, completely specifies the linear metric perturbations $h_{ab}^{(1)}$ in the internal scheme. For example, Wald's theorem (Wald 1973) shows that if $\Psi_0^{(1)}$ for linear vacuum perturbations of the Kerr geometry, and if the geometry is regular at the future event horizon, then the linear perturbation $h_{ab}^{(1)}$ is zero up to a gauge transformation $h_{ab}^{(1)} \to h_{ab}^{(1)} + \xi_{a;b} + \xi_{b;a}$ and a change in the mass and spin parameters. Further (Chrzanowski 1975; Wald 1978), as described in Section 2.7, the linear internal perturbations $h_{ab}^{(1)}$ can be computed from $\Psi_0^{(1)}$, up to gauge and a shift in mass and spin. Hence it is appropriate to study $\Psi_0^{(1)}$.

There is a decoupled field equation for the lowest-order part of Ψ_0. As in Section 2.7, this equation is completely separable with respect to T, R, θ and ϕ. For stationary vacuum perturbations the angular dependence in any mode is of the form $e^{im\phi}S(\theta)$, where

$$\frac{1}{\sin\theta}\frac{d}{d\theta}\left(\sin\theta\frac{dS}{d\theta}\right) + \left(\lambda + 2 - \frac{m^2}{\sin^2\theta} - \frac{4m\cos\theta}{\sin^2\theta} - 4\cot^2\theta\right)S = 0. \tag{3.3.4}$$

This equation has eigenvalues $\lambda = (L-2)(L+3)$ for $L \geq 2, L \geq |m|$, and eigenfunctions $S(\theta) = {}_2Y_L^m(\theta)$, the spherical harmonics of spin weight 2 (Goldberg et al. 1967). The radial equation is

$$\Delta\frac{d^2W}{dR^2} + 6(R-1)\frac{dW}{dR} + \{[\chi^2 m^2 + 4i\chi m(R-1)]\Delta^{-1} - \lambda\}W = 0. \tag{3.3.5}$$

This is a hypergeometric equation, with regular singular points at $R = R_- = 1(1-\chi^2)^{\frac{1}{2}}, R_+,$ and ∞. As $R \to \infty$, a solution of the radial equation behaves either as R^{L-2} or as R^{-L-3}.

We apply this to the perturbations at $O(M)$ in the QS scheme. As follows from the remarks on matching near the end of the previous section, the $O(M)$ QS metric perturbations cannot grow any faster than R as $R \to \infty$, when referred to coordinates $T, R \sin \theta \cos \phi = X, R \sin \theta \sin \phi = Y, R \cos \theta = Z$. Moreover, the part growing as R is determined by the $M^0 r^1$ terms in the external scheme, i.e. by the first derivatives of the background metric on l_0. Provided that the functions involved are sufficiently regular, the $O(M)$ part of Ψ_0, due to the lowest-order QS perturbations, will then be $O(R^{-1})$ as $R \to \infty$. Hence in each mode specified by (L, m) the radial function $W_L^m(R)$ must be $O(R^{-L-3})$ as $R \to \infty$.

The argument now takes a different course according as $m \neq 0$ or $m = 0$. Suppose first that $m \neq 0$. Then it may be verified that

$$
W_L^m(R) = \text{const.} \times (R - R_-)^{-2-\gamma_m} (R - R_+)^{-L-1+\gamma_m} \times
$$
$$
\times F\left[L - 1, L + 1 - 2\gamma_m; 2L + 2; \left(\frac{R_- - R_+}{R - R_+}\right)\right], \qquad (3.3.6)
$$

where

$$
\gamma_m = im\chi/(R_+ - R_-)
$$

and $F[,;;]$ is the standard notation for the hypergeometric function (Erdélyi 1953). Before imposing the boundary condition that the perturbations be regular at the future event horizon, and hence at $R = R_+$, we remark that the Kinnersley tetrad becomes singular as $R \to R_+$. A tetrad which is regular near the future event horizon $R = R_+$ has been described by Hartle (1974); in particular,

$$
l_H^a = \frac{\Delta}{2(R^2 + \chi^2)} l^a, \qquad m_H^a = m^a - \frac{i\chi \sin \theta}{\sqrt{2}(R + i\chi \cos \theta)} l^a, \qquad (3.3.7)
$$

where $l_H^a, n_H^a, m_H^a, \bar{m}_H^a$ is the Hartle tetrad. We transform the $O(M)$ part of Ψ_0 to this tetrad, and find (Hartle 1974) that in the (L, m) mode it behaves as

$$
e^{im\phi'} {}_2Y_L^m(\theta) \left(\frac{\Delta}{4R_+}\right)^2 \left(\frac{R - R_-}{R - R_+}\right)^{\gamma_m} W_L^m(R) \qquad \text{as } R \to R_+,
$$

where (θ, ϕ') are Kerr angular coordinates which are well behaved near $R = R_+$. To find the behaviour of $W_L^m(R)$ as $R \to R_+$, we use the relation (Erdélyi 1953)

$$
F[A, B; C; \zeta]/\Gamma(C)
$$
$$
= \frac{\Gamma(B - A)}{\Gamma(B)\Gamma(C - A)}(-\zeta)^{-A} F[A, 1 - C + A; 1 - B + A; \zeta^{-1}]
$$
$$
+ \frac{\Gamma(A - B)}{\Gamma(A)\Gamma(C - B)}(-\zeta)^{-B} F[B, 1 - C + B; 1 - A + B; \zeta^{-1}],
$$
$$
\tag{3.3.8}
$$

which is valid for application to eqn (3.3.6) as $R \to R_+$. Altogether, this shows that the $O(M)$ part of Ψ_0 in this regular tetrad contains a term which behaves as $(R - R_+)^{2-2\gamma_m}$ in the (L, m) mode as $R \to R_+$, unless $\Psi_0 \equiv 0$ in this mode.

Since the QS perturbation must be regular at $R = R_+$, and $m \neq 0$, we must have $\Psi_0 \equiv 0$ in this mode at $O(M)$.

Now suppose that $m = 0$ and that the perturbation is regular at the future event horizon $R = R_+$. The boundary condition as $R \to \infty$ again implies that $W_L^0(R)$ is given by eqn (3.3.6), where now $\gamma_m = 0$. Suppose that the constant in (3.3.6) is non-zero. Then, instead of (3.3.8) one may use a more complicated formula (Erdélyi 1953) to show that Ψ_0 will be non-zero on the horizon at $O(M)$ in the Hartle tetrad. But arguments which will be elaborated in Section 3.5 (see eqns (3.5.8)–(3.5.22)) show that the horizon shear and hence Ψ_0 on the horizon must be zero for stationary axisymmetric perturbations. We conclude that $\Psi_0 \equiv 0$ at $O(M)$ is the axisymmetric modes also.

Hence $\Psi_0 \equiv 0$ at $O(M)$ in the QS scheme. By a slight modification of Wald's theorem (Wald 1973), mentioned above, which states that Ψ_0 gives an almost complete description of vacuum perturbations of the Kerr metric, we find that the $O(M)$ QS metric perturbations can be described, after a coordinate transformation, merely by changes in the mass (unperturbed value 1) and angular momentum (unperturbed value χ, directed along $\theta = 0$) of the Kerr solution. We shall suppose that these coordinate transformations, one of which exists for each background time τ_0 labelling a QS scheme, can be fitted together so as to define a coordinate transformation $(\tau, x, y, z, M) \to (\tau', x', y', z', M)$ on the five-dimensional manifold $\mathcal{M}$, which preserves the properties of the asymptotic expansions that we have so far assumed. It will follow from the considerations in Section 3.5 that the $O(M)$ perturbations in the QS mass and QS scalar angular momentum are independent of the background time τ' along the worldline l_0. We can now choose to ignore these perturbations, since in a physical context they would correspond to a small error in our estimate of mass and angular momentum, which can easily be reabsorbed.

Thus we can assume that (τ, x, y, z, M) have been chosen so that all $O(M)$ QS perturbations have been eliminated. The QS perturbations are now

$$M^2 j_{ab}^{(2)}(0, R, \theta, \phi) + O(M^3);$$

we shall see in Section 3.5 that $j_{ab}^{(2)}$ in general describes non-trivial internal perturbations.

3.4 Equations of motion

The elimination of the $O(M)$ QS perturbations in the previous section also has the merit of removing all terms in the external scheme which behave as $M^k r^l g(\tau, \theta, \phi)$, where $k + l = 1$. In particular, we see that the background metric is $\mathrm{diag}(-1, 1, 1, 1) + O(r^2)$ near l_0, so that l_0 must be a timelike geodesic in $(\mathcal{M}_0, g_{ab}^{(0)})$, i.e. the lowest-order approximation to the path of the small black hole in the background is always a geodesic, regardless of the black hole's rotation. The argument in Section 3.3 can be regarded as showing that were the worldline non-geodesic, then the structure of the gravitational field near the unperturbed event horizon $\{R = R_+\}$ would be disastrously altered by the perturbing effects of the background.

We mention two other types of argument which might be used to show that l_0 is a geodesic. In the first of these, which does not apply to black holes, one might assume that the space–times $(\mathcal{M}_M, g_{ab}(M))$ each contain matter in a region $\{R < \text{some constant}\}$, such that the energy–momentum tensor $T^{ab}(x^c)$ can be approximated by a distribution $\int d\tau M u^a(\tau) u^b(\tau) \delta(x^c, y^c, (\tau))$, where $l_0 = \{y^c(\tau)\}$ and $u^b(\tau) = (d/d\tau) y^b(\tau)$. The geodesic property then follows on using the conservation equation $T^{ab}{}_{|b} = 0$, where the subscript vertical bar denotes covariant differentiation with respect to the background; this argument was given by Robertson (1937) in a discussion of the motion of test particles.

Alternatively, one may work with the empty space field equations for the external scheme. If one examines those parts of the $O(M)$ linearized external field equations which are most singular as $r \to \infty$, one again finds that l_0 must be a geodesic. An argument of this type was first given by Infeld and Schild (1949). In Kates (1980a), it is also shown by a matching argument how a compact object, not necessarily a black hole, but possibly with strong internal gravity, moves on a geodesic in a background geometry.

We remark that, in order to produce a 'pole–dipole' test particle, which deviates from geodesic motion as a result of spin forces, according to the Papapetrou equations (Papapetrou 1951; Rüdinger 1983), one expects to have to scale angular momentum as $O(M)$ rather than $O(M^2)$ as here in the limit $M \to 0$. This scaling is not appropriate to rotating black holes, since the Kerr parameter a (in the usual notation) should remain less than M. In some sense, the rotation of the black hole will affect its motion with respect to the background at $O(M)$ and higher orders. But at these orders the internal structure of the black hole cannot be ignored, and it does not seem feasible to give precise local geometrical expression to the statement that 'the black hole deviates from geodesic motion'. These are, however, cases as with a binary black hole system (Chapter 4), or with a single black hole in a nearly stationary background, where one can examine the black hole over very long time-scales, of order M^{-1} or longer, in which case the cumulative effect of non-geodesic $O(M)$ and higher terms in the equations of motion may in some cases have built up to $O(1)$ corrections to the black hole's position.

So far we have not considered the time dependence of the three-vector χ^i which represents the spin of the black hole in relation to the background. We now show that the four-vector χ^a, with components $(0, \chi^i)$ in coordinates (τ, x, y, z), is parallel propagated along the geodesic l_0 in the background. Consider the empty space Einstein equations, linearized up to order M^2 in the external scheme:

$$0 = R_{ab} = R_{ab}^{(0)} + (1/2)M[g_a^{(1)d}{}_{|bd} + g_b^{(1)d}{}_{|ad} - g_{ab}^{(1)|d}{}_{|d} - g^{(1)d}{}_{d|ab}]$$

$$+ M^2\{(1/2)[g_a^{(2)d}{}_{|bd} + g_b^{(2)d}{}_{|ad} - g_{ab}^{(2)|d}{}_{|d} - g^{(2)d}{}_{d|ab}]$$

$$+ (1/2)[g^{(1)de}g_{ab|de}^{(1)} + g^{(1)de}g_{de|ab}^{(1)} - g^{(1)de}g_{ad|be}^{(1)} - g^{(1)de}g_{bd|ae}^{(1)}]$$

$$+ (1/4)[2g^{(1)de}{}_{|d}g^{(1)}_{ab|e} - 2g^{(1)de}{}_{|d}g^{(1)}_{ae|b} - 2g^{(1)de}{}_{|d}g^{(1)}_{be|a}$$
$$- g^{(1)|d}_{ab}g^{(1)e}_e{}_{|d} + g^{(1)d}_a{}_{|b}g^{(1)e}_e{}_{|d} + g^{(1)d}_b{}_{|a}g^{(1)e}_e{}_{|d} + g^{(1)}_{de|a}g^{(1)de}{}_{|b}$$
$$- 2g^{(1)|d}_{ae}g^{(1)|e}_{bd} + 2g^{(1)|d}_{ae}g^{(1)e}_b{}_{|d}]\} + O(M^3). \tag{3.4.1}$$

Here $R^{(0)}_{ab}$ is the background Ricci tensor ($= 0$ locally) and we raise indices of varied quantities with the background metric. Thus

$$g^{(1)d}_a = g^{(0)de}g^{(1)}_{ae}, \qquad \text{etc.} \tag{3.4.2}$$

This expression for variations in the Ricci tensor can be derived using eqns (7.3) and (7.4) of Hawking and Ellis (1973).

Referring to our eqn (3.2.7), we see that the most singular terms in the $O(M^2)$ external field equations behave as r^{-4} as $r \to 0$. The $O(r^{-4})$ part of these equations is already satisfied, since covariant differentiation may be replaced by partial differentiation in considering these terms, which only involve the most singular parts of $g^{(1)}_{ab}, g^{(2)}_{ab}$. So for these purposes we may replace the background by Minkowski space, and take the space–times $(\mathcal{M}_M, g_{ab}(M))$ to be exactly Kerr solutions, while preserving eqn (3.2.7); i.e. our $O(M^2)O(R^{-4})$ external equations agree with those for an exact Kerr family where they are satisfied.

In considering the $O(M^2)O(r^{-3})$ part of the external field equations, we use the fact that the $O(M)$ QS perturbations have been removed. This shows that the next terms beyond eqn (3.2.7) in asymptotic expansions of $g^{(0)}_{ab}, g^{(1)}_{ab}, g^{(2)}_{ab}$, as $r \to 0$ are respectively $O(r^2)$, $O(r)$, $O(1)$. Then the $O(M^2)O(r^{-3})$ part of eqn (3.4.1), with $a = 0$, $b = i$, gives

$$0 = r^{-6}(-2\dot\chi^j\chi^j x^i r^2 - \dot\chi^j x^j \chi^i r^2 - \chi^j x^j \dot\chi^i r^2 + 8\dot\chi^j x^j \chi^k x^k x^i) \tag{3.4.3}$$

$\forall x^i$, with $\dot\chi^i = (d/d\tau)\chi^i(\tau)$. Hence

$$\dot\chi^i = 0. \tag{3.4.4}$$

Thus χ^a has constant components, say $(0,0,0,\chi)$, in coordinates (τ, x, y, z), and is parallel propagated along l_0 with respect to the background. The remaining $O(M^2)O(r^{-3})$ external equations are now satisfied by an argument like that in the previous paragraph.

3.5   Effect of background on internal structure

We now examine the lowest-order QS perturbations (i.e. the quasi-stationary internal perturbations) given by $M^2 j^{(2)}_{ab}(0, R, \theta, \phi)$. Since we have removed the $O(M)$ QS terms, $j^{(2)}_{ab}$ describes the largest deviations of the structure of the gravitational field near the event horizon from that of a Kerr solution, due to the distorting effect of the background universe. We shall see that $j^{(2)}_{ab}$ is essentially determined by the Riemann tensor of the background on l_0.

Let us decompose the background Riemann tensor on l_0 into its electric and magnetic parts (Ellis 1971) with respect to u^a, where u^a is the velocity vector of l_0, $\equiv (l, 0, 0, 0)$ in (τ, x, y, z) coordinates on $\mathcal{M}_0$. Thus we define the electric tensor

$$E_{ac} = C^{(0)}_{abcd} u^b u^d = E_{ca} \tag{3.5.1}$$

and magnetic tensor

$$H_{ac} = \frac{1}{2} \eta_{ab}{}^{gh} C^{(0)}_{g\,hcd} u^b u^d = H_{ca}, \tag{3.5.2}$$

where $C^{(0)}_{abcd}$ is the background Weyl tensor on l_0, and η^{abcd} is the totally anti-symmetric tensor such that $\eta^{0123} = (-g^{(0)})^{-1/2}$, where $g^{(0)} = \det(g^{(0)}_{ab})$. Then

$$E_{ac} u^c = H_{ac} u^c = 0, \tag{3.5.3}$$

so that E_{ac}, H_{ac} define three-dimensional symmetric tensors $E_{\alpha\gamma}, H_{\alpha\gamma}$ which give a complete description of the background Riemann tensor (or Weyl tensor) on l_0 *in vacuo*. We now decompose $C^{(0)}_{abcd}$ into parts with convenient rotational properties. First write $E_{\alpha\gamma}$ in the form

$$E_{\alpha\gamma} = \alpha_0 \begin{pmatrix} -1 & 0 & 0 \\ 0 & -1 & 0 \\ 0 & 0 & 2 \end{pmatrix} + \alpha_1 (\frac{3}{2})^{\frac{1}{2}} \begin{pmatrix} 0 & 0 & 1 \\ 0 & 0 & i \\ 1 & i & 0 \end{pmatrix} + \alpha_2 (\frac{3}{2})^{\frac{1}{2}} \begin{pmatrix} 1 & i & 0 \\ i & -1 & 0 \\ 0 & 0 & 0 \end{pmatrix}$$

$$+ \alpha_{-1} (\frac{3}{2})^{\frac{1}{2}} \begin{pmatrix} 0 & 0 & 1 \\ 0 & 0 & -i \\ 1 & -i & 0 \end{pmatrix} + \alpha_{-2} (\frac{3}{2})^{\frac{1}{2}} \begin{pmatrix} 1 & -i & 0 \\ -i & -1 & 0 \\ 0 & 0 & 0 \end{pmatrix}, \tag{3.5.4}$$

where $\alpha_{-1} = \bar{\alpha}_1$, $\alpha_{-2} = \bar{\alpha}_2$, and the α_m are, of course, functions of τ. Similarly, we decompose $H_{\alpha\gamma}$ in terms of coefficients $\beta_m (-2 \le m \le 2)$.

Now the arguments of Section 3.2 show that the $O(M^2)$ QS metric perturbations are $O(R^2)$ as $R \to \infty$ when referred to coordinates (T, X, Y, Z), and that the matching determines the $M^2 R^2$ parts in terms of those parts of the background metric which behave as r^2 as $r^2 \to 0$. We then find that at $O(M^2)$ in the QS scheme

$$\Psi_0 = \sum_{m=-2}^{+2} \frac{1}{2} (\alpha_m + i\beta_m)_2 Y_2^m(\theta, \phi) + 0(R^{-1}) \tag{3.5.5}$$

as $R \to \infty$, where

$$_2Y_2^0(\theta, \phi) = 6 \sin^2 \theta,$$
$$_2Y_2^{\pm 1}(\theta, \phi) = -2\sqrt{6} \sin \theta (\cos \theta \pm 1) e^{\pm i\phi},$$
$$_2Y_2^{\pm 2}(\theta, \phi) = 2\sqrt{6}(2 - \sin^2 \theta \pm 2 \cos \theta) e^{\pm 2i\phi}$$

are spin-weight-2 harmonics (Goldberg *et al.* 1967). Note that the limiting value $\sum_{m=-2}^{+2} \frac{1}{2}(\alpha_m + i\beta_m)_2 Y_2^m(\theta, \phi)$ is just what would be found by 'matching out' the Kinnersley tetrad to large R, giving in (τ, x, y, z) coordinates

$$l^a \equiv (1 \sin\theta \cos\phi, \sin\theta \sin\phi, \cos\theta),$$

$$n^a \equiv \frac{1}{2}(1, -\sin\theta \cos\phi, -\sin\theta \sin\phi, -\cos\theta),$$

$$m^a = \frac{1}{\sqrt{2}}(0, \cos\theta \cos\phi, \cos\theta \sin\phi, -\sin\theta)$$

$$+ \frac{i}{\sqrt{2}}(0, -\sin\phi, \cos\phi, 0),$$

(3.5.6)

and computing Ψ_0 for the background.

Hence the $L = 2$ modes will in general be present in Ψ_0 at $O(M^2)$. Now when $m \neq 0$, it may be verified that a solution of the radial equation (3.3.5) for $L = 2$ which satisfies the boundary conditions of Section 3.3 at the future event horizon $R = R_+$ is

$$W_2^m(R) = (R - R_-)^{-3-\gamma_m}(R - R_+)^{-2+\gamma_m} \times$$

$$\times F\left[1, 3 + 2\gamma_m; -1 + 2\gamma_m; \left(\frac{R - R_+}{R - R_-}\right)\right].$$

(3.5.7)

When $m = 0$, two linearly independent solutions of the radial equation are

$$W \equiv 1,$$

(3.5.8)

$$W = \int_R^\infty \frac{d\omega}{(\omega^2 - 2\omega + \chi^2)^3}.$$

(3.5.9)

When one examines Ψ_0 at $O(M^2)$ in the Hartle tetrad of Section 3.3, on the unperturbed future horizon $R = R_+$, one finds for the first solution (3.5.8) that

$$\Psi_0 = 0$$

(3.5.10)

on the horizon, and for the second solution (3.5.9),

$$\Psi_0 = \text{const.} \times \sin^2\theta.$$

(3.5.11)

In fact, the boundary conditions on the horizon imply that we only need consider the first solution. To find the boundary conditions, we follow Hawking and Hartle (1972) and consider the Newman–Penrose equation (Newman and Penrose 1962)

$$\frac{d\sigma}{ds} = 2\rho\sigma + (3\epsilon - \bar{\epsilon})\sigma + \Psi_0.$$

(3.5.12)

Here

$$\rho = -l_{a;b}m^a\bar{m}^b \tag{3.5.13}$$

is the convergence of the null geodesic generators of the event horizon, with tangent vector $l^a = dx^a/ds$. Also

$$\sigma = -l_{a;b}m^a m^b \tag{3.5.14}$$

is the shear of the horizon, and

$$\epsilon = -\frac{1}{2}(l_{a;b}n^a l^b - m_{a;b}\bar{m}^a l^b). \tag{3.5.15}$$

The vectors l^a, n^a, m^a, $\bar{m}^a$ form a null tetrad. We assume that m^a and $\bar{m}^a$ are parallel propagated along the horizon generators, so that $\epsilon = \bar{\epsilon}$. These conditions are satisfied by the Hartle tetrad for the unperturbed solution, so that we may take

$$l^a = l_H^a + O(M^2), \qquad n^a = n_H^a + O(M^2), \qquad m^a = m_H^a + O(M^2). \tag{3.5.16}$$

This implies that

$$s = V + O(M^2) \tag{3.5.17}$$

(apart from a constant), where V is the null Kerr coordinate which measures 'group time' according to the Kerr isometry group along the null generators of the unperturbed Killing horizon. We write

$$\rho = \rho^{(0)} + M\rho^{(1)} + M^2\rho^{(2)} + \dots \tag{3.5.18}$$

for the perturbed convergence, and similarly for σ, ϵ, Ψ_0. Then (Hawking and Hartle 1972)

$$\rho^{(0)} = 0, \qquad \sigma^{(0)} = 0, \qquad \epsilon^{(0)} = (1-\chi^2)^{\frac{1}{2}}/4R_+. \tag{3.5.19}$$

Since the perturbations are $O(M^2)$, we have

$$\rho^{(1)} = \sigma^{(1)} = \epsilon^{(1)} = 0. \tag{3.5.20}$$

Then eqn (3.5.12) gives

$$\frac{d\sigma^{(2)}}{ds} = 2\epsilon^{(0)}\sigma^{(2)} + \Psi_0^{(2)}. \tag{3.5.21}$$

Now it is already implicit in our approximations that the position and shape of the event horizon are in some sense differentiable functions of the space–time metric. We shall further suppose that the structure of the perturbed event horizon in the QS scheme can be determined from the purely local metric perturbations. This is plausible at least at order M^2, where the perturbations are stationary, since in a stationary metric an event horizon has a local characterization as a stationary null surface. At order $M^i(i > 2)$, the properties of the event horizon should not vary at a rate faster that V^{i-2} as $V \to \infty$, in order to comply with the requirement of quasi-stationarity, and so should also be determined locally. Hence for an $L = 2, m = 0$ mode at $O(M^2)$, the horizon should undergo a stationary, axisymmetric perturbation. But the null generator L^a is given by

$$l^a = K^a + \omega \tilde{K}^a + O(M^2), \tag{3.5.22}$$

where $\omega = \chi/2R_+$ is the angular velocity of the Kerr horizon, K^a is the time-translation Killing vector, and $\tilde{K}^a$ is the rotational Killing vector of the Kerr solution. Since ρ and σ represent the variations in the intrinsic horizon metric as the generator l^a winds around horizon, $\rho^{(2)}$ and $\sigma^{(2)}$ must be zero for our $m = 0$ mode. Equation (3.5.21) then shows that $\Psi_0^{(2)} = 0$ on the horizon. Thus only the solution $W \equiv$ const. of the $L = 2, m = 0$ radial equation is acceptable.

We can now write the $L = 2$ parts of Ψ_0 in the Kinnersley tetrad at $O(M^2)$ in the QS scheme in the form

$$
\begin{aligned}
\Psi_0 = &\frac{1}{2}(\alpha_0 + i\beta_0)_2 Y_2^0(\theta, \phi) \\
&+ \sum_{\{m=-2,-1,1,2\}} \frac{(\alpha_m + i\beta_m)\Gamma(3 + 2\gamma_m)(R_+ - R_-)^5}{48\Gamma(-1 + 2\gamma_m)} \times \\
&\times {}_2 Y_2^m(\theta, \phi)(R - R_-)^{-3-\gamma_m}(R - R_+)^{-2+\gamma_m} \\
&\times F\left[1, 3 + 2\gamma_m; -1 + 2\gamma_m; \left(\frac{(R - R_+)}{(R - R_-)}\right)\right].
\end{aligned}
\tag{3.5.23}
$$

Here we have used Erdélyi (1953) to find the asymptotic behaviour as $R \to \infty$ of the hypergeometric function in eqn (3.5.7), together with the boundary conditions (3.5.5). Further, eqn (3.5.5) shows that the $L > 2$ parts of Ψ_0 at $O(M^2)$ are $O(R^{-1})$ as $R \to \infty$. By an argument as in Section 3.3, these $L > 2$ parts must be identically zero. Thus eqn (3.5.23), by Wald's uniqueness theorem (Wald 1973), gives almost all the information about the $O(M^2)$ QS perturbations. We see that the largest internal perturbations due to the background are caused by conformal curvature on l_0, and are of a quadrupole character.

We may now argue as in Hawking and Hartle (1972) to find the lowest-order rates of change of area and angular momentum of the black hole. First, eqn (3.5.23) and the remarks in Section 3.3 show that the lowest-order Ψ_0 on the horizon is

$$
\Psi_0^{(2)} = \frac{(R_+ - R_-)^4}{768(R_+)^2} \times
$$
$$
\times \sum_{m=-2}^{+2} (\alpha_m + i\beta_m)\frac{\Gamma(3 + 2\gamma_m)}{\Gamma(-1 + 2\gamma_m)} {}_2 Y_2^m(\theta, \phi'). \tag{3.5.24}
$$

Then eqns (3.5.21) and (3.5.22) show that when $\Psi_0^{(2)}$ has $e^{im\phi'}$ dependence on ϕ', one has on the horizon

$$\sigma^{(2)} = \frac{\Psi_0^{(2)}}{im\omega - 2\epsilon^{(0)}}. \tag{3.5.25}$$

Hence, on the horizon

$$\sigma^{(2)} = \frac{(R_+ - R_-)^3}{192 R_+} \times$$

$$\times \sum_{m=-2}^{+2} (\alpha_m + i\beta_m) \frac{\Gamma(3 + 2\gamma_m)}{\Gamma(2\gamma_m)} {}_2 Y_2^m(\theta, \phi'). \tag{3.5.26}$$

We also need the Newman–Penrose equation (*in vacuo*)

$$\frac{d\rho}{ds} = \rho^2 + \sigma\bar{\sigma} + (\epsilon + \bar{\epsilon})\rho. \tag{3.5.27}$$

Hence

$$\frac{d\rho^{(2)}}{ds} = 2\epsilon^{(0)}\rho^{(2)}. \tag{3.5.28}$$

But $\rho^{(2)}$ will be periodic along a generator as it winds around the horizon. Thus

$$\rho^{(2)} = 0. \tag{3.5.29}$$

Similarly,

$$\frac{d\rho^{(3)}}{ds} = 2\epsilon^{(0)}\rho^{(3)}. \tag{3.5.30}$$

But we expect that $\rho^{(3)}$ will be $O(s)$ as $s \to \infty$, since the $O(M^3)$ QS perturbations are $O(V)$ as $V \to \infty$. Hence also

$$\rho^{(3)} = 0. \tag{3.5.31}$$

Then eqn (3.5.27) shows

$$\frac{d\rho^{(4)}}{ds} = \sigma^{(2)}\bar{\sigma}^{(2)} + \epsilon^{(0)}\rho^{(4)}. \tag{3.5.32}$$

If we consider the area $A(V)$ of the instantaneous horizon at constant V, then

$$\frac{dA(V)}{dV} = -2 \int \rho\, dA, \tag{3.5.33}$$

where the integral is taken over the instantaneous 2-surface. When eqn (3.5.32) is integrated over this surface, we find

$$\frac{dA(V)}{dV} = \frac{M^4}{\epsilon^{(0)}} \int \sigma^{(2)} \bar{\sigma}^{(2)} dA + O(M^5), \qquad (3.5.34)$$

where the derivative term has vanished since the perturbation is quasi-stationary. Here $\int dA(\ )$ refers to the unperturbed instantaneous horizon metric (Smarr 1973) (see subsection 2.5.2)

$$ds^2 = (R_+^2 + \chi^2 \cos^2 \theta) d\theta^2 + \frac{(R_+^2 + \chi^2) \sin^2 \theta}{(R_+^2 + \chi^2 \cos^2 \theta)} d\phi'^2. \qquad (3.5.35)$$

Using eqn (3.5.26), we find in the QS scheme

$$\frac{dA(V)}{dV} = \frac{16}{15} \pi M^4 (1 - \chi^2)^{-1/2} \chi^2$$
$$\times \sum_{m=-2}^{+2} (\mid \alpha^m \mid^2 + \mid \beta_m \mid^2)[1 + \chi^2(m^2 - 1)] \times$$
$$\times [4 + \chi^2(m^2 - 4)] + O(M^5). \qquad (3.5.36)$$

To find the real increase of area, we suppose that the hypersurfaces of constant τ were chosen to coincide with hypersurfaces of constant V near the horizon. The only differences between the real and QS rates of change of area are caused by the time rescaling (by factor M) and and the conformal factor M^2 removed from the internal metric perturbations. Hence if $^R A(\tau)$ is the real area at time τ, then

$$\frac{d^R A(\tau)}{d\tau} = M \frac{dA(V)}{dV}. \qquad (3.5.37)$$

As written, the rate of area increase depends on the α_m, β_m, which were given a somewhat arbitrary normalization. We can rewrite the expression in terms of invariants, and so remove any such arbitrariness. Recall that u^a is the unit tangent vector to the geodesic l_0 in the background, and let z^a be the unit spacelike vector in the background into which the spin direction ties. Define

$$B_1 = R^{(0)abcd} R^{(0)}_{abcd},$$
$$B_2 = R^{(0)abcd} u_b u_d R^{(0)}_{aecf} u^e u^f,$$
$$B_3 = R^{(0)abcd} z_c u_d R^{(0)}_{abef} z^e u^f, \qquad (3.5.38)$$
$$B_4 = R^{(0)abcd} u_b z_c u_d R^{(0)}_{aefg} u^e z^f u^g,$$
$$B_5 = R^{(0)abcd} z_b u_c z_d R^{(0)}_{aefg} z^e u^f z^g,$$

where $R^{(0)}_{abcd}$ is the background Riemann tensor on l_0. Then

$$\frac{dA(V)}{dV} = \frac{16}{45}\pi M^4 \chi^2 (1 - \chi^2)^{-1/2} \times$$
$$\times [-(1 + 3\chi^2)B_1 + 16(1 + 3\chi^2)B_2 \qquad\qquad (3.5.39)$$
$$- 6(1 + 3\chi^2)B_3 - 3(8 + 29\chi^2)B_4 - 15\chi^2 B_5] + O(M^5).$$

Since the Kerr solution is stable (Whiting 1989), then secular effects such as area increase must correspond to variations in the basic parameters of the Kerr solution over a long time-scale. For a Kerr solution with mass μ and the usual spin parameter a, the area of the instantaneous event horizon is

$$A = 8\pi\mu[\mu + (\mu^2 - a^2)^{\frac{1}{2}}]. \qquad\qquad (3.5.40)$$

So if we know the lowest-order time dependence of A and μ, say, we also know how a varies at the lowest order. An argument is given in Hawking and Hartle (1972), which shows that the mass is constant at $O(M^4)$ for QS perturbations with our parametrization, provided that the perturbations are asymptotically flat, being caused by a distant matter distribution. Clearly, our QS perturbations are not asymptotically flat (see eqn (3.5.5)), so that the argument cannot be used as it stands. But the mass and angular momentum changes are being caused by purely local effects near the horizon, for which the far field is irrelevant. Since these local effects can presumably be simulated by asymptotically flat perturbations, the mass is constant at $O(M^4)$ in the QS scheme. Hence all the area change at $O(M^4)$ is caused by the change in a. Hence, in the QS scheme

$$\frac{da}{dV} = -\frac{(1 - \chi^2)^{1/2}}{8\pi\chi}\frac{dA}{dV} + O(M^5). \qquad\qquad (3.5.41)$$

We see that the presence of the background always forces the black hole to lose angular momentum. Also the mass of the black hole changes by a fraction of $O(1)$ on a background time-scale of $O(M^{-4})$, while the angular momentum changes by a fraction of $O(1)$ on a τ scale of $O(M^{-3})$; this justifies our assertions in Sections 3.2 and 3.3 about the time dependence of low-order mass and spin terms.

The secular changes on the horizon at $O(M^4)$ in the QS scheme are determined completely by $\rho^{(4)}$ and $\sigma^{(4)}$. We expect the parameters $(\theta, \phi' - \omega V)$ labelling an unperturbed null generator to change on a V-time-scale of order M^{-4}, as the horizon evolves. Because the perturbations are quasi-stationary, secular effects in $\rho^{(4)}, \sigma^{(4)}$ are averages of some function of ϕ', taken over many revolutions of the null generator around the horizon. Thus, secular effects depend only on θ (at the lowest order). The 'rings' of null geodesics at constant θ will slowly be moved through new values of θ. In particular, the polar geodesics at $\theta = 0, \pi$ will remain polar at the lowest order. This makes it hard from the point of view of the internal scheme to assign a direction to the angular momentum change of the black hole due to this effect. One can nevertheless examine the precession of the black hole's spin-axis due to other effects, in the external scheme, as in Thorne and Hartle (1985)—see Section 3.9. In particular, in the case in which the 'external universe' is due to a second black hole, the precession can be found at low post-Newtonian orders. That is, one observes the precession over time-scales comparable with negative powers of M. The situation is similar to that in Section 3.4 respecting the deviation from geodesic motion. In this case of two black holes, this can be evaluated using the external scheme (see Chapter 4).

Our knowledge of $\rho^{(2)}$ and $\sigma^{(2)}$ allows us to write down the perturbed intrinsic horizon metric at $O(M^2)$ in the QS scheme corresponding to the non-axisymmetric modes. Let us use coordinates $x^1 = \theta, x^2 = \tilde{\phi}, x^3 = V$ on the horizon, and assume (making a gauge transformation if necessary) that the generator l^a is given by $l^a = (\partial/\partial V)^a$. For the unperturbed horizon, we would have $\tilde{\phi} = \phi - \omega V$. Then the perturbed intrinsic degenerate metric can be written

$$
ds^2 = \left\{ f_{11} + M^2 \left[{}_0n_{11} + \sum_{\{m=1,2\}} \left({}_mn_{11}e^{im(\tilde{\phi}+\omega V)} + {}_m\bar{n}_{11}e^{-im(\tilde{\phi}+\omega V)} \right) \right] \right\} d\theta^2
$$

$$
+ 2M^2 \left[{}_0n_{12} + \sum_{\{m=1,2\}} \left({}_mn_{12}e^{im(\tilde{\phi}+\omega V)} + {}_m\bar{n}_{12}e^{-im(\tilde{\phi}+\omega V)} \right) \right] d\theta d\tilde{\phi}
$$

$$
+ \left\{ f_{22} + M^2 \left[{}_0n_{22} + \sum_{\{m=1,2\}} \left({}_mn_{22}e^{im(\tilde{\phi}+\omega V)} \right. \right. \right.
$$

$$
\left. \left. \left. + {}_m\bar{n}_{22}e^{-im(\tilde{\phi}+\omega V)} \right) \right] \right\} d\tilde{\phi}^2 + O(M^3). \tag{3.5.42}
$$

Here the unperturbed terms

$$
f_{11} = R_+^2 + \chi^2 \cos^2 \theta,
$$
$$
f_{22} = \frac{(R_+^2 + \chi^2)^2 \sin^2 \theta}{(R_+^2 + \chi^2 \cos^2 \theta)}, \tag{3.5.43}
$$

(eqn (3.5.35)) and ${}_mn_{11}, {}_mn_{12}, {}_mn_{22}(m = 0, 1, 2)$ are functions of θ, with

$$
{}_0n_{11}, {}_0n_{12}, {}_0n_{22}
$$

real. To find ${}_mn_{11}, {}_mn_{12}, {}_mn_{22}$ for $m = 1, 2$, we recall that ρ and σ on the horizon are quantities depending only on the intrinsic geometry of the null surface (Penrose 1968) and can be computed from eqn (3.5.42) by embedding the null surface in a space–time with an extra null coordinate V' such that the space–time metric is given by the addition of an extra term $-2dV dV'$ to eqn (3.5.42). Examining the Hartle tetrad on the horizon (Hartle 1974), we see that

$$m^\theta = \frac{1}{\sqrt{2}(R_+ + i\chi\cos\theta)} + O(M^2),$$

$$m^{\tilde{\phi}} = \frac{i(R_+ - i\chi\cos\theta)}{\sqrt{2}\sin\theta(R_+^2 + \chi^2)}, \tag{3.5.44}$$

$$m^V = O(M^2).$$

Write

$$\sigma^{(1)} = \sum_{\{m=-2,-1,1,2\}} Z_m(\theta)e^{im\phi'}. \tag{3.5.45}$$

Then by computing ρ and σ as above from eqn (3.5.42), and considering modes with $m = \pm 1, m = \pm 2$ in isolation, we find that

$$\rho^{(2)} = 0$$

$$\Rightarrow (R_+^2 + \chi^2)^2 \sin^2\theta\,_m n_{11} + (R_+^2 + \chi^2\cos^2\theta)^2\,_m n_{22} = 0, m = 1, 2. \tag{3.5.46}$$

Also

$$-\frac{im\omega}{4(R_+ + i\chi\cos\theta)^2}\,_m n_{11} + \frac{m\omega(R_+ - i\chi\cos\theta)}{2(R_+ + i\chi\cos\theta)(R_+^2 + \chi^2)\sin\theta}\,_m n_{12}$$

$$+\frac{im\omega(R_+ - i\chi\cos\theta)^2}{4(R_+^2 + \chi^2)^2\sin^2\theta}\,_m n_{22} = Z_m, \qquad m = 1, 2, \tag{3.5.47}$$

$$\frac{im\omega}{(4R_+ + i\chi\cos\theta)^2}\,_m\bar{n}_{11} - \frac{m\omega(R_+ - i\chi\cos\theta)}{2(R_+ + i\chi\cos\theta)(R_+^2 + \chi^2)\sin\theta}\,_m\bar{n}_{12}$$

$$-\frac{im\omega(R_+ - i\chi\cos\theta)^2}{4(R_+^2 = \chi^2)^2\sin^2\theta}\,_m\bar{n}_{22} = Z_{-m}, \qquad m = 1, 2. \tag{3.5.48}$$

Hence, for $m = 1, 2$,

$$_m n_{11} = \frac{i(R_+ + i\chi\cos\theta)^2}{m\omega}Z_m + \frac{i(R_+ - i\chi\cos\theta)^2}{m\omega}\bar{Z}_{-m},$$

$$_m n_{12} = \frac{(R_+^2 + \chi^2)(R_+ + i\chi\cos\theta)\sin\theta}{m\omega(R_+ - i\chi\cos\theta)}Z_m$$

$$-\frac{(R_+^2 + \chi^2)(R_+ + i\chi\cos\theta)\sin\theta}{m\omega(R_+ + i\chi\cos\theta)}\bar{Z}_{-m}, \tag{3.5.49}$$

$$_m n_{22} = -\frac{i(R_+^2 + \chi^2)\sin^2\theta}{m\omega(R_+ - i\chi\cos\theta)^2}Z_m - \frac{i(R_+^2 + \chi^2)\sin^2\theta}{m\omega(R_+ + i\chi\cos\theta)^2}\bar{Z}_{-m}.$$

One may then use eqn (3.5.26) for $\sigma^{(2)}$ to find (slightly lengthy) expressions for $_m n_{11}, _m n_{12}, _m n_{22}(m = 1, 2)$. This approach does not give us $_0 n_{11}, _0 n_{12}, _0 n_{22}$. The general perturbation of the four-metric, and hence of the intrinsic horizon geometry, including the case $m = 0$, may be found by taking the solution Ψ_0 of the Teukolsky equation (Teukolsky 1973) and deducing the perturbed four-metric from the work of Chrzanowski (1975) and Wald (1978), following the description of Section 2.7.

3.6 The non-rotating case

In this section we examine the internal structure in the case in which the small black hole is Schwarzschild-like. More precisely, we assume that the black hole is rotating so slowly that $\chi = 0$ and that the $O(M)$ QS perturbations of the Schwarzschild solution zero (so that there is no perturbation among the Kerr family at $O(M)$). We may then apply well-known results on Schwarzschild perturbations to find the QS metric terms $M^2 j_{ab}^{(2)}(0, R, \theta, \phi)$.

If we adopt the gauge condition of Regge and Wheeler (1957), then in Schwarzschild coordinates (T, R, θ, ϕ), stationary perturbations can be classified by angular numbers (L, m) and into modes of even or odd parity, where in an even (L, m) mode

$$j_{ab}^{(2)} = \begin{pmatrix} \left(1 - \frac{2}{R}\right) H_0 Y_L^m & H_1 Y_L^m & 0 & 0 \\ H_1 Y_L^m & \left(\frac{R}{R-2}\right) H_2 Y_L^m & 0 & 0 \\ 0 & 0 & R^2 K Y_L^m & 0 \\ 0 & 0 & 0 & R^2 \sin^2\theta K Y_L^m \end{pmatrix}, \quad (3.6.1)$$

and in an odd (L, m) mode

$$j_{ab}^{(2)} = \begin{pmatrix} 0 & 0 & -\frac{1}{\sin\theta} h_0 \frac{\partial}{\partial\phi} Y_L^m & \sin\theta h_0 \frac{\partial}{\partial\theta} Y_L^m \\ 0 & 0 & -\frac{1}{\sin\theta} h_1 \frac{\partial}{\partial\phi} Y_L^m & \sin\theta h_1 \frac{\partial}{\partial\theta} Y_L^m \\ -\frac{1}{\sin\theta} h_0 \frac{\partial}{\partial\phi} Y_L^m & -\frac{1}{\sin\theta} h_1 \frac{\partial}{\partial\phi} Y_L^m & 0 & 0 \\ \sin\theta h_0 \frac{\partial}{\partial\theta} Y_L^m & \sin\theta h_1 \frac{\partial}{\partial\theta} Y_L^m & 0 & 0 \end{pmatrix}.$$

$$(3.6.2)$$

Here $Y_L^m(\theta, \phi)$ is the usual spherical harmonic, and $H_0, H_1, H_2, K, h_0, h_1$ are all functions of R. We shall only be interested in the $L = 2$ modes, since they give the leading internal perturbations (Section 3.5).

Stationary Schwarzschild perturbations have been discussed by Vishveshwara (1970); in the even case the field equations imply $H_0 = H_2 = H$, $H_1 \equiv 0$, where

$$\frac{d^2 H}{dR*^2} + \frac{2}{R}\left(1 - \frac{2}{R}\right)\frac{dH}{dR*} - \left[\frac{4}{R^4} + \frac{L(L+1)}{R^2}\left(1 - \frac{2}{r}\right)\right] H = 0 \quad (3.6.3)$$

and $R* = R + 2\log(R - 2)$. The boundary conditions at the horizon require

$$H \to 0 \qquad \text{as } R \to 2. \quad (3.6.4)$$

Hence

$$H = C_m R(R - 2) \quad (3.6.5)$$

for some constant C_m, when $L = 2$. We find K from two of the field equations
(Edelstein and Vishveshwara 1970)

$$\frac{dK}{dR} - \frac{dH}{dR} - \frac{2}{R(R-2)}H = 0,$$

$$(R - 1)\frac{dK}{dR} - (R - 2)\frac{dH}{dR} + 2(H - K) = 0.$$

(3.6.6)

Hence, when $L = 2$,

$$K = C_m(R^2 - 2).$$

(3.6.7)

For odd stationary perturbations, $h_1 \equiv 0$ and

$$\frac{d^2 h_0}{dR^{*2}} - \frac{2}{R^2}\frac{dh_0}{dR*} \left[\frac{L(L+1)}{R^2} - \frac{4}{R^3}\right]\left(1 - \frac{2}{R}\right)h_0 = 0.$$

(3.6.8)

The boundary conditions at the horizon require

$$h_0 \to 0 \qquad \text{as } R \to 2.$$

(3.6.9)

Hence, when $L = 2$,

$$h_0 = D_m R^2(R - 2)$$

(3.6.10)

for some constant D_m.

To compare the metric perturbations with the work of the previous section,
we first specify our normalization of spherical harmonics:

$$Y_2^0(\theta, \phi) = 3\cos^2\theta - 1,$$

$$Y_2^{+1}(\theta, \phi) = \sqrt{6}\sin\theta\cos\theta e^{\pm i\phi},$$

$$Y_2^{+2}(\theta, \phi) = \sqrt{6}\sin^2\theta e^{\pm 2i\phi}.$$

(3.6.11)

Then for an $(L = 2, m)$ even mode, in the Kinnersley tetrad

$$\Psi_0 = -\frac{1}{2}M^2 \frac{H}{R(R-2)} {}_2Y_2^m(\theta, \phi) + O(M^3),$$

(3.6.12)

and for an $(L = 2, m)$ odd mode,

$$\Psi_0 = \frac{i}{2}M^2 \left[\frac{1}{R(R-2)}\frac{dh_0}{dr} - \frac{2}{R^2(R-2)^2}h_0\right] {}_2Y_2^m(\theta, \phi) + O(M^3).$$

(3.6.13)

Altogether,

$$\Psi_0 = -\frac{M^2}{2} \sum_{m=-2}^{+2} (C_m - 3iD_m) {}_2Y_2^m(\theta, \phi) + O(M^3).$$

(3.6.14)

Comparing this with the asymptotic behaviour of Ψ_0 as $R \to \infty$ in eqn (3.5.5), we find

$$C_m = -\alpha_m, \qquad D_m = \frac{1}{3}\beta_m. \qquad (3.6.15)$$

Thus the electric part of the background Weyl tensor produces even internal perturbations, and the magnetic part produces odd perturbations.

By Wald's theorem, we have already given a complete description of the $O(M^2)$ QS perturbations, apart from the $L = 0, 1$ modes which represent changes in mass and angular momentum respectively among the Kerr family (Vishveshvara 1970; Zerilli 1970; Wald 1973). The arguments of Section 3.5 show that the $L = 0, 1$ perturbations will be constant over a background time-scale, as measured by τ.

We see that the $O(M^2)$ QS metric perturbations, when referred to coordinates $(T, R \sin\theta \cos\phi, R \sin\theta \sin\phi, R \cos\theta)$ are $O(R^2)$ as $R \to \infty$; this is consistent with our matching conditions in Section 3.2. The $M^2 R^2$ QS terms match onto those background terms which behave as R^2 when $r \to 0$. Of course, the Regge–Wheeler gauge condition on the internal scheme enforces a coordinate condition on these $O(r^2)$ background terms.

From the explicit metric perturbations in the internal scheme, one can find the event horizon. In the case of one large and one small Schwarzschild black hole, with masses M and m, the event horizon has been found approximately (Bishop 1988) in the case either of time-symmetric initial data, or that the small black hole falls from rest at infinity.

3.7 Solution of external problem

The Green function approach to general relativity developed by Sciama *et al.* (1969) allows us to give a useful representation of the lowest-order external perturbation $g_{ab}^{(1)}$, when the external scheme refers to the vacuum Einstein equations. To show this, we first define a coordinate transformation which simplifies the form of the perturbations $g_{ab}^{(1)}$ of the external metric. This transformation is given by $(\tau, x, y, z, M) \to (\tau, x', y', z', M)$ on a subset of $\mathcal{N}$ such that $x'r = r'x$, $y'r = r'y$, $z'r = r'z$, and

$$r = r'\left(1 + \frac{M}{2r'}\right)^2 \qquad (3.7.1)$$

for $r <$ some constant, $r/M >$ some constant; outside these regions the coordinate transformation should be defined so as to be regular. We may assume that the transformation has been defined so that the components of $g_{ab}^{(0)}$ are unchanged. We note that eqn (3.2.7) for the most singular of $g_{ab}^{(1)}$ near l_0 has been replaced by

$$g_{ab}^{(1)} \sim \frac{2}{r'}\text{diag}(1, 1, 1, 1) \qquad (3.7.2)$$

as $r' \to 0$, in (τ, x', y', z') coordinates. In fact the coordinate transformation $r = r'(1 + M/2r')^2$ just gives the transformation from Schwarzschild to isotropic coordinates in the Schwarzschild metric. Now let us make a gauge transformation which puts $g_{ab}^{(1)}$ in the harmonic gauge (Hawking and Ellis 1973). This will simplify the linearized Einstein field equation in the external scheme.

Following Sciama *et al.* (1969), we consider

$$\delta g^{ab} = -g^{(0)ac} g^{(0)bd} g_{cd}^{(1)}, \tag{3.7.3}$$

which gives the first-order variation in g^{ab} in the external scheme. Define

$$\psi_a = \frac{1}{2} g_{bc}^{(0)} \delta g^{bc}{}_{|a} - g_{ab}^{(0)} \delta g^{bc}{}_{|c}, \tag{3.7.4}$$

where we recall that the subscripted vertical bar denotes covariant differentiation with respect to the background. Because of our coordinate conditions on $g_{ab}^{(0)}$ and $g_{ab}^{(1)}$ in (τ, x, y, z) coordinates, we find that ψ_a is $O(1)$ as $r' \to 0$, in (τ, x', y', z') coordinates. By a further coordinate transformation, ψ_a may be removed entirely. Let us solve the equation

$$g^{(0)ab} \xi_{c|ab} = \psi_c \tag{3.7.5}$$

to find a vector field ξ_c on $\mathcal{M}_0$. Then perform a coordinate transformation on $\mathcal{N}$ to coordinates $(\hat{\tau}, \hat{x}, \hat{y}, \hat{x}, M)$, where

$$(\hat{\tau}, \hat{x}, \hat{y}, \hat{z}) = (\tau, x', y', z') + M \xi^a + O(M^2) \tag{3.7.6}$$

as $M \to 0$, with τ, x', y', z' fixed. This transformation has the effect of adding $\xi^{a|b} + \xi^{b|a}$ to δg^{ab}, so that $\psi_a = 0$ in $(\hat{\tau}, \hat{x}, \hat{y}, \hat{z})$ coordinates. We still have

$$g_{ab}^{(1)} \sim \frac{2}{\hat{r}} \mathrm{diag}(1,1,1,1) \tag{3.7.7}$$

as $\hat{r} \to 0$, in $(\hat{\tau}, \hat{x}, \hat{y}, \hat{z})$ coordinates, where $\hat{r}^2 = \hat{x}^2 + \hat{y}^2 + \hat{z}^2$.

The harmonic gauge condition $\psi_a = 0$ simplifies the linearized empty-space Einstein equations, to give

$$g^{(0)cd} \delta g^{ab}{}_{|cd} + 2R^{(0)a}{}_c{}^b{}_d \, \delta g^{cd} = 0. \tag{3.7.8}$$

Let $E^{a'b'}{}_{cd}(x', x)$ be the retarded Green function for this equation (Friedlander 1975) such that

$$g^{(0)ef} E^{a'b'}{}_{cd|ef} + 2R^{(0)e}{}_c{}^f{}_d E^{a'b'}{}_{ef}$$
$$= g^{(0)a'}{}_{(c} g^{(0)b'}{}_{d)} [g^{(0)}(x') g^{(0)}(x)]^{-1/4} \delta(x, x'). \tag{3.7.9}$$

Here $E^{a'b'}{}_{cd} = E^{(a'b')}{}_{(cd)}(x', x)$ is a two-point tensor, that is, it has tensorial properties at x' with respect to the indices a', b' and at x with respect to the indices c, d. Also $g^{(0)}(x) = \det(g_{ab}^{(0)}(x))$, $\delta(x, x')$ is the Dirac distribution, and

$g^{(0)a'}{}_c(x, x')$ denotes the two-point vector of geodesic parallel transport (which is only defined locally). Use of Green's identity leads, as in Sciama *et al.* (1969), to a Kirchhoff representation

$$\delta g^{a'b'}(x') = \int_{\partial\Omega} g^{(0)cd}[E^{a'b'}{}_{ef|c}\delta g^{ef} - E^{a'b'}{}_{ef}\delta g^{ef}{}_{|c}] \times$$
$$\times [-g^{(0)}(x)]^{\frac{1}{2}} dS_d, \qquad (3.7.10)$$

where Ω is a volume containing x', and dS_d is the outward-directed coordinate surface element on the boundary $\partial\Omega$.

We choose Ω to be bounded by a small tube $\{\hat{r} = \epsilon\}$, with $l_0 \cap \Omega \neq \emptyset$, and by a surface 'near infinity'. We assume that the contribution from the large surface tends to zero as Ω increases to fill the whole of $\mathcal{M}_0 - l_0$. This is in some sense a requirement that the linearized external solution should be completely determined by the black hole, without any additional gravitational radiation incoming from infinity. Assuming that the relevant functions are sufficiently well behaved, our conditions on δg^{ab} near l_0 imply

$$\delta g^{a'b'}(x') = 8\pi \int d\tau u^e(\tau)u^f(\tau)E^{a'b'}{}_{ef}(x', x(\tau))[-g^{(0)}(x(\tau))]^{\frac{1}{2}}, \qquad (3.7.11)$$

which is the desired integral representation. When translated back into our original coordinate system (τ, x, y, z), these results show that δg^{ab} is defined via such an integral representation up to the addition of a gauge term $\xi^{a|b} + \xi^{b|a}$.

In general, $E^{a'b'}{}_{ef}$ will include tail effects (DeWitt and Brehme 1960; Price 1972; Friedlander 1975), i.e. its support will include points x with timelike separation from x'. This implies that the $O(M^3)$ QS perturbations may depend, through the matching, on the whole past history of the black hole, whereas the $O(M^2)$ QS perturbations are caused only by background curvature at one time τ. Roughly speaking, outgoing information about the black-hole field is partially reflected back into the black hole off the conformal curvature of the background.

It might be possible to extend this Green function approach in order to solve higher-order external equations. One would need to subtract off the most singular parts of $g^{(n)}_{ab}$ as $r \to 0$, so that the integrals involved would converge near l_0; these singular parts would presumably be known already from the matching.

One could also try to consider the field equations with matter in the external scheme, rather than with empty space as in this section. The situation becomes more complicated since effects due to the black hole will change the energy-momentum tensor T_{ab} by $O(M)$, and so will produce extra $O(M)$ changes in the external scheme. A Green function would have to be found for the gravity-matter coupled system.

3.8 A small black hole in the Brans–Dicke theory

In general, if one has a relativistic theory of gravity different from general relativity, black holes will move on paths which differ from the paths taken by weak-field 'test' matter (Will 1993). From this one may possibly rule the non-Einsteinian theory out observationally. The same applies to the motion (say) of neutron stars with strong internal gravity. For example, in the binary pulsar PSR 1913+16, one has very accurate tests of general relativity, because the timing measurments are so accurate (Damour and Taylor 1992; Will 1993).

To take an example, the procedure which we have set up in this chapter for analyzing the behaviour of a small black hole in a background, under the Einstein field equations, can also be applied to the same problem in the Brans–Dicke theory of gravitation (Brans and Dicke 1961). The Brans–Dicke field equations were originally formulated in terms of a metric g^{ab} and scalar field Φ which satisfy

$$R_{ab} - \frac{1}{2}g_{ab}R = 8\pi\Phi^{-1}T_{ab}$$
$$+ \xi\Phi^{-2}(\Phi_{;a}\Phi_{;b} - \frac{1}{2}g_{ab}g^{cd}\Phi_{;c}\Phi_{;d})$$
$$+ \Phi^{-1}(\Phi_{;ab} - g_{ab}g^{cd}\Phi_{;cd}), \tag{3.8.1}$$
$$g^{cd}\Phi_{;cd} = 8\pi(3 + 2\xi)^{-1}T. \tag{3.8.2}$$

Here R_{ab} and R are the Ricci tensor and scalar of g^{ab}, and a subscripted semi-colon denotes covariant differentiation with respect to g^{ab}. Also T_{ab} is the usual energy–momentum tensor which satisfies $T_{ab;c}g^{bc} = 0$, and ξ is a constant. In this formulation, we say that the theory is viewed in the Brans–Dicke frame. Alternatively, one may make the conformal transformation $\bar{g}_{ab} = \Phi g_{ab}, \bar{T}_{ab} = \Phi^{-1}T_{ab}$. Then the field equations become

$$\bar{R}_{ab} - \frac{1}{2}\bar{g}_{ab}\bar{R} = 8\pi\bar{T}_{ab} + (3 + 2\xi)\Phi^{-2}(16\pi)^{-1} \times$$
$$\times (\Phi_{|a}\Phi_{|b} - \frac{1}{2}\bar{g}_{ab}\bar{g}^{cd}\Phi_{|c}\Phi_{|d}), \tag{3.8.3}$$
$$\bar{g}^{cd}(\log\Phi)_{|cd} = 8\pi(3 + 2\xi)^{-1}\bar{T}. \tag{3.8.4}$$

Here $\bar{R}_{ab}$ and $\bar{R}$ are the Ricci scalar and tensor of $\bar{g}^{ab}$, and covariant differentiation with respect to $\bar{g}^{ab}$ is denoted by a stroke. We say that the theory is viewed in the Einstein frame when written in this second way.

A normal test body, with negligible self-gravitation, can be shown to move on a geodesic in the Brans–Dicke frame (Will 1993). But a body with significant gravitational binding energy will, in general, deviate from geodesic motion in the Brans–Dicke frame. A black hole is an example of a body with binding energy of the same order as its rest-mass energy; moreover, if the black hole is exactly stationary, then it has no scalar monopole moment (Hawking 1972b), unlike an ordinary test body. By arguing that there is no flux of the scalar field through a small world tube surrounding the black hole, and hence no force on the black hole due to the scalar field, Hawking was led to conjecture that a small black hole moves along a geodesic in the Einstein frame with respect to the background universe.

We are able to verify this result by means of our approximation method. Before setting up an internal approximation scheme for the small black hole, we recall the result of Hawking (1972b): that the stationary vacuum black-hole solutions in the Brans–Dicke theory are identical to the stationary vacuum black-hole solutions of the Einstein equations and have $\Phi \equiv$ const. So we again take the internal metric to be that of a Kerr solution, perturbed by the presence of the background. In the Einstein frame, the internal scheme can be expressed by

$$
\begin{aligned}
\bar{g}_{ab} =& M^2 \bar{j}^{(0)}_{ab}(R,\theta) + M^3 \bar{j}^{(1)}_{ab}(MT,R,\theta,\phi) \\
&+ M^4 \bar{j}^{(2)}_{ab}(MT,R,\theta,\phi) + \cdots,
\end{aligned}
\tag{3.8.5}
$$

$$
\begin{aligned}
\Phi =& \Phi^{(0)I}(MT) + M\Phi^{(1)}(MT,R,\theta,\phi) \\
&+ M^2 \Phi^{(2)I}(MT,R,\theta,\phi) + \cdots.
\end{aligned}
\tag{3.8.6}
$$

Here $\bar{j}^{(0)}_{ab}$ is a unit-mass Kerr metric in Boyer–Lindquist coordinates (T,R,θ,ϕ). The form of the expansion for Φ is dictated by the requirement that the lowest-order internal solution, found by letting $M \to 0$ with T,R,θ,ϕ constant, should have $\Phi \equiv$ const., as in the exactly stationary solutions. This condition permits $\Phi^{(0)I}$ to have slow time dependence, but no spatial dependence.

We use matching coordinates (τ,x,y,z,M) which are related to the internal coordinates (T,R,θ,ϕ) just as before. The external scheme can be written

$$
\begin{aligned}
\bar{g}_{ab} =& \bar{g}^{(0)}_{ab}(\tau,x,y,z) + M\bar{g}^{(1)}_{ab}(\tau,x,y,z) \\
&+ M^2 \bar{g}^{(2)}_{ab}(\tau,x,y,z) + \cdots,
\end{aligned}
\tag{3.8.7}
$$

$$
\begin{aligned}
\Phi =& \Phi^{(0)E}(\tau,x,y,z) + M\Phi^{(1)E}(\tau,x,y,z) \\
&+ M^2 \Phi^{(2)E}(\tau,x,y,z) + \cdots,
\end{aligned}
\tag{3.8.8}
$$

where $\bar{g}^{(0)}_{ab}(\tau,x,y,z)$ and $\Phi^{(0)E}(\tau,x,y,z)$ are the background metric and scalar field in the Einstein frame. The labels I and E have been attached to the expression $\Phi^{(n)}$ to distinguish internal from external terms.

Since Φ acts as a source in eqn (3.8.3) through products of its first derivatives, it only generates curvature of $O(1)$ in $\bar{g}_{ab}$ in the internal region. Hence the lowest-order QS perturbations of $\bar{j}^{(0)}_{ab}$ satisfy the vacuum Einstein equations, linearized about the Kerr solution. The boundary conditions for $\bar{j}^{(1)}_{ab}(0,R,\theta,\phi)$ as $R \to \infty$ are determined by the $O(r)$ part of $\bar{g}^{(0)}_{ab}$ as $r \to 0$, and one may argue just as before to find that $\bar{j}^{(1)}_{ab}$ can be eliminated by choice of coordinates. Then, from the matching,

$$
\bar{g}^{(0)}_{ab}(\tau,x,y,z) = \mathrm{diag}(-1,1,1,1) + O(r^2)
\tag{3.8.9}
$$

as $r \to 0$, so that the lowest-order worldline of the black hole is indeed a geodesic with respect to $\bar{g}_{ab}^{(0)}$.

Further inferences can be made by examining the matching conditions. From the matching for Φ, we find

$$\Phi^{(0)I}(MT) = \Phi^{(0)E}(\tau, 0, 0, 0), \qquad (3.8.10)$$

where $\tau = \tau_0 + MT$ is an appropriate background time. Moreover, the fact that $\Phi^{(0)I}$ depends only on time implies that

$$\Phi^{(1)E} = O(1), \qquad \Phi^{(2)E} = O(r^{-1}), \qquad \Phi^{(3)E} = O(r^{-2}), \ldots \qquad (3.8.11)$$

as $r \to 0$. The condition that $\Phi^{(1)E}$ is bounded near the worldline can be regarded as a statement that the black hole has no scalar multipole moments at the lowest order.

The conditions (3.8.11) allow us to prove that the spin vector of the black hole is parallel transported along the geodesic, with respect to $\bar{g}_{ab}^{(0)}$. We may recall that the corresponding property in the Einstein theory was found from the $M^2 r^{-3}$ part of the field equations in the external scheme. Here the parts of $\bar{g}_{ab}^{(0)}$, $\bar{g}_{ab}^{(1)}$, and $\bar{g}_{ab}^{(2)}$ which contribute to this part of the field equations have the same behaviour as in $g_{ab}^{(0)}$, $g_{ab}^{(1)}$, and $g_{ab}^{(2)}$ of Section 3.4. Also eqns (3.8.3) and (3.8.11) show that the scalar field does not contribute to this part of the field equations. The argument then proceeds as in Section 3.4.

The largest internal perturbations of the scalar field and conformal metric can also be analysed. The lowest-order QS perturbation in Φ is

$$M\left[T\frac{d}{d(MT)}\Phi^{(0)I}(MT)\,|_{MT=0} + \phi^{(1)I}(0, R, \theta, \phi)\right]$$
$$= MN(T, R, \theta, \phi), \qquad (3.8.12)$$

say. The field equation (3.8.4) for Φ shows that, *in vacuo*,

$$\bar{j}^{(0)cd}N_{|cd} = 0, \qquad (3.8.13)$$

where the covariant differentiation is with respect to the Kerr background $\bar{j}_{ab}^{(0)}$. Now the part $T[d/d(MT)]\Phi^{(0)I}(MT)\,|_{MT=0}$ of n is already harmonic in the Kerr background. Thus $\Phi^{(1)I}(0, R, \theta, \phi)$ separately is harmonic in the Kerr background. If one wished, one could compute $\Phi^{(1)I}(0, R, \theta, \phi)$ in terms of hypergeometric functions; as $R \to \infty$, $\Phi^{(1)I}(0, r, \theta, \phi) \sim \alpha X + \beta Y + \gamma Z$ due to the spatial gradient of $\Phi^{(0)E}$ near the black hole, while the boundary condition at the horizon is that $\Phi^{(1)I}(MT, R, \theta, \phi)$ should be regular there. Then one could obtain a Teukolsky equation for the metric perturbation $\bar{j}_{ab}^{(2)}$. This would include source terms arising from products of gradients of N. The boundary conditions on $\bar{j}_{ab}^{(2)}(0, R, \theta, \phi)$ as $R \to \infty$ will be due to the r^2 parts of $\bar{g}_{ab}^{(0)}$ near the worldline.

We remark that the close resemblance of the analysis in this section to the earlier arguments about black holes in the pure Einstein theory has only been possible because, in the Einstein frame, Φ appears on the right-hand side of eqn (3.8.3) in a form involving products of first derivatives. Had we instead tried to repeat the earlier arguments in the Brans–Dicke frame, we should have found that Φ produces curvature of $O(M^{-1})$ in g_{ab} in the internal region, due to the second derivatives of Φ in eqn (3.8.1); the $O(M)$ QS metric perturbations could not then be zero. More details on the motion of black holes in generic gravitational theories are given in Will (1993), in the chapter on structure and motion of compact objects in alternative theories of gravity. In general, the treatment will be analogous to that in the present section, taking matched asymptotic expansions of all fields involved.

3.9 Conclusions

We have seen that a technique involving matched asymptotic expansions can be brought to bear on the problem of a small black hole in a background universe, in the context of the Einstein field equations. In particular, it helps us to understand how the black hole moves with respect to the background, and also how the background gravitational field distorts the internal structure of the black hole. Under our assumptions, the black hole moves approximately along a timelike geodesic in the background, and its angular momentum vector is approximately parallel transported along the geodesic in the background metric. The largest deviations from the Kerr geometry near the black hole are of a quadrupole nature, caused by local background curvature. The technique can also be used in order to understand the behaviour of a small black hole in the Brans–Dicke theory. One would similarly like to investigate the system of black holes in other theories of gravity.

These results can be extended to the case of an (electrically or magnetically) charged black hole in a background universe containing electromagnetic fields. It is shown in Kates (1980c) that the black hole obeys the Lorentz force law, as a consequence of the electrovac field equations and matching. An extension of the treatment given here might also deal with the situation in which the background $(M_0, g_{ab}^{(0)})$ contains matter on the zeroth-order worldline l_0. One would expect to be able to treat the black hole's swallowing of matter supplied by the external region as part of the internal scheme. This would be useful astrophysically, since black holes near the centres of galaxies are typically expected to have accretion discs around them and rotate with quite large a/M but $a/M < 1$ (Blandford and Thorne 1979).

Another possible astrophysical application concerns the leading correction to the equation of spin propagation of the black hole. As shown by Thorne and Hartle (1985), this is due to coupling between the black hole and the electric part E_{ij} of the Weyl tensor (3.5.1). The leading correction is found by matching to be

$$\frac{d\Sigma}{dt} = \Omega_T \times \Sigma, \tag{3.9.1}$$

where

$$\Omega_T{}^i = -E^i{}_j M \chi s^j. \tag{3.9.2}$$

Here, Σ is the angular momentum vector Mas of the black hole, with s being a unit vector in the direction of Σ. We recall that $|\Sigma| = \chi M^2 = aM$. Thus the black hole precesses with angular velocity Ω_T proportional to the rotation parameter χ and to the external electric-type curvature. This tidal torque might conceivably be detected by its influence on the shapes of jets that emerge from the nuclei of some galaxies and quasars (Blandford and Thorne 1979; Begelman et al. 1980, 1984; Thorne et al. 1986).

In several models for jet production, a supermassive black hole occurs at the origin of the jet. Moreover, the directions of the two jets are tied to the rotation axis of the black hole. Hence, if the hole precesses, then the jets will develop an S-like shape. (Such jets have been reported by radio astronomers.) The length scales of the S-shapes are $\sim 10^3$–10^7 light years. These correspond to the time scales for the precession of $\sim 10^3$–10^9 years, depending on the jet speed. In addition, the time scale $1/|\Omega_T|$ for the torqued precession is shown in the example below to be in the observed range.

Let us examine a supermassive black hole at the centre of a quite dense elliptical star cluster. In order of magnitude E_{jk} will be $aG\rho$. Here a is a dimensionless measure of the deviations of the cluster from sphericity ($a \simeq 1$ for a pancake-shaped cluster), G is Newton's gravitation constant, and ρ is the mean mass density in the cluster. The resulting torqued precession rate will be

$$\Omega_T \sim aG\rho M\chi = a\left[\frac{1}{4\times 10^7 \text{yr}}\right]\left[\frac{\rho}{10^9 M_\odot/\text{ly}^3}\right]\left[\frac{M}{10^9 M_\odot}\right]\chi. \tag{3.9.3}$$

The torqued precession, then, for a massive black surrounded by a dense and non-spherical cluster, could be large. For more typical situations, the precession would turn out to be negligible.

However, when one studies other examples of precession (Thorne and Hartle 1985), such as a two-body system (Chapter 4), one is reminded that the dominant effect tends to be geodetic precession, in which the parallel transport of the black hole's spin along a geodesic for the two-body problem leads to a net precession as viewed from a distant nearly flat coordinate system. Further, one needs to take account of the 'gravitomagnetic' torque on a black hole due to the angular momentum of another body. The details are discussed in Sections 4.6 and 4.7. For example, for a black hole in a circular orbit of radius b around an external body of mass $M_E \gg M$, let the geodetic precession of the spin be given by $d\Sigma/dt = \Omega_{\text{geod}} \times \Sigma$, relative to inertial frames far away from the system (at $r \gg b$). Similarly, the gravitomagnetic precession is given by Ω_{GM}, and the 'torqued' precession (3.9.1) is given by Ω_T. One finds (from Chapter 4 or from Thorne and Hartle (1985)) that the relative magnitudes of these three precessions are

$$| \, \Omega_T \, | : | \, \Omega_{GM} \, | : | \, \Omega\text{geod} \, | \simeq \chi \left(\frac{M}{M_E} \frac{M}{b} \right)^{\frac{1}{2}} : \left(\frac{R_E}{b} \right)^{\frac{1}{2}} : 1, \qquad (3.9.4)$$

where R_E is the external body's radius, and we have assumed that the body is rotating as rapidly as possible (the centrifugal force being of the order of the self-gravity force) so as to make the gravitomagnetic precession as large as possible. Note that, for this orbital situation, $\Omega_T \ll \Omega_{GM} \leq \Omega_{\text{geod}}$.

Thus the torqued precession is in general expected to be much less than the geodetic and gravitomagnetic precessions. In Begelman *et al.* (1980), geodetic precession has been identified as astrophysically the most important precession that a black hole can undergo, and it was suggested to be the most likely cause for the S-shaped distortions of jets in galactic nuclei and quasars. This raises an interesting question: In what external geometry is the black hole moving? One possibility is that one may often find two supermassive objects occupying the same galactic nucleus. Then gravitational drag by normal stars, i.e. 'tidal friction', will cause the two black holes to sink to the centre of the galactic nucleus and find each other, forming a binary. If one of the black holes M_1 produces a jet along its spin axis, then the jet will precess with an angular velocity given exactly in the following Chapter 4, approximated here as

$$\Omega_{\text{geod}} + \Omega_{GM} \simeq \frac{3M_2}{2b} \left(\frac{M_1 + M_2}{b^3} \right)^{\frac{1}{2}}$$

$$\simeq \left(\frac{1}{10^4 \text{yr}} \right) \left(\frac{M_1}{10^8 M_\odot} \right) \left(\frac{M_1 + M_2}{10^8 M_\odot} \right)^{\frac{1}{2}} \left(\frac{10^{17} \text{cm}}{b} \right)^{\frac{5}{2}}, \quad (3.9.5)$$

where M_2 is the second black hole's mass, and $M_\odot$ is the solar mass. This effect may possibly account for the S-shape of some jets, and certainly encourages one to study the dynamics of two black holes in a bound orbit, as in the following Chapter 4.

4

INTERACTION OF TWO BLACK HOLES IN THE SLOW-MOTION LIMIT

4.1 Introduction

Following the ideas of Chapter 3, in this chapter we shall describe a method of analysing the interaction of two black holes in a wide variety of physical situations (D'Eath 1975b). The black holes may be rotating and are permitted in our approach to have comparable masses, subject to two types of restriction on the interaction. First, it is assumed that the spatial separation of the black holes is large compared to the radii of their event horizons, so that each black hole is in the far field of the other. Second, we suppose that the relative velocity of the black holes is small compared to the speed of light, so that their combined gravitational field can be described in terms of a slow-motion approximation scheme. Then we should expect that the dominant interaction between them is just that between a pair of Newtonian point particles; it is further assumed in a slow-motion scheme that the Newtonian potential energy of the interaction is comparable to the relative kinetic energy. This allows one to consider a bound system of two black holes moving at non-relativistic speeds. It also allows one to study a quasi-Newtonian scattering of two black holes at non-relativistic speeds in which the scattering angle may be quite large (Ruffini and Wheeler 1971; Hansen 1972). These two types of interaction are described by the same equations. The aim is then to find more and more accurate approximations to the dynamical space–time geometry, subject to the vacuum Einstein field equations, and from this description of the geometry to find more and more accurate approximation to the equations of motion and spin propagation of the black holes.

While a slow-motion perturbation of Minkowski space will represent the gravitational field throughout most of the space, it will break down in regions near the black holes, where the field strength becomes large. In Chapter 3, this question was considered for the interaction of a black hole with an external 'background' universe. In that chapter it was seen how the gravitational field can be described in terms of two approximation schemes, valid in rather different regions of space–time, provided that a characteristic mass or length scale M of the black hole is much smaller than the length- scales on which the background geometry varies. In one of the perturbation expansions, the external scheme, the effect of the black hole is to produce a small perturbation of the background geometry; this expansion is valid in the far field of the black hole. In the present context, it corresponds to the slow-motion perturbation of flat space by the sum of the far fields of the two black holes, together with radiation effects and non-linear effects of higher order, which is valid away from the strong-field regions of each black hole. The other perturbation expansion used in Chapter 3, the internal scheme, is valid in the strong-field region of the black hole, and describes the structure of the field on length scales of order M. The background only produces small distortions of the basic Kerr black-hole geometry in this region, since the curvature of the background is much less than the curvature M^{-2} of the black hole on these length-scales. In the present context there will be two such internal approximation schemes, one for each black hole. One hole feels only the far field of the other, which is weak since it can be described by a slow-motion perturbation of flat space; as a result the 'background' curvature acting on one of the black holes is much less than the curvature in the near-field region, so that the interaction again produces only a small perturbation in the basic Kerr geometry near the black hole. The internal and external asymptotic expansions for the gravitational field must then be related by matching in some intermediate region near the relevant black hole in which both schemes are valid; this matching procedure was discussed in some detail in Chapter 3.

Now that we have summarized our approach to the description of the gravitational field, let us turn our attention to the motion of the black holes. Over short periods of time, the motion will be dominated by the Newtonian interaction. But in order to follow the behaviour of the black holes over longer periods, we must understand the higher-order relativistic corrections to the Newtonian motion. For example, if the black holes are rotating, there should be modifications due to spin–orbit and spin–spin forces. There will also be effects due to the emission of gravitational waves, and to the capture of gravitational radiation by the holes. Moreover, the presence of a second black hole will cause the direction of the spin of the first to precess slowly. It is the main aim of this chapter to calculate the most significant corrections to the equations of motion and the lowest-order equations of spin propagation. Before discussing our method further, we mention two previous approaches relevant to this problem, which will suggest the nature of the equations of motion and spin propagation which we seek.

Wald (1972) has examined the interaction of spinning bodies in general relativity by studying the behaviour of a spinning test particle (not a black hole) in the far field of a rotating gravitational source. The test particle obeys the Papapetrou equations (Papapetrou 1951) which are derived from the conservation equations

$$T^{ab}{}_{;b} = 0 \qquad (4.1.1)$$

by making certain assumptions about the matter in the test particle, and describe the deviation of the particle's motion from a geodesic due to spin effects, together with the variation of the spin along the particle worldline. Suppose that M and $\mathbf{J}$ are the mass and angular momentum of the source, which is at the origin of the spatial coordinates, and $\mathbf{r}$, $\mathbf{v}$, and $\boldsymbol{\Sigma}$ are the position, velocity, and spin of the test body. We use geometrical units here such that Newton's constant G and the speed of light c are 1. Then, as shown in Wald (1972), the active gravitational effect $\mathbf{J}$ gives the test body an acceleration

$$\boldsymbol{\alpha} = r^{-5}\mathbf{v} \times [2r^2\mathbf{J} - 6\mathbf{r}(\mathbf{r}\cdot\mathbf{J})]. \tag{4.1.2}$$

The lowest-order spin–orbit force, due to deviation from geodesic motion in the dominant Schwarzschild part of the far field, is given as

$$\mathbf{F} = -3Mr^{-5}\{r^2\mathbf{v} \times \boldsymbol{\Sigma} + 2\mathbf{r}[\mathbf{v}\cdot(\mathbf{r} \times \boldsymbol{\Sigma})] - (\mathbf{r}\cdot\mathbf{v})(\mathbf{r} \times \boldsymbol{\Sigma})\}. \tag{4.1.3}$$

The lowest-order spin–spin force, again due to deviation from geodesic motion in a far field which takes account of $\mathbf{J}$, is

$$\mathbf{F} = -\nabla\{r^{-5}[-r^2\boldsymbol{\Sigma}\cdot\mathbf{J} + 3(\mathbf{r}\cdot\mathbf{J})(\mathbf{r}\cdot\boldsymbol{\Sigma})]\}. \tag{4.1.4}$$

The spin propagation equations show that the precession of the spin vector of the test body due to the Schwarzschild part of the far field is essentially

$$\frac{d\boldsymbol{\Sigma}}{dt} = \frac{3}{2}Mr^{-3}(\mathbf{r} \times \mathbf{v}) \times \boldsymbol{\Sigma}. \tag{4.1.5}$$

The largest effect on the spin precession due to the active gravitational effect of $\mathbf{J}$ is

$$\frac{d\boldsymbol{\Sigma}}{dt} = r^{-5}\boldsymbol{\Sigma} \times [r^2\mathbf{J} - 3\mathbf{r}(\mathbf{r}\cdot\mathbf{J})]. \tag{4.1.6}$$

These equations describe most of the dominant effects to be found in the distant interaction of two normal spinning bodies. If the Papapetrou equations also give the leading effects in the motion and spin propagation of black holes, then eqns (4.1.2)–(4.1.6) should apply to the situation in which a small rotating black hole moves in the far field of a much larger black hole. We might hope that any equations which we derive for the interaction of two black holes of comparable mass should enable us to recover the effects of eqns (4.1.2)–(4.1.6) by taking one of the masses to be zero, after making allowance (if necessary) for any differences in the choice of coordinates used to describe the gravitational field.

Another approach to the problem of equations of motion in general relativity is through the method of Einstein, Infeld, and Hoffmann (which we refer to as the EIH method) (Einstein et al. 1938; Einstein and Infeld 1940, 1948). EIH show that the empty-space field equations alone determine the motion of the type of particles which they consider. We encountered a similar phenomenon in Chapter 3, where we used only empty-space equations and found that a small black hole moves approximately on a geodesic of the background universe. The EIH method is a slow-motion approximation scheme, in which a finite number of low-mass, slowly moving particles interact by Newtonian forces together with small relativistic corrections. The gravitational field away from the particles is treated as a slow-motion perturbation of flat space. Just as we found for the metric terms $g_{ab}^{(1)}$, $g_{ab}^{(2)}$, ... of Chapter 3, which describe the first-, second-, ... order perturbations induced on the background by the black hole, the EIH metric perturbations of flat space become singular at the worldlines of the particles. The type of singularity occurring is restricted by EIH so that the particles should in some sense be as spherical as possible and represent non-rotating bodies. When the Einstein equations are solved order by order in the perturbation scheme, one finds that certain integrability conditions must be satisfied by the metric components up to a given perturbation order, before the field equations can be solved to find the next-order metric components. The EIH equations of motion are integrability conditions of this type, which state that certain integrals taken over two-surfaces surrounding the particle worldlines must vanish. EIH showed by this means that the lowest-order equations of motion are indeed Newtonian, and also computed the lowest-order corrections to the equations of motion for the two-body problem. The corrections are of post-Newtonian order, i.e. of order GM/Rc^2 smaller than Newtonian terms, where R is a typical distance between the particles, M is a typical mass, and we have written the G and c dependence explicitly. When applied to a bound two-body system, these equations show that the post-Newtonian perihelion precession in $6\pi(M_1 + M_2)Gp^{-1}c^{-2}$ radians per revolution (Robertson 1938), where M_1 and M_2 are the particle masses and p is the semi-latus rectum of the relative orbit.

Our approximation scheme should also allow us to recover EIH's post-Newtonian effects when we specialize to the case of two non-rotating black holes, after making allowance for differences in the choice of coordinates. This will provide an independent check on our analysis, quite apart from the comparison with eqns (4.1.2)–(4.1.6) (see Anderson 1987).

It would be possible to treat the interaction of two black holes by a combination of the matching technique of Chapter 3 with the EIH method. A small modification of EIH's surface integral method allows one to extract equations of spin propagation as well as equations of motion from the vacuum field equations. This approach can in principle be taken; unfortunately, it leads to algebra much worse than that encountered by EIH, since it is hard to combine the matching technique with convenient coordinate conditions on the slow-motion perturbations of flat space. Fortunately, an alternative approach is available, since (Chrzanowski 1975; Wald 1978) one may compute vacuum metric perturbations of the Kerr geometry explicitly from the Riemann tensor perturbations analyzed by Teukolsky (1973). In the latter approach, to be described in this chapter, we find equations of motion and spin propagation for the black holes by computing certain parts of the space–time metric in two different ways. These parts of the metric are computed in one way in the course of the slow-motion perturbation of flat space; in this computation one imposes boundary conditions at infinity. They are also computed in terms of one of the internal schemes describing perturbations of a Kerr metric, which rely on the formulations of Teukolsky, Chrzanowski, and Wald; in this computation one has imposed boundary conditions at the future horizon of the black hole. When one relates the internal scheme to the slow-motion scheme by matching, the equations of motion and spin propagation follow. This method is considerably simpler to apply than EIH's, and again uses only the empty-space field equations. Ideally, of course, one would like to compare the results of such a perturbation approach with the results of a fully non-linear numerical calculation

(Smarr *et al.* 1976). In the past few years, a number of authors have pursued such numerical calculations with a view to understanding black-hole interactions better (Anninos *et al.* 1993, 1995*a*, *b*).

As in the EIH method, metric perturbations of flat space which describe the combined far fields of the two black holes in the matching approach become singular near the black-hole worldlines. Whereas EIH determine the singular behaviour of the metric perturbations by solving the field equations and using rules about the addition of terms which represent extra multipoles in their particles, here we find the singular terms near a worldline by matching to the appropriate internal approximation scheme. This should allow us to express unambiguously the condition that the 'particles' in our slow-motion scheme are really Kerr-like black holes. One expects that the entire physical system is defined by a small number of parameters describing the orbits, masses, and spins of the black holes, together with the boundary condition that no gravitational radiation is incoming from past null infinity. There should be no freely specifiable multipoles for the black holes beyond the basic Kerr parameters. For example, by the arguments of Chapter 3 (see Section 3.5), the largest distortion of the geometry near one of the black holes is of a quadrupole character, determined by the strongest part of the 'background' Riemann tensor, which will arise here from the Newtonian field of the other black hole. At a sufficiently high order in the approximation scheme, the quadrupole bulge will produce active gravitational effects via the matching. These effects are determined entirely by the 'background' Riemann tensor, together with the boundary condition of regularity on the future event horizon, as we found in Chapter 3; higher-order internal distortions are similarly determined by the 'external' field and horizon boundary conditions. This is in contrast to the EIH papers, where the particles could be given extra multipole structure, but high multipoles are suppressed in order to produce a unique approximation scheme. An approach to the motion of a system of compact objects, which might be black holes, neutron stars, or other objects, based on the ideas described here is given by Damour (1983, 1987).

We now explain the scaling which is used in ordering the various small parameters in the problem. The combined gravitational field of the two bodies, regarded as a slow-motion perturbation of flat space, is described in terms of time and space coordinates (u, w^i), where i runs from 1 to 3 and u is regarded as the zeroth coordinate. All metric perturbations depend on time through the variable (ϵu), where ϵ is a small parameter, and on space through w^i; hence spatial gradients are much larger than time derivatives. The black holes move approximately on orbits

$$w^i = r_1^i(\epsilon u), \qquad w^i = r_2^i(\epsilon u), \qquad (4.1.7)$$

where r_1^i and r_2^i are certain position functions which we shall show later to obey Newtonian equations of motion for point particles. Thus as $\epsilon \to 0$, we fix the spatial positions of the orbits but lengthen the time needed to traverse them. The masses M_1, M_2 of the black holes scale as

$$M_1 = \epsilon^2 \mu_1, \qquad M_2 = \epsilon^2 \mu^2. \tag{4.1.8}$$

The oriented Kerr spin parameters a_1^i, a_2^i (where $|\mathbf{a}_1|$ would be the Kerr parameter a_1 of the first black hole in the usual notation) scale as

$$\mathbf{a}_1 = \chi_1 M_1, \qquad \mathbf{a}_2 = \chi_2 M_2, \tag{4.1.9}$$

where $|\chi_1| < 1$, $|\chi_2| < 1$. The indices $i, j, k, l, \ldots$ will always run from 1 to 3, and we treat quantities such as a_1^k, B^{kl}, labelled by these indices, as Cartesian vectors or tensors, to which a Cartesian summation convention applies. A dot over a quantity indicates differentiation with respect to (ϵu). For example, $\dot{r}_1^i = [d/d(\epsilon u)]r_1^i(\epsilon u)$.

This scaling ensures that the Newtonian potential energy $M_1 M_2 \mid \mathbf{r}_1 - \mathbf{r}_2 \mid^{-1}$ and kinetic energy $\frac{1}{2}\epsilon^2 M_1 \mid \dot{\mathbf{r}}_1 \mid^2 + \frac{1}{2}\epsilon^2 M_2 \mid \dot{\mathbf{r}}_2 \mid^2$ are both of order ϵ^4. It is applicable to a bound two-body system, but could also be used to describe a scattering orbit, provided that the velocities were small compared to c, and that the kinetic energy was not too much greater than the potential energy.

We may now classify the post-Newtonian corrections which we expect to find in the equations of motion and spin propagation, according to their order in the small parameter ϵ. We first outline our attitude towards the intepretation of these corrections.

The higher-order equations of motion or spin propagation will be written as time-evolution equations for certain correction terms to the lowest-order position or angular momentum. For example, the largest correction to the equation of motion for the first black hole will appear as an equation for $\ddot{r}_{12}^i(\epsilon u)$, where $\epsilon^2 r^i{}_{12}(\epsilon u)$ is a position displacement such that $r_1^i(\epsilon u) + \epsilon^2 r^i{}_{12}(\epsilon u)$ might be regarded as the position of the black hole, allowing for corrections up to $O(\epsilon^2)$. We prefer not to regard $r_{12}^i(\epsilon u)$ in this way, since the black hole is itself spread out over a length-scale of $O(M_1) = O(\epsilon^2)$. Instead, we remark that, in the case of a bound system of two black holes, the relativistic corrections will gradually accumulate to produce effects of $O(1)$ over a sufficiently long time-scale of $O(\epsilon^{-n})$, for some $n > 0$, and that the black-hole orbits should be well approximated by a sequence of Newtonian orbits in which the Newtonian orbit parameters change over a time-scale $\gg O(\epsilon^{-1})$, as measured by the time u. Thus, although it will be convenient to refer to $\epsilon^2 r_{12}^i(\epsilon u)$ as the post-Newtonian position shift of the first hole, we prefer to treat it as describing part of a change in the orbit and parameters which is taking place over a time scale of $O(\epsilon^{-3})$. Similarly, the spin-propagation equations will be regarded as describing fractional changes of $O(1)$ in the Kerr parameters $\mathbf{a}_1, \mathbf{a}_2$ over long time-scales.

We shall say that an effect is n post-Newtonian if (for a bound system) it produces $O(1)$ changes in the orbit or $O(1)$ fractional changes in $\mathbf{a}_1, \mathbf{a}_2$ over a time-scale of $O(\epsilon^{-2n-1})$. Then the largest deviations from Newtonian motion are post-Newtonian deviations found by EIH. In particular, it should be possible to recover the EIH perihelion shift for a bound orbit. The largest precession of the spin vectors will also be a post-Newtonian effect, given by some generalization

of eqn (4.1.5). Effects such as those given by eqns (4.1.2) and (4.1.3), in which
a moving body is affected by the spin of another body, or in which a rotating
moving body is affected by the Schwarzschild far field of another body, produce
$1\frac{1}{2}$ post-Newtonian changes in the orbit. The precession of one spin vector due
to the rotation of the other body will also be a $1\frac{1}{2}$ post-Newtonian effect, as in
eqn (4.1.6). The second post-Newtonian effects will be quite complicated, but
among them one should find the largest spin–spin force, as in eqn (4.1.4).

For quasi-Newtonian scattering, one can consider the same equations of motion and spin propagation. During the scattering of the two black holes, the
direction of asymptotic motion and direction of asymptotic spin will have been
changed by an amount given by asymptotic series in ϵ. By observing the changes
in motion and spin over very long time-scales, one would again be able to detect
the corrections in motion and spin at post-Newtonian and higher orders.

It has been demonstrated with the help of matched asymptotic expansions,
both for weakly and strongly self-gravitating bodies, that gravitational radiation effects first appear at the $2\frac{1}{2}$ post-Newtonian order, and that quadrupole
radiation is emitted in agreement with the linearized theory (Burke 1971; Kates
1980b). Gravitational radiation with strong-field slow-motion sources is discussed
further by Thorne (1980b) and Futamase (1985, 1987). A review of the generation of gravitational waves is given by Thorne (1977). Further suggestions for
improvement are given by Anderson (1985). The strong-field argument involves
matching as in this chapter between the internal 'body' zone and the external
nearby-Newtonian zone; then one must further match (as in the weak-field argument) between the external zone and the wave zone, which describes gravitational
waves propagating out to future null infinity $\mathcal{I}^+$ (with no input from $\mathcal{I}^-$)—these
gravitational waves have the low frequency corresponding to the Newtonian time-
scale $O(\epsilon^{-1})$. In the present case of two black holes, say, in a quasi-Newtonian
bound state, one can follow Kates (1980b). If properly formulated and completed,
the matching approach should meet the objections of Ehlers et al. (1976). The
$2\frac{1}{2}$ post-Newtonian gravitational radiation means that the black holes will spiral
towards each other and collide on a time-scale of $O(\epsilon^{-6})$; some details of the
radiation pattern and of the spiralling process are given by Peters and Mathews (1963) and Peters (1964). Post-Newtonian corrections to the gravitational
radiation are given by Wagoner and Will (1976).

For simplicity, set the Newtonian centre of mass at the origin. Define the
tracefree Newtonian quadrupole moment

$$I_{ij}^T(\epsilon t) = (M_1 x_1{}^i x_1{}^j + M_2 x_2{}^i x_2{}^j) - (1/3)\delta_{ij}(M_1 x_1{}^k x_1{}^k + M_2 x_2{}^k x_2{}^k). \quad (4.1.10)$$

Following Peters (1964) and Misner et al. (1973), the rate of emission of energy
at lowest ($2\frac{1}{2}$ post-Newtonian) order is

$$L = - < \frac{dE}{dt} >= \frac{1}{5} < \dddot{I}_{ij}^T \dddot{I}_{ij}^T > . \quad (4.1.11)$$

Here the brackets $< \ >$ indicate that the right-hand side of eqn (4.1.11) has been averaged over several periods of the black holes' motion. Momentum is also radiated in gravitational waves for an eccentric binary—see Fitchett (1981, 1984) and Fitchett and Detweiler (1984). For an elliptical relative orbit with semi-major axis a and eccentricity e, eqn (4.1.11) gives

$$< \frac{dE}{dt} > = \frac{-32}{5} \frac{G^4 M_1{}^2 M_2{}^2 (M_1 + M_2)}{c^5 a^5 (1 - e^2)^{7/2}} \left(1 + \frac{73}{24} e^2 + \frac{37}{96} e^4\right). \tag{4.1.12}$$

The corresponding rate of decrease of the period P of the binary system is then given by

$$P^{-1} \frac{dP}{dt} = -\frac{3}{2} E^{-1} \frac{dE}{dt} = \frac{-96}{5} \frac{M_1 M_2 (M_1 + M_2)}{c^5 a^4} \left(1 + \frac{73}{24} e^2 + \frac{37}{96} e^4\right). \tag{4.1.13}$$

From this and from the equation of emission of angular momentum, one finds (Peters 1964) the $2 - \frac{1}{2}$ post-Newtonian evolution of a and e. In the case of a circular orbit, the $2 - \frac{1}{2}$ post-Newtonian estimate for the time to collision of the black holes is

$$T_c(a_0) = (a_0)^4 / (4\beta), \tag{4.1.14}$$

where a_0 is the initial value of a, and

$$\beta = \frac{64}{5} \frac{G^3 M_1 M_2 (M_1 + M_2)}{c^5}. \tag{4.1.15}$$

The corresponding equations for a hyperbolic quasi-Newtonian encounter are given in Hansen (1972), and will be described in Chapter 5 in the context of the general hyperbolic encounter.

In the case of the binary pulsar PSR 1913+16 (Hulse and Taylor 1975), the rate of change $\dot{P}$ of the period has been measured to be in accordance with eqn (4.1.13), when one takes values for the masses M_1, M_2 and semi-major axis a given by Newtonian and post-Newtonian effects (Will 1993). This has provided the first observational test of the emission of gravitational waves in accordance with general relativity.

Momentum, as well as energy, is radiated in this quasi-Newtonian process. The centre of mass of a binary pair can acquire considerable velocity during the emission of gravitational waves (Fitchett 1981, 1984; Fitchett and Detweiler 1984).

Secular changes in the masses and scalar angular momenta of the black holes only appear at the $4\frac{1}{2}$ post-Newtonian order and higher. For, by the arguments of Chapter 3, the largest internal perturbations of (say) black hole 1 will be due to the dominant part of the 'external' Riemann tensor; as mentioned already, this dominant part arises from the Newtonian field of black hole 2, and will be of order $M_2 \mid \mathbf{r}_1 - \mathbf{r}_2 \mid^{-3}$, i.e., of $O(\epsilon^2)$. Now, in Chapter 3, we found it convenient to describe the gravitational field near a black hole of small mass M_1 in terms of 'internal' time and space coordinates scaled by a factor M_1, so that the black hole has structure on a coordinate scale of $O(1)$, which corresponds to distances in space–time of $O(M_1)$. When a conformal factor $(M_1)^2$ is removed from the metric, we are left with perturbations of a unit-mass Kerr metric (with some angular momentum χ_1). In our present case, the 'external' Riemann tensor which perturbs black hole 1 is some function of (ϵu). A rescaled time coordinate appropriate for describing the structure of black hole 1 (provided that we are not too close to the event horizon) is $U = u/M_1 = u/(\epsilon^2 \mu_1)$. Hence, when we view the internal scheme in terms of perturbations of a unit-mass Kerr metric, the perturbations will depend on time through functions of $(\epsilon^3 U)$. Viewed in this way, with the conformal factor $(M_1)^2$ removed, the internal scheme will be called the QS (quasistationary) scheme, as in Chapter 3, since the perturbations change so slowly with internal time U; a discussion of the various viewpoints on the geometry of a black hole in an external field is given in Section 3.2. We further found in Chapter 3 that the largest QS perturbations will be of order $(M_1)^2$ times the external Riemann tensor, as measured in some orthonormal frame. Hence in our case the QS perturbations will be of order $(M_1)^2 \epsilon^2 = O(\epsilon^6)$. Since the area change of the event horizon is quadratic in the perturbations (Section 3.5), $[d(\text{area})/dU]/\text{area} = O(\epsilon^{12})$. Hence $[d(\text{area})/du]/\text{area} = O(\epsilon^{10})$, so that secular changes in the black hole's Kerr parameters are $4 - \frac{1}{2}$ post-Newtonian, as claimed above.

We have used the scaling of small parameters described by eqns (4.1.7)–(4.1.9) since it would apply to a large class of interactions between two black holes, and in particular to the interesting case in which the black holes form a bound system and will eventually collide. But it is not necessary to combine our matching technique with a slow-motion expansion about flat space. One could instead use a fast-motion scheme in which the masses and spins scale as before, but now the lowest-order orbits are $r_1^i(u)$, $r_2^i(u)$, and all metric components in the perturbation of flat space depend on time through u rather than (ϵu). This would correspond to a distant relativistic scattering in which the two bodies are only slightly deflected from straight-line motion. This scaling would have the merit of exhibiting gravitational radiation at much lower orders than in the slow-motion scheme (see, for example, the work of Thorne and Kovács (1975), Crowley and Thorne (1977), and Kovács and Thorne (1977, 1978)—see also D'Eath (1979b)). This leads to bremsstrahlung gravitational radiation at second order in the particle masses (Peters 1970). For quite relativistic encounters, the radiation is concentrated near the forward and backward directions—see also Westpfahl (1985). Yet another régime is described by letting the relative velocity of the holes tend to the speed of light; in this case the black holes behave almost like two colliding impulsive plane-fronted gravitational waves (Penrose 1974—see Chapters 5–8 below).

In Section 4.2 we shall discuss the basic assumptions of the method, and illustrate the matching at low orders in the slow-motion scheme. In Sections 4.3 and 4.4, certain first and $1\frac{1}{2}$ post-Newtonian metric terms, which are needed for finding equations of motion and spin propagation, are computed by using the post-linearized Einstein equations together with the matching conditions.

Section 4.5 describes the computation of certain parts of the largest internal perturbations of the Kerr metric for either black hole, where we find boundary conditions on the perturbations by matching in towards the black holes and then apply the work of Chrzanowski (1975) and Wald (1978). The matching arguments which lead to the equations of motion and spin propagation up to $1\frac{1}{2}$ post-Newtonian order are given in Section 4.6, together with the calculation of the largest spin–spin force, and in Section 4.7 we comment on the results.

4.2 Basic method—lowest orders

We begin the exposition of the method by making some basic assumptions about the approximation scheme. In the coordinate system (u, w^i) which is used to describe the combined gravitational far field of the two black holes, we expand the metric as

$$
\begin{aligned}
g_{00} &= -1 + \epsilon Q^1_{00}(\epsilon u, w^k) + \epsilon^2 Q^2_{00}(\epsilon u, w^k) + \cdots, \\
g_{0i} &= \epsilon Q^1_{0i}(\epsilon u, w^k) + \epsilon^2 Q^2_{0i}(\epsilon u, w^k) + \cdots, \\
g_{ij} &= \delta_{ij} + \epsilon Q^1_{ij}(\epsilon u, w^k) + \epsilon^2 Q^2_{ij}(\epsilon u, w^k) + \cdots.
\end{aligned}
\tag{4.2.1}
$$

Traditionally, a metric term Q^n_{00} is known as $(n/2 - 1)$ post-Newtonian, Q^n_{0i} as $n/2 - 1/2$ post-Newtonian, and Q^n_{ij} as $n/2$ (Chandrasekar and Esposito 1970). These names are explained by the properties that a particle moving on a geodesic with a velocity of $O(\epsilon)$ is given a coordinate acceleration of $O(\epsilon^{2n+2})$ by an nth post-Newtonian term, that in the Newtonian limit particle accelerations are $O(\epsilon^2)$, and that the largest deviations from Newtonian dynamics ('post-Newtonian effects') are always $O(\epsilon^2)$ smaller than Newtonian effects.

We assume (for motivation see below) that

$$
\begin{aligned}
Q^1_{00} &= Q^3_{00} = 0, \\
Q^2_{00} &= 2\mu_1 \left| \mathbf{w} - \mathbf{r}_1 \right|^{-1} + 2\mu_2 \left| \mathbf{w} - \mathbf{r}_2 \right|^{-1}, \\
Q^1_{0i} &= Q^2_{0i} = 0, \\
Q^3_{0i} &= - 2\mu_1 \dot{r}^i_1 \left| \mathbf{w} - \mathbf{r}_1 \right|^{-1} \\
&\quad - 2\mu_1 \dot{r}^k_1 (w^k - r^k_1)(w^i - r^i_1) \left| \mathbf{w} - \mathbf{r}_1 \right|^{-3} \\
&\quad - 2\mu_2 \dot{r}^i_2 \left| \mathbf{w} - \mathbf{r}_2 \right|^{-1} \\
&\quad - 2\mu_2 \dot{r}^k_2 (w^k - r^k_2)(w^i - r^i_2) \left| \mathbf{w} - \mathbf{r}_2 \right|^{-3}, \\
Q^1_{ij} &= Q^3_{ij} = 0, \\
Q^2_{ij} &= 2\mu_1 (w^i - r^i_1)(w^j - r^j_1) \left| \mathbf{w} - \mathbf{r}_1 \right|^{-3} \\
&\quad + 2\mu_2 (w^i - r^i_2)(w^j - r^j_2) \left| \mathbf{w} - \mathbf{r}_2 \right|^{-3}.
\end{aligned}
\tag{4.2.2}
$$

These terms should describe, at the lowest orders, the conditions that the black holes have 'masses' $\epsilon^2 \mu_1, \epsilon^2 \mu_2$ and move approximately along worldlines $\mathbf{w} = \mathbf{r}_1(\epsilon u), \mathbf{r}_2(\epsilon u)$. For if we put $\mu_2 = 0$ (say), then the metric at low orders is that of a Kerr black hole with mass $\epsilon^2\mu_1$, and oriented Kerr parameter $\mathbf{a}_1$, which has been given a translation and a slow Lorentz boost by velocity $\epsilon\dot{\mathbf{r}}_1$, so that it moves along a straight line $\mathbf{w} = \mathbf{r}_1(\epsilon u)$. Thus the low-order metric for $\mu_2 = 0$ is obtained from a low-mass Kerr solution, expressed in 'Cartesian' coordinates based on Boyer–Lindquist coordinates as

$$ds^2 = -dt^2 + dr^k dr^k + 2\epsilon^2\mu_1 r^{-1} dt^2$$
$$+ 2\epsilon^2\mu_1 r^k r^l r^{-3} dr^k dr^l + O(\epsilon^4), \qquad (4.2.3)$$

by a Poincaré transformation

$$t = (1 - \epsilon^2|\dot{\mathbf{r}}_1|^2)^{-\frac{1}{2}}(u - \epsilon\dot{r}_1^k w^k),$$
$$r^i = w^i - \epsilon(1 - \epsilon^2|\dot{\mathbf{r}}_1|^2)^{-\frac{1}{2}}\dot{r}_1^i u - r_1^i(0) \qquad (4.2.4)$$
$$+ |\dot{\mathbf{r}}_1|^{-2}[(1 - \epsilon^2|\dot{\mathbf{r}}_1|^2)^{-\frac{1}{2}} - 1]\dot{r}_1^k w^k \dot{r}_1^i,$$

corresponding to a spatial translation by $r_1^i(0)$, and a boost by the velocity $\epsilon\dot{\mathbf{r}}_1$. The terms representing the two moving black holes may be superposed at low orders to give eqn (4.2.2), since non-linear terms are not present in the lowest-order Einstein equations

$$Q^2_{00,kk} = 0,$$
$$Q^2_{ij,kk} + Q^2_{kk,ij} - Q^2_{ik,jk} - Q^2_{jk,ik} = Q^2_{00,ij}, \qquad (4.2.5)$$
$$Q^3_{0i,kk} - Q^3_{0k,ik} = \epsilon^{-1}Q^2_{ik,0k} - \epsilon^{-1}Q^2_{kk,0i}.$$

When matching the gravitational field far outside the black holes to the field near (say) black hole 1, we shall work in terms of matching coordinates (t, r^i), which will shortly be related to (u, w^i). The 'non-interacting' transformation (4.2.4) must be part of the full transformation (4.2.10) between the matching coordinates (t, r^i) and the coordinates (u, w^i), which cover the whole external region. In the internal approximation scheme for black hole 1, the metric is that of a Kerr solution with mass $\epsilon^2\mu_1$ and spin parameter $\mathbf{a}_1$, perturbed by the interaction with the second black hole. We have already pointed out in the Introduction that when we remove a conformal factor $(\epsilon^2\mu_1)^2$ from the internal metric to produce a QS scheme, the perturbations will be of $O(\epsilon^6)$ and higher; we suppose that 'internal' coordinates have been chosen so that no gauge terms of order less than ϵ^6 have been left in the QS scheme. The matching coordinates (t, r^i) are related to the Boyer–Lindquist coordinates (T, R, θ, ϕ) used to describe the internal scheme, just as in Section 3.2 for the matching coordinates (τ, x, y, z). For example, if the spin of the black hole is in the r^3 direction, then

$$t = t_0 + M_1 T,$$
$$r^1 = M_1 R \sin\theta \cos\phi,$$
$$r^2 = M_1 R \sin\theta \sin\phi,$$
$$r^3 = M_1 R \cos\theta,$$

(4.2.6)

provided that (T, R, θ, ϕ) is outside the ergosurface of the unperturbed Kerr metric, where t_0 is some 'external time'. If the spin of the black hole is in some other direction, then eqn (4.2.6) should be modified by the obvious rotation of spatial coordinates. Note the rescaling of time and distance by the factor $M_1 = \epsilon_1 \mu^2$, which we mentioned in Section 4.1. As a result, when we match out a QS perturbation of order ϵ^I, it produces metric terms in the matching coordinates which behave as $\epsilon^I (r\epsilon^{-2})^J$ near $r = 0$, apart from time and angular dependence; here J depends on the radial behaviour of the QS perturbation, and $r = |\mathbf{r}|$. When we match out the dominant Kerr part of the internal metric, we obtain a post-linearized Kerr metric (i.e. a series for the Kerr metric in negative power of r).

Combining all these terms, and making one extra assumption, we take the metric in (t, r^i) coordinates to be

$$g_{00} = -1 + \epsilon P_{00}^1(\epsilon t, r^k) + \epsilon^2 P_{00}^2(\epsilon t, r^k) + \cdots,$$
$$g_{0i} = \epsilon P_{0i}^1(\epsilon t, r^k) + \epsilon^2 P_{0i}^2(\epsilon t, r^k) + \cdots,$$
$$g_{ij} = \delta_{ij} + \epsilon P_{ij}^1(\epsilon t, r^k) + \epsilon^2 P_{ij}^2(\epsilon t, r^k) + \cdots,$$

(4.2.7)

where

$$P_{00}^1 = 0, \qquad P_{00}^2 = 2\mu_1 r^{-1} + O(r^2), \qquad P_{00}^3 = O(r^2),$$
$$P_{00}^4 = O(r), \qquad P_{00}^5 = O(r), \ldots,$$
$$P_{0i}^1 = 0, \qquad P_{0i}^2 = O(r^2), \qquad P_{0i}^3 = O(r^2),$$
$$P_{0i}^4 = -2(\mu_1)^2 \epsilon_{ijk} \chi_1^j r^k r^{-3} + O(r), \qquad P_{0i}^5 = O(r), \ldots,$$
$$P_{ij}^1 = 0, \qquad P_{ij}^2 = 2\mu_1 r^i r^j r^{-3} + O(r^2), \qquad P_{ij}^3 = O(r^2),$$
$$P_{ij}^4 = (\mu_1)^2 [(4 - 2\,|\,\chi_1\,|^2) r^{-4} r^i r^j - r^{-2} \chi_1^i \chi_1^j$$
$$\qquad + r^{-4} \chi_1^k r^k (r^i \chi_1^j + \chi_1^i r^j) + |\,\chi_1\,|^2\, r^{-2} \delta_{ij}] + O(r),$$
$$P_{ij}^5 = O(r), \cdots$$

(4.2.8)

as $r \to 0$. Here the terms given as $O(r^n)$ arise from matching out the internal perturbations, while the remaining explicit terms are the lowest-order parts of the linearized Kerr metric. The extra assumption is that the terms due to internal perturbations appear only in $P_{00}^2, P_{0i}^2, P_{ij}^2$, and higher-order components, so that $I - 2J \geq 2$ in the expression in the previous paragraph. This assumption is plausible since the internal perturbations are caused by 'background' curvature, which is here $O(\epsilon^2)$, and so should admit a description in (t, r^i) coordinates by metric perturbations of $O(\epsilon^2)$ or higher; no inconsistencies are apparent between the assumption and the rest of the approximation scheme. The $O(r^n)$ bounds on radial behaviour follow from the matching. For example, the largest $O(\epsilon^6)$ QS perturbations produce terms behaving as r^2 in P_{ab}^2, as r^1 in P_{ab}^4, as r^0 in P_{ab}^6, etc. The $O(\epsilon^7)$ QS perturbations produce terms behaving as r^2 in P_{ab}^3, as r^1 in P_{ab}^5, as r^0 in P_{ab}^7, etc., and the $O(\epsilon^8)$ QS perturbations produce terms behaving as r^3 in P_{ab}^2, as r^2 in P_{ab}^4, as r^1 in P_{ab}^6, etc.

The transformation between the matching coordinates (t, r^i) and the coordinates (u, w^i) which cover the whole external region is assumed to be of the form

$$
\begin{aligned}
t &= u + \epsilon f_1(\epsilon u, w^k) + \epsilon^2 f_2(\epsilon u, w^k) + \cdots, \\
r^i &= w^i - r_1^i(\epsilon u) + \epsilon^2 h_2^i(\epsilon u, w^k) + \epsilon^3 h_3^i(\epsilon u, w^k) + \cdots,
\end{aligned}
\tag{4.2.9}
$$

where $f_1, f_2, \ldots, h_2^i, h_3^i, \ldots$ are regular functions of $(\epsilon u, w^k)$. We only look for one such transformation that is consistent with our assumptions about the P_{ab}^n and Q_{ab}^n. In fact, there is a large variety of these transformations, and we choose the one which apparently corresponds to the simplest perturbation schemes; any alternative transformation should merely lead to a description of the same physical system (i.e. the same gravitational field) in different coordinates.

The transformation chosen has the form

$$
\begin{aligned}
t ={}& u - \epsilon \dot{r}_1^k w^k + \epsilon A + \frac{1}{2}\epsilon(\epsilon u)\dot{r}_1^k \dot{r}_1^k - \frac{1}{2}\epsilon^3 \dot{r}_1^k \dot{r}_1^k \dot{r}_1^l w^l \\
&+ \frac{3}{8}\epsilon^3 (\epsilon u)(\dot{r}_1^k \dot{r}_1^k)^2 - \epsilon^3 \dot{r}_{12}^k w^k + \epsilon^3 B \\
&+ \epsilon^3 B^k (w^k - r_1^k) + \epsilon^3 B^{kl}(w^k - r_1^k)(w^l - r_1^l) \\
&+ \epsilon^3 B^{klm}(w^k - r_1^k)(w^l - r_1^l)(w^m - r_1^m) \\
&- \epsilon^4 \dot{r}_{13}^k w^k + \epsilon^4 G + \epsilon^4 G^k (w^k - r_1^k) \\
&+ \epsilon^4 G^{kl}(w^k - r_1^k)(w^l - r_1^l) + O(\epsilon^5), \\
r^i ={}& w^i - r_1^i + \frac{1}{2}\epsilon^2 \dot{r}_1^k w^k \dot{r}_1^i - \frac{1}{2}\epsilon^2 \dot{r}_1^k \dot{r}_1^k r_1^i - \epsilon^2 r_{12}^i \\
&+ \epsilon^2 F_{12}^{ik}(w^k - r_1^k) + \epsilon^2 D^{ik}(w^k - r_1^k) \\
&+ \epsilon^2 D^{ikl}(w^k - r_1^k)(w^l - r_1^l) + \epsilon^2 D^{iklm}(w^k - r_1^k) \times \\
&\times (w^l - r_1^l)(w^m - r_1^m) - \epsilon^3 r_{13}^i + \epsilon^3 F_{13}^{ik}(w^k - r_1^k) + O(\epsilon^4).
\end{aligned}
\tag{4.2.10}
$$

Here we can separate out the terms which would effect the transformation if black hole 2 were absent. These are

$$
\begin{aligned}
t &= u - \epsilon \dot{r}_1^k w^k + \frac{1}{2}\epsilon(\epsilon u)\dot{r}_1^k \dot{r}_1^k - \frac{1}{2}\epsilon^3 \dot{r}_1^k \dot{r}_1^k \dot{r}_1^l w^l + \frac{3}{8}\epsilon^3 (\epsilon u)(\dot{r}_1^k \dot{r}_1^k)^2 + O(\epsilon^5), \\
r^i &= w^i - r_1^i + \frac{1}{2}\epsilon^2 \dot{r}_1^k w^k \dot{r}_1^i - \frac{1}{2}\epsilon^2 \dot{r}_1^k \dot{r}_1^k r_1^i + O(\epsilon^4)
\end{aligned}
\tag{4.2.11}
$$

and arise from the slow-motion Poincaré transformation of eqn (4.2.4). The terms

$$A, B, B^k, B^{kl}, B^{klm}, D^{ik}, D^{ikl}, D^{iklm}, G, G^k, G^{kl}$$

are functions of (ϵu), where

$$B^{kl} = B^{(kl)}, \qquad B^{klm} = B^{(klm)},$$
$$D^{ik} = D^{(ik)}, \qquad D^{ikl} = D^{i(kl)}, \qquad D^{iklm} = D^{i(klm)} \qquad (4.2.12)$$

and are needed in the matching at low orders. For in the QS scheme we have already chosen that no gauge terms of order less than ϵ^6 should be present in the metric, and we shall also impose coordinate restrictions on the QS perturbations at orders ϵ^6 and ϵ^7. These coordinate restrictions dictate part of the form of the $P^n_{ab}(\epsilon t, r^i)$ in eqn (4.2.8). But it will also be convenient to impose certain coordinate conditions on the $Q^n_{ab}(\epsilon u, w^i)$ found in the course of a slow-motion perturbation of flat space. The terms $A, B, \ldots$ are required to make the two types of coordinate condition compatible. Coordinate transformations analogous to eqn (4.2.10) have also been studied in celestial mechanics by Kopejkin (1988), Brumberg and Kopejkin (1989), and Klioner and Voinov (1993).

The terms $r^i_{12}, r^i_{13}, F^{ik}_{12}, F^{ik}_{13}$ are also functions of (ϵu), where

$$F^{ik}_{12} = F^{[ik]}_{12}, \qquad F^{ik}_{13} = F^{[ik]}_{13}. \qquad (4.2.13)$$

They are again determined by the matching conditions at low orders, but have a more direct physical significance than $A, B, \ldots$. We shall later be able to interpret r^i_{12} by regarding $\epsilon^2 r^i_{12}$ as the post-Newtonian position shift of the first black hole, under the interpretation given in the Introduction. Similarly, $\epsilon^3 r^i_{13}$ will be a $1 - \frac{1}{2}$ post-Newtonian position shift. Were we to write down the transformation equation (4.2.10) to higher orders, we should include $n/2$ post-Newtonian terms r^i_{1n} for $n \geq 4$ in the same way. At a lower order, r^i_1 is effectively r^i_{10}; the Newtonian interpretation for r^i_1, r^i_2 will shortly be justified when we show that r^i_1, r^i_2 obey the Newtonian equations for point particles. Note that we have not included a $\frac{1}{2}$ post-Newtonian term r^i_{11} in the transformation. Roughly speaking, this is because there are no $\frac{1}{2}$ post-Newtonian metric terms Q^n_{ab} to produce $\frac{1}{2}$ post-Newtonian deflections. Had we included terms r^i_{11} and r^i_{21} in the transformations to matching coordinates (where r^i_{2n} are introduced for the second black hole just as r^i_{1n} for the first), we should have found that r^i_{11}, r^i_{21} obey the Newtonian equations describing small displacements $\epsilon r^i_{11}, \epsilon r^i_{21}$ from the lowest-order orbits r^i_1, r^i_2; that is, that the deviations corresponded to Newtonian rather than post-Newtonian effects.

We shall also be able to interpret F^{ij}_{12} and F^{ij}_{13} as describing the precession of the black-hole spin axis on post-Newtonian and $1 - \frac{1}{2}$ post-Newtonian time scales: $\delta_{ij} + \epsilon^2 F^{ij}_{12} + \epsilon^3 F^{ij}_{13}$ is essentially a rotation matrix close to the identity. At higher orders in the transformation of eqn (4.2.10), we would include $n/2$ post-Newtonian terms F^{ij}_{1n} for $n \geq 4$ in the same way. Lower-order rotational terms can be shown to be excluded from the transformation by the matching conditions. Corresponding terms F^{ij}_{2n} will arise in the transformation from (u, w^i)

to matching coordinates for the second black hole.

We now show how some of the functions A, B, etc., are detemined by the matching. For example, by transforming the (t, r^i) form of the metric (eqn (4.2.8)) to (u, w^i) coordinates, we find that near the first black hole

$$
\begin{aligned}
g_{uu} &= -1 + 2\epsilon^2 \mu_1 r^{-1} + 2\epsilon^2 \ddot{r}_1^k w^k - 2\epsilon^2 \dot{A} \\
&\quad - 2\epsilon^2 (\epsilon u) \ddot{r}_1^k \dot{r}_1^k + \epsilon^2 O(r^2) + O(\epsilon^3) \\
&= -1 + 2\epsilon^2 \mu_1 |\mathbf{w} - \mathbf{r}_1|^{-1} + 2\epsilon^2 \ddot{r}_1^k w^k \\
&\quad - 2\epsilon^2 \dot{A} - 2\epsilon^2 (\epsilon u) \ddot{r}_1^k \dot{r}_1^k + \epsilon^2 O(|\mathbf{w} - \mathbf{r}_1|^2) + O(\epsilon^3).
\end{aligned}
\tag{4.2.14}
$$

But from our assumptions (eqn (4.2.2)) about the low-order Q_{ab}^n,

$$
\begin{aligned}
g_{uu} &= -1 + 2\epsilon^2 \mu_1 |\mathbf{w} - \mathbf{r}_1|^{-1} + 2\epsilon^2 \mu_2 |\mathbf{r}_1 - \mathbf{r}_2|^{-1} \\
&\quad - 2\epsilon^2 \mu_2 (w^k - r_1^k)(r_1^k - r_2^k) |\mathbf{r}_1 - \mathbf{r}_2|^{-3} + \epsilon^2 O(|\mathbf{w} - \mathbf{r}_1|^2) \\
&\quad + O(\epsilon^3).
\end{aligned}
\tag{4.2.15}
$$

Hence,

$$
\dot{A} = \ddot{r}_1^k \dot{r}_1^k - \mu_2 |\mathbf{r}_1 - \mathbf{r}_2|^{-1} - (\epsilon u) \ddot{r}_1^k \dot{r}_1^k, \tag{4.2.16}
$$

$$
\ddot{r}_1^i = -\mu_2 (r_1^i - r_2^i) |\mathbf{r}_1 - \mathbf{r}_2|^{-3}. \tag{4.2.17}
$$

Similarly, for the second black hole,

$$
\ddot{r}_2^i = \mu_1 (r_1^i - r_2^i) |\mathbf{r}_1 - \mathbf{r}_2|^{-3}. \tag{4.2.18}
$$

Consistently with eqns (4.2.17) and (4.2.18), we may work in the centre-of-mass frame, in which

$$
\mu_1 r_1^i + \mu_2 r_2^i = 0. \tag{4.2.19}
$$

Thus, the two black holes do indeed move approximately on Newtonian orbits in each other's far field. This is related to the property established in Chapter 3, that a small black hole moves (at lowest order) on a geodesic of the background universe which surrounds it. In the present case the 'background' for one black hole is essentially the Newtonian field of the other, contained in Q_{00}^n, and eqns (4.2.17) and (4.2.18) are just the lowest-order geodesic equations in these backgrounds. Our derivation of the Newtonian equations has used only the vacuum Einstein equations in conjunction with the matching approach; the higher-order equations of motion will again follow from matching Q_{00}^n, for $n \geq 4$.

Continuing with our determination of A, B, etc., we next compute D^{ik}, D^{ikl} by examining Q_{ij}^2, for by transforming to (u, w^i) coordinates, given the (t, r^i) metric form of eqn (4.2.8), we find

$$
Q_{ij}^2 = 2\mu_1 (w^i - r_1^i)(w^j - r_1^j) |\mathbf{w} - \mathbf{r}_1|^{-3}
$$

$$+ 2D^{ij} + (2D^{ijk} + 2D^{jik})(w^k - r_1^k) + O(|\mathbf{w} - \mathbf{r}_1|^2) \qquad (4.2.20)$$

as $|\mathbf{w} - \mathbf{r}_1| \to 0$. But, by eqn (4.2.2),

$$
\begin{aligned}
Q_{ij}^2 = {} & 2\mu_1(w^i - r_1^i)(w^j - r_1^j)\,|\mathbf{w} - \mathbf{r}_1|^{-3} \\
& + 2\mu_2(r_1^i - r_2^i)(r_1^j - r_2^j)\,|\mathbf{r}_1 - \mathbf{r}_2|^{-3} \\
& + 2\mu_2[\delta_{ik}(r_1^j - r_2^j)\,|\mathbf{r}_1 - \mathbf{r}_2|^{-3} \\
& + (r_1^i - r_2^i)\delta_{jk}\,|\mathbf{r}_1 - \mathbf{r}_2|^{-3} \\
& - 3(r_1^i - r_2^i)(r_1^j - r_2^j)(r_1^k - r_2^k)\,|\mathbf{r}_1 - \mathbf{r}_2|^{-5}](w^k - r_1^k) \\
& + O(|\mathbf{w} - \mathbf{r}_1|^2). \qquad (4.2.21)
\end{aligned}
$$

Hence

$$D^{ij} = \mu_2(r_1^i - r_2^i)(r_1^j - r_2^j)\,|\mathbf{r}_1 - \mathbf{r}_2|^{-3}, \qquad (4.2.22)$$

$$
\begin{aligned}
D^{ijk} = {} & \mu_2(r_1^i - r_2^i)\delta_{jk}\,|\mathbf{r}_1 - \mathbf{r}_2|^{-3} \\
& - \frac{3}{2}\mu_2(r_1^i - r_2^i)(r_1^j - r_2^j)(r_1^k - r_2^k)\,|\mathbf{r}_1 - \mathbf{r}_2|^{-5}. \qquad (4.2.23)
\end{aligned}
$$

Then, by examining the matching for Q_{0i}^3 in a similar way, we find

$$
\begin{aligned}
B^i = {} & -\mu_2(\mu_1 + \mu_2)^{-2}[2(\mu_1)^2 + 3\mu_1\mu_2 + \frac{3}{2}(\mu_2)^2]\dot{r}_1^i\,|\mathbf{r}_1|^{-1} \\
& -\mu_2(\mu_1 + \mu_2)^{-2}[2(\mu_1)^2 + 4\mu_1\mu_2 + \frac{3}{2}(\mu_2)^2]\dot{r}_1^k r_1^k r_1^i\,|\mathbf{r}_1|^{-3}, \qquad (4.2.24)
\end{aligned}
$$

$$
\begin{aligned}
B^{ij} = {} & \frac{1}{2}\mu_1(\mu_2)^2(\mu_1 + \mu_2)^{-2}(\dot{r}_1^i r_1^j + r_1^i \dot{r}_1^j)\,|\mathbf{r}_1|^{-3} \\
& + (\mu_2)^2(\mu_1 + \mu_2)^{-1}\dot{r}_1^k r_1^d(\frac{3}{2}r_1^i r_1^j\,|\mathbf{r}_1|^{-5} - \delta_{ij}\,|\mathbf{r}_1|^{-3}) \qquad (4.2.25)
\end{aligned}
$$

$$\dot{F}_{12}^{ij} = (\mu_2)^2(\mu_1 + \mu_2)^{-2}(2\mu_1 + \frac{3}{2}\mu_2)(r_1^i \dot{r}_1^j - \dot{r}_1^i r_1^j)\,|\mathbf{r}_1|^3. \qquad (4.2.26)$$

Here the knowledge of $\dot{F}_{12}^{ij}$ already gives the lowest-order (post-Newtonian) spin precession of each black hole—see Section 4.6.

If we take the metric given by eqns (4.2.1) and (4.2.2) and transform it to the form of eqn (4.2.7), using the transformation equation (4.2.10), then the resulting P_{ab}^n now have the form of eqn (4.2.8) for $n \le 3$. We must await more detailed information about the internal perturbations and about the higher-order Q_{ab}^n before we can determine the remaining functions $B, B^{klm}, D^{iklm}, G, G^k, G^{kl}$, or their time derivatives.

4.3 Computation of Q_{00}^4

We are now in a position to calculate some higher-order metric terms which will be needed later in finding equations of motion and spin propagation, and we begin with the remaining post-Newtonian term Q_{00}^4. This satisfies the field equation

$$Q_{00,kk}^4 = 2\epsilon^{-1}Q_{0k,0k}^3 - \epsilon^{-2}Q_{kk,00}^2 + Q_{kl}^2 Q_{00,kl}^2 + Q_{kl,l}^2 Q_{00,k}^2$$
$$- \frac{1}{2}Q_{00,k}^2 Q_{ll,k}^2 - \frac{1}{2}Q_{00,k}^2 Q_{00,k}^2, \qquad (4.3.1)$$

which follows from writing out the Einstein equations to second order in small perturbations as in eqn (3.4.1). If we write

$$R_1^i = w^i - r_1^i, \qquad R_2^i = w^i - r_2^i, \qquad (4.3.2)$$

then, using eqn (4.2.2),

$$Q_{00,kk}^4 = \sum_{\alpha=1,2} [-2\mu_\alpha \ddot{r}_\alpha^k R_\alpha^k |\mathbf{R}_\alpha|^{-3} + 2\mu_\alpha \dot{r}_\alpha^k \dot{r}_\alpha^k |\mathbf{R}_\alpha|^{-3}$$
$$- 6\mu_\alpha (\dot{r}_\alpha^k R_\alpha^k)^2 |\mathbf{R}_\alpha|^{-5}] - 4\mu_1\mu_2(|\mathbf{R}_1|^{-1}|\mathbf{R}_2|^{-3} + |\mathbf{R}_1|^{-3}|\mathbf{R}_2|^{-1})$$
$$+ 12\mu_1\mu_2 (R_1^k R_2^k)^2(|\mathbf{R}_1|^{-3}|\mathbf{R}_2|^{-5} + |\mathbf{R}_1|^{-5}|\mathbf{R}_2|^{-3})$$
$$- 16\mu_1\mu_2 R_1^k R_2^k |\mathbf{R}_1|^{-3}|\mathbf{R}_2|^{-3}.$$
$$(4.3.3)$$

By transforming the (t, r^i) form of the metric given in eqn (4.2.8) to (u, w^i) coordinates, we find

$$Q_{00}^4 = \mu_1(\dot{r}_1^k \dot{r}_1^k r_1^l - \dot{r}_1^k r_1^k \dot{r}_1^l + 2r_{12}^l)R_1^l |\mathbf{R}_1|^{-3}$$
$$\mu_1(\dot{r}_1^k R_1^k)^2 |\mathbf{R}_1|^{-3} + 2\mu_1(\dot{r}_1^k \dot{r}_1^k - 2\mu_2 |\mathbf{r}_1 - \mathbf{r}_2|^{-1})|\mathbf{R}_1|^{-1} + O(1)$$
$$(4.3.4)$$

as $|\mathbf{R}_1| \to 0$. Solving eqn (4.3.3) subject to eqn (4.3.4), a similar condition as $|\mathbf{R}_2| \to 0$, and the condition that $Q_{00}^4 \to 0$ as $|\mathbf{w}| \to \infty$, we find

$$Q_{00}^4 = \sum_{\alpha=1,2} [\mu_\alpha \ddot{r}_\alpha^k R_\alpha^k |\mathbf{R}_\alpha|^{-1} + 2\mu_\alpha \dot{r}_\alpha^k \dot{r}_\alpha^k |\mathbf{R}_\alpha|^{-1}$$
$$- 2\mu_1\mu_2 |\mathbf{r}_1 - \mathbf{r}_2|^{-1} |\mathbf{R}_\alpha|^{-1}$$
$$+ \mu_\alpha (\dot{r}_\alpha^k R_\alpha^k)^2 |\mathbf{R}_\alpha|^{-3} - \mu_\alpha \dot{r}_\alpha^k r_\alpha^k \dot{r}_\alpha^l R_\alpha^l |\mathbf{R}_\alpha|^{-3}$$
$$+ \mu_\alpha \dot{r}_\alpha^k \dot{r}_\alpha^k r_\alpha^l R_\alpha^l |\mathbf{R}_\alpha|^{-3}$$
$$+ 2\mu_\alpha r_{\alpha 2}^k R_\alpha^k |\mathbf{R}_\alpha|^{-3} - 2\mu_1\mu_2 r_\alpha^k |\mathbf{r}_\alpha|^{-1} R_\alpha^k |\mathbf{R}_\alpha|^{-3}]$$
$$+ 2\mu_1\mu_2 R_1^k R_2^k(|\mathbf{R}_1|^{-1}|\mathbf{R}_2|^{-3} + |\mathbf{R}_1|^{-3}|\mathbf{R}_2|^{-1})$$
$$- 4\mu_1\mu_2 |\mathbf{R}_1|^{-1}|\mathbf{R}_2|^{-1}. \qquad (4.3.5)$$

We remark on the usefulness of the matching method in this calculation, since it has easily given us the information in eqn (4.3.4) on the singular parts of Q_{00}^4 near the black-hole worldline. Without the use of matching, we might have had some difficulty in establishing which solution of the Poisson equation (4.3.3) to take.

One can then calculate $\dot{B}$, where $B(\epsilon u)$ is defined in eqn (4.2.10), by examining the condition that the terms appearing in P_{00}^4 due to internal perturbations are $O(r)$ as $r \to 0$. We omit the calculation, since a knowledge of $\dot{B}$ will not be of use in the rest of this work.

Our expression for Q_{00}^4 now allows us to interpret $\epsilon^2 r_{12}^i$ and $\epsilon^2 r_{22}^i$ as the post-Newtonian position shifts of the black holes, according to the interpretation given in the Introduction, because slow changes in the Newtonian orbit parameters over a time-scale of $O(\epsilon^{-3})$ will be expressed in the post-Newtonian metric components by terms which may grow to the size of Newtonian metric components over this time-scale. The only such terms are

$$2\mu_1 r_{12}^l R_1^l \, |\mathbf{R}_1|^{-3} + 2\mu_2 r_{22}^l R_2^l \, |\mathbf{R}_2|^{-3}$$

in Q_{00}^4. These would be produced by replacing r_1^i, r_2^i in Q_{00}^2 by $r_1^i + \epsilon^2 r_{12}^i$, $r_2^i + \epsilon^2 r_{22}^i$, since

$$Q_{00}^2 = 2\mu_1 \, |\mathbf{w} - \mathbf{r}_1|^{-1} + 2\mu_2 \, |\mathbf{w} - \mathbf{r}_2|^{-1} \, ,$$

and Q_{00}^4 shifts by

$$2\mu_1 r_{12}^l R_1^l \, |\mathbf{R}_1|^{-3} + 2\mu_2 r_{22}^l R_2^l \, |\mathbf{R}_2|^{-3}$$

under the replacement $r_1^i \to r_1^i + \epsilon^2 r_{12}^i$, $r_2^i \to r_2^i + \epsilon^2 r_{22}^i$. Thus r_{12}^i, r_{22}^i determine the changes in Newtonian orbit parameters over a post-Newtonian time-scale.

4.4 Computation of $1\frac{1}{2}$ post-Newtonian metric terms

Usually one does not encounter any metric terms at $1\frac{1}{2}$ post-Newtonian order in a slow-motion perturbation of flat space, but in the present context we are forced to include them. This occurs because we are scaling the angular momentum of each black hole as mass^2, whereas in an ordinary slow-motion scheme for fluid bodies the angular momentum scales as mass$^{3/2}$.

We have already taken the $1\frac{1}{2}$ post-Newtonian term Q_{ij}^3 to be zero. Then Q_{0i}^4 must satisfy the field equation

$$Q_{0i,kk}^4 - Q_{0k,ik}^4 = 0. \tag{4.4.1}$$

By transforming from the (t, r^i) form of the metric in eqn (4.2.8), we find

$$Q_{0i}^4 = -2(\mu_1)^2 \epsilon_{ikl} \chi_1^k R_1^l \, |\mathbf{R}_1|^{-3} + O(1) \tag{4.4.2}$$

as $|\mathbf{R}_1| \to 0$, with a similar condition as $|\mathbf{R}_2| \to 0$. We take the solution

$$\begin{aligned} Q_{0i}^4 = &- 2(\mu_1)^2 \epsilon_{ikl} \chi_1^k R_1^l \, |\mathbf{R}_1|^{-3} \\ &- 2(\mu_2)^2 \epsilon_{ikl} \chi_2^k R_2^l \, |\mathbf{R}_2|^{-3} \end{aligned} \tag{4.4.3}$$

of eqn (4.4.1), that is, the sum of the two source fields, which satisfies the boundary conditions at the holes and tends to zero as $|\mathbf{w}| \to \infty$.

The field equation for Q_{00}^5 gives

$$Q_{00,kk}^5 = 2\epsilon^{-1}Q_{0k,0k}^4 = 0. \tag{4.4.4}$$

When we transform from eqn (4.2.8), we find

$$
\begin{aligned}
Q_{00}^5 = \ & 2\mu_1 r_{13}^k R_1^k \left|\mathbf{R}_1\right|^{-3} \\
& + 4(\mu_1)^2 \epsilon_{klm} r_1^{ik} \chi_1^l R_1^m \left|\mathbf{R}_1\right|^{-3} \\
& + O(1)
\end{aligned}
\tag{4.4.5}
$$

as $\left|\mathbf{R}_1\right| \to 0$, with a similar condition as $\left|\mathbf{R}_2\right| \to 0$. If we solve eqn (4.4.4), subject to these boundary conditions and the requirement that $Q_{00}^5 \to 0$ as $\left|\mathbf{w}\right| \to \infty$, we obtain simply

$$
\begin{aligned}
Q_{00}^5 = \ & 2\mu_1 r_{13}^k R_1^k \left|\mathbf{R}_1\right|^{-3} \\
& + 4(\mu_1)^2 \epsilon_{klm} r_1^{ik} \chi_1^l R_1^m \left|\mathbf{R}_1\right|^{-3} \\
& + 2\mu_2 r_{23}^k R_2^k \left|\mathbf{R}_2\right|^{-3} \\
& + 4(\mu_2)^2 \epsilon_{klm} r_2^{ik} \chi_2^l R_2^m \left|\mathbf{R}_2\right|^{-3},
\end{aligned}
\tag{4.4.6}
$$

i.e., again the sum of the two source fields.

Given these expressions for Q_{0i}^4 and Q_{00}^5, one can now calculate the functions G^k and $\dot{G}$, where G^k and G appear in eqn (4.2.10). We find G^k by transforming the Q_{ab}^n to (t, r^i) coordinates and requiring that the terms in P_{0i}^4 due to internal perturbations be $O(r)$ as $r \to 0$, as in eqn (4.2.8). This gives

$$G^k = 2(\mu_2)^2 \epsilon_{klm} \chi_2^l (r_1^m - r_2^m) \left|\mathbf{r}_1 - \mathbf{r}_2\right|^3, \tag{4.4.7}$$

which is needed in calculating the $1\frac{1}{2}$ post-Newtonian equation of motion. Similarly, one finds $\dot{G}$ by transforming coordinates and requiring that $P_{00}^5 = O(r)$ as $r \to 0$; the expression for $\dot{G}$ is not used later and we do not give it.

The appearance of r_{13}^i and r_{23}^i in Q_{00}^5 shows that they may also be interpreted as describing slow changes in the Newtonian orbit parameters over a $1\frac{1}{2}$ post-Newtonian time-scale, just as r_{12}^i and r_{22}^i determine post-Newtonian orbital effects. The remaining terms in Q_{00}^5 of eqn (4.4.6) arise from applying a slow-motion boost to the matched-out Kerr field of eqn (4.2.8).

4.5   Explicit internal perturbations

Before we can proceed to find equations of motion and spin propagation at higher orders by the matching method, we need some more explicit information about the internal perturbations at the lowest orders in the QS scheme. As pointed out in Section 4.1, the lowest-order QS perturbations of black hole 1 will be at $O(\epsilon^6)$, due to the curvature produced by the Newtonian field of the other black hole, and can be given with the help of Section 3.5. Since the time dependence of the slow-motion perturbations is slowed by a factor ϵ, and since another factor ϵ^2 is gained in the course of matching to the internal scheme (eqn (4.2.6)), the QS perturbations of a unit-mass Kerr metric depend on 'internal time' T through functions of $(\epsilon^3 T)$. This implies that the QS perturbations at orders ϵ^6, ϵ^7, and ϵ^8 separately obey the field equations describing stationary linearized vacuum gravitational perturbations of a unit-mass Kerr metric. Only at order ϵ^9 do time derivatives start to appear in the field equations for the QS scheme, while non-linear effects only become important there at order ϵ^{12}.

As seen in Section 2.7, for vacuum perturbations of the Kerr metric, the perturbed geometry may be described in terms of the Newman–Penrose quantity $\Psi_0 = C_{abcd}l^a m^b l^c m^d$ (eqn (3.3.3)), where C_{abcd} is the Weyl tensor of the perturbed space–time, taking the Kinnersley null tetrad $l^a, n^a, m^a, \bar{m}^a$ of eqn (3.3.2) for a unit-mass Kerr solution with angular momentum χ (Teukolsky 1973). The unperturbed value of Ψ_0 is zero, and the part of Ψ_0 corresponding to linearized perturbations is independent of infinitesimal coordinate transformations (gauge transformations) and of infinitesimal tetrad rotations. Teukolsky derived an equation for Ψ_0 alone, not involving other perturbed field quantities, from the field equations linearized about the Kerr metric. He further showed that the equation for Ψ_0 is completely separable in Boyer–Lindquist coordinates, with solutions of the form $W(R)S(\theta)e^{i\omega T}e^{im\phi}$. As noted in Section 3.3, it has been shown by Wald (1973) that under reasonable conditions, Ψ_0 uniquely determines the geometry for linearized perturbations, apart from the addition of gauge terms and of perturbations which merely change the mass and angular momentum of the black hole. Chrzanowski (1975) has given constructive expression to Wald's theorem by providing explicit formulas for the metric perturbations in terms of the perturbed Ψ_0. In turn, Wald (1978) justified Chrzanowski's work. These ideas have been further elaborated by Chandrasekhar (1979, 1983), who has found the general linear metric perturbation of the Kerr metric, obeying the linearized Einstein equation.

The matching conditions reveal that the $O(\epsilon^6)$ QS metric perturbations are $O(R^2)$ as $R \to \infty$, where the fastest growing R^2 parts correspond to the r^2 parts of P_{ab}^2 in eqn (4.2.8). Similarly, the $O(\epsilon^7)$ QS metric perturbations are again $O(R^2)$ as $R \to \infty$, with the R^2 parts corresponding to the r^2 parts of P_{ab}^3. The $O(\epsilon^8)$ QS metric perturbations are $O(R^3)$ as $R \to \infty$, where the R^3 parts arise from the r^3 parts of P_{ab}^2. It follows from arguments given in Section 3.5, where we use the Teukolsky equation for Ψ_0 together with boundary conditions of regularity at the future event horizon of the black hole, that the $O(\epsilon^6)$ and $O(\epsilon^7)$ QS perturbations each consist only of modes with angular quantum number $L = 2$. By a slight extension of these arguments, it follows that the $O(\epsilon^8)$ QS perturbations consist of a sum of modes which only have $L = 3$ or 2. It will be particularly important to have explicit information about the $L = 2$ stationary modes for these lowest-order QS perturbations. Only in that case will we be able to complete the matching model to obtain, e.g., the post-Newtonian equation of motion and $1\frac{1}{2}$ post-Newtonian equation of spin precession given in Section 4.6.

In the context of the two-black-hole problem, we see that the lowest-order

QS metric perturbations can be found explicitly, provided that we have a means of computing metric perturbations from Ψ_0 and that we can identify the electric and magnetic curvature tensors E_{ij}, H_{ij}, which act as 'background' curvature, producing these low-order $L = 2$ modes.

It is shown in Chrzanowski (1975) and Wald (1978), as described in Section 2.7, that if Ψ_0 satisfies Teukolsky's equation for vacuum perturbations about the Kerr metric, then there is a complex metric perturbation h_{ab} constructed from Ψ_0 such that both $\mathrm{Re}(h_{ab})$ and $i\mathrm{Im}(h_{ab})$ satisfy the linearized Einstein equations and lead back to (complex) multiples of the original Ψ_0, where we use the defining equation $\Psi_0 = C_{abcd}l^a m^b l^c m^d$. Explicitly,

$$
\begin{aligned}
h_{ab} = \rho^{-4}\{ & n_a n_b(2\pi + 2\bar\beta + \bar\delta)(3\pi - 4\bar\beta - \bar\delta) \\
& + \bar m_a \bar m_b(\bar\mu + 4\mu - 2\gamma + \Delta)(4\gamma - \mu - \Delta) \\
& - n_{((a}\bar m_{b)}[2\bar\gamma + 2\gamma - 4\mu - \Delta)(4\bar\beta - 3\pi + \bar\delta) \\
& + (2\pi + 4\bar\beta + \bar\tau + \bar\delta)(4\gamma - \mu - \Delta)]\}\Psi_0.
\end{aligned}
\tag{4.5.1}
$$

Here the Newman–Penrose operators $\nabla, \bar\delta$ act on functions f according to $\nabla f = n^a f_{,a}$ and $\bar\delta f = \bar m^a f_{,a}$. The quantities $\rho, \beta, \pi, \tau, \mu, \gamma$ are Newman–Penrose spin coefficients relating to the Kinnersley tetrad. In Boyer–Lindquist coordinates for a unit-mass Kerr metric with angular momentum χ, they take the values

$$
\begin{aligned}
\rho &= -1/(R - i\chi\cos\theta), & \beta &= -\bar\rho(\cot\theta)/2\sqrt{2}, \\
\pi &= i\chi\rho^2(\sin\theta)/\sqrt{2}, & \tau &= -i\chi\rho\bar\rho(\sin\theta)\sqrt{2}, \\
\mu &= \rho^2\bar\rho\Delta/2, & \gamma &= \mu + \rho\bar\rho(R - 1)/2.
\end{aligned}
\tag{4.5.2}
$$

We can now use eqn (4.5.1) to find explicitly the Kerr $L = 2$ stationary metric perturbations corresponding to the Ψ_0 of eqn (3.5.23), which gives the general stationary $L = 2$ solution for Ψ_0. Since we shall only need the first two terms in an asymptotic expansion of the metric perturbations about $R = \infty$, we expand eqn (3.5.23) as

$$
\Psi_0 = \sum_{m=-2}^{2} \frac{1}{2}(\alpha_m + i\beta_m)_2 Y_2^m(\theta, \phi)[1 + (im\chi/R) + O(R^{-2})]
\tag{4.5.3}
$$

near $R = \infty$. One proceeds by taking in turn the only non-zero member of $\{\alpha_m, \beta_m : 0 \leq m \leq 2\}$ to be $\alpha_0, \alpha_1, \alpha_2$ with α_1, α_2 real, and computing h_{ab} in eqn (4.5.1); the expressions corresponding to α_1 or α_2 imaginary can be found by performing rotations about the axis $\theta = 0$. Since only the largest terms of h_{ab} near $R = \infty$ are required, one may also use expansions of the Kinnersley tetrad components (eqn (3.2.2)) and spin coefficients (eqn (4.5.2)) about $R = \infty$ to simplify the calculation.

In the case of $L = 2$ stationary modes caused only by an electric tensor E_{ij}, so that all $\beta_m = 0$, we find the metric perturbation

$$'j_{ab} = -\frac{2}{3}\mathrm{Re}(h_{ab}). \qquad \qquad \text{·(4.5.4)}$$

To see this, note that both $\mathrm{Re}(h_{ab})$ and $i\mathrm{Im}(h_{ab})$ are vacuum metric perturbations regular at the future event horizon which lead back to complex multiples of the original Ψ_0 via $\Psi_0 = C_{abcd}l^a m^b l^c m^d$. By explicit computation, we find that the ratio of Ψ_0 for $\mathrm{Re}(h_{ab})$ and Ψ_0 for $i\mathrm{Im}(h_{ab})$ is purely imaginary. Hence there is a unique linear combination of $\mathrm{Re}(h_{ab})$ and $i\mathrm{Im}(h_{ab})$, which leads back to the original Ψ_0. By Wald's uniqueness theorem (1973), this is precisely the metric perturbation corresponding to the given Ψ_0, modulo gauge transformations, and the addition of angular momentum and mass to the black hole. The correct linear combination of eqn (4.5.4) is found by computing $\lim_{R\to\infty}\Psi_0$ and comparing with eqn (4.5.3). We remark that it follows from Chrzanowski (1975) that the perturbations caused by an electric tensor are even under the parity transformation $(T, R, \theta, \phi) \to (T, R, \pi - \theta, \pi + \phi)$.

We write down the $L = 2$ metric perturbations in terms of coordinates (T, R^1, R^2, R^3), where $R^1 = R\sin\theta\cos\phi, R^2 = R\sin\theta\sin\phi, R^3 = R\cos\theta$ and the labels $i, j, \ldots$ refer to $R^i, R^j, \ldots (i = 1, 2, 3, \text{etc.})$. In the gauge of eqn (4.5.1), we find

$$'j_{TT} = E_{kl}R^k R^l - 4R^{-1}E_{kl}R^k R^l + O(1),$$

$$'j_{Ti} = \frac{1}{3}R^{-1}R^i E_{jk}R^j R^k + \frac{2}{3}RE_{ij}R^j - \frac{2}{3}R^{-2}R^i E_{jk}R^j R^k$$
$$- \frac{4}{3}E_{ij}R^j - \frac{2}{3}E_{ij}\epsilon_{jkl}\chi^k R^l + \frac{4}{3}\epsilon_{ijk}E_{jl}\chi^k R^l + O(1),$$

$$'j_{ij} = \frac{2}{3}E_{ij}R^k R^k + \frac{1}{3}\delta_{ij}E_{kl}R^k R^l$$
$$+ (\frac{2}{3}R^{-1}E_{km}\chi^l R^m + \frac{1}{3}R^{-3}E_{mn}\chi^l R^k R^m R^n)(R^i\epsilon_{jkl} + R^j\epsilon_{ikl}) \qquad (4.5.5)$$
$$+ \frac{2}{3}R^{-3}R^i R^j\epsilon_{klm}E_{kl}\chi^m R^l R^n$$
$$- \frac{2}{3}R^{-1}\delta_{ij}\epsilon_{kmn}E_{kl}\chi^m R^l R^n$$
$$+ \frac{2}{3}R^{-1}R^l\chi^m R^m(E_{ki}\epsilon_{jkl} + E_{kj}\epsilon_{ikl}) + O(1)$$

as $R \to \infty$. By a gauge transformation, this perturbation can be put in the simpler form

$$j_{TT} = E_{kl}R^k R^l - 4R^{-1}E_{kl}R^k R^l + O(1),$$

$$j_{Ti} = -\frac{2}{3}E_{ij}\epsilon_{jkl}\chi^k R^l + \frac{4}{3}\epsilon_{ijk}E_{jl}\chi^k R^l + O(1), \qquad (4.5.6)$$

$$j_{ij} = -\frac{2}{3}E_{ij}R^k R^k + \frac{1}{3}\delta_{ij}E_{kl}R^k R^l + O(1).$$

Similarly, for $L = 2$ stationary modes caused only by a magnetic tensor H_{ij}, we find the metric perturbation

$$'k_{ab} = \frac{2}{3}\text{Re}(h_{ab}), \qquad (4.5.7)$$

where now all coefficients α_m are zero. These perturbations are odd under the parity transformation. In the gauge of eqn (4.5.1),

$$'k_{TT} = -2H_{kl}R^k\chi^l + O(1),$$

$$'k_{Ti} = \frac{2}{3}\epsilon_{ijk}H_{jl}R^kR^l - \frac{4}{3}R^{-1}\epsilon_{ijk}H_{jl}R^kR^l + \frac{1}{3}R^{-3}R^iH_{jk}\chi^lR^jR^kR^l$$

$$- \frac{2}{3}R^{-1}H_{ij}\chi^kR^jR^k - \frac{1}{3}R^{-1}\chi^iH_{jk}R^jR^k - \frac{2}{3}R^{-1}R^iH_{jk}\chi^jR^k$$

$$- \frac{2}{3}RH_{ij}\chi^j + O(1), \qquad (4.5.8)$$

$$'k_{ij} = \frac{1}{3}R^{-1}H_{km}R^lR^m(\epsilon_{ikl}R^j + \epsilon_{jkl}R^i) + \frac{1}{3}RR^l(H_{ik}\epsilon_{jkl} + H_{jk}\epsilon_{ikl})$$

$$- \frac{4}{3}H_{ij}\chi^kR^k - \frac{2}{3}\delta_{ij}H_{kl}R^k\chi^l + O(1)$$

as $R \to \infty$. This perturbation can be simplified by a gauge transformation to give

$$k_{TT} = -2H_{kl}R^k\chi^l + O(1),$$

$$k_{Ti} = \frac{2}{3}\epsilon_{ijk}H_{jl}R^kR^l - \frac{4}{3}R^{-1}\epsilon_{ijk}H_{jl}R^kR^l + O(1), \qquad (4.5.9)$$

$$k_{ij} = -\frac{4}{3}H_{ij}\chi^kR^k - \frac{2}{3}\delta_{ij}H_{kl}R^k\chi^l + O(1).$$

Armed with eqns (4.5.6) and (4.5.9), we can now find the parts of the QS metric perturbations which are relevant for computing equations of motion and spin propagation. Since we know that the $O(\epsilon^6)$ and $O(\epsilon^7)$ QS perturbations for each black hole are separately $L = 2$ stationary, we may assume that they are in the gauges of eqns (4.5.4) and (4.5.9). For example, the $O(\epsilon^6)$ QS perturbations of black hole 1 are taken to be $\epsilon^6(\mu_1)^2(j_{ab} + k_{ab})$, where we have reinstated the perturbation factors $\epsilon^6(\mu_1)^2$ which mass^2, and where χ^i, E_{kl}, H_{kl} appearing in eqns (4.5.4) and (4.5.9) are replaced by $\chi_1^i, E_{kl}^{16}, H_{kl}^{16}$; the label '16' appearing in E_{kl}^{16}, H_{kl}^{16} refers to the $O(\epsilon^6)$ perturbations of black hole 1.

We find E_{kl}^{16} and H_{kl}^{16} by examining the matching conditions for the metric perturbations Q_{ab}^2 of flat space. Recall that we have already dealt with the matching for Q_{ab}^2 up to terms behaving as $|\mathbf{w} - \mathbf{r}_1|$ near the first black hole. Our assumption of a specific form based on eqns (4.5.6) and (4.5.9) for the largest $O(\epsilon^6)$ QS perturbations now allows us to deal with all terms in Q_{ab}^2 behaving as $|\mathbf{w} - \mathbf{r}_1|^2$ near the first black hole. First, by examining the matching for Q_{00}^2 at $O(|\mathbf{w} - \mathbf{r}_1|^2)$, we find

$$E_{kl}^{16} = 3\mu_2(r_1^k - r_2^k)(r_1^l - r_2^l)|\mathbf{r}_1 - \mathbf{r}_2|^{-5} - \mu_2\delta_{kl}|\mathbf{r}_1 - \mathbf{r}_2|^{-3}. \qquad (4.5.10)$$

This expression just arises from making a Taylor expansion of the part $2\mu_2 |\mathbf{w} - \mathbf{r}_2|^{-1}$ of Q_{00}^2 due to the second black hole. From the matching for Q_{0i}^2 at $O(|\mathbf{w} - \mathbf{r}_2|^2)$ we find

$$H_{kl}^{16} = 0. \tag{4.5.11}$$

Thus the dominant internal perturbations of black hole 1 are purely even, due to the electric tensor E_{kl}^{16} which represents the Newtonian curvature caused by black hole 2. Matching the $|\mathbf{w} - \mathbf{r}_1|^2$ part of Q_{ij}^2 gives us the quantity D^{ijkl} appearing in the transformation eqn (4.2.10); a knowledge of D^{ijkl} is necessary in computing the post-Newtonian equation of motion. We have

$$D^{ijkl} = \frac{5}{2}\mu_2 S^i S^j S^k S^l S^{-7} + \frac{1}{6}\mu_2 S^{-3}(\delta_{ij}\delta_{kl} + \delta_{ik}\delta_{jl} + \delta_{il}\delta_{jk})$$
$$- \frac{1}{6}\mu_2 S^{-5}(\delta_{ij}S^k S^l + \delta_{ik}S^j S^l + \delta_{il}S^j S^k)$$
$$- \frac{5}{6}\mu_2 S^i S^{-5}(\delta_{jk}S^l + \delta_{jl}S^k + \delta_{kl}S^j), \tag{4.5.12}$$

where

$$S^i = r_1^i - r_2^i, \qquad S = |\mathbf{S}|. \tag{4.5.13}$$

The $O(\epsilon^7)$ QS perturbations of black hole 1 are taken to be of the form $\epsilon^7 (\mu_1)^2 (j_{ab} + k_{ab})$, where χ^i, E_{kl}, H_{kl} appearing in eqns (4.5.6) and (4.5.9) for j_{ab}, k_{ab} are replaced by $\chi_1^i, E_{kl}^{17}, H_{kl}^{17}$. By analogy with our calculation of E_{kl}^{16} and H_{kl}^{16}, we compute E_{kl}^{17} and H_{kl}^{17} from matching the Q_{ab}^3. Again, we have already dealt with matching for Q_{ab}^3 up to $O(|\mathbf{w} - \mathbf{r}_1|)$ near black hole 1, and our assumption about the gauge for the $O(\epsilon^7)$ QS perturbations now leads to complete information about the matching for Q_{ab}^3 up to $O(|\mathbf{w} - \mathbf{r}_1|^2)$. By matching the $|\mathbf{w} - \mathbf{r}_1|^2$ part of Q_{00}^3 we obtain

$$E_{kl}^{17} = 0. \tag{4.5.14}$$

From the $|\mathbf{w} - \mathbf{r}_1|^2$ part of Q_{0i}^3 we find

$$H_{kl}^{17} = -3(\mu_1 + \mu_2)S^{-5}\dot{r}_1^n S^m (\epsilon_{knm}S^l + \epsilon_{lnm}S^k). \tag{4.5.15}$$

We see that the second-largest QS perturbations are odd under parity change, and are due to a magnetic tensor H_{kl}^{17} which represents mass current effects. Examination of the $|\mathbf{w} - \mathbf{r}_1|^2$ part of Q_{0i}^3 also gives us the matching term B^{klm} of eqn (4.2.10); this term is not needed later and we do not write it down. Matching for Q_{ij}^3 at $O(|\mathbf{w} - \mathbf{r}_1|^2)$ does not give any new information, but merely acts as a check on the consistency of our treatment so far.

By using Chrzanowski (1973) and Wald (1978) together with our matching method in this section, we have found enough explicit information on the $O(\epsilon^6)$ and $O(\epsilon^7)$ QS perturbations to enable us to derive all equations of motion and spin propagation up to and including $1 - \frac{1}{2}$ post-Newtonian order; the derivation of these equations will be described in the next section. We also leave to the next section any discussion of the explicit metric perturbations at $O(\epsilon^8)$ in the QS scheme, which are needed in computing the dominant second post-Newtonian part of the spin-spin force.

As mentioned in Chapter 3, weak perturbations of a large Schwarzschild black hole of mass M in the presence of a small Schwarzschild black hole of mass m have been studied, leading to an approximate calculation of the distorted event horizon, in the two cases of time-symmetric initial data and infall of the small black hole from rest at infinity (Bishop 1988). In the case of time-symmetric initial data (Misner 1960; Brill and Lindquist 1963; Bowen and York 1980), the emitted gravitational waves have been studied, together with the distortion of the event horizon (Tomimatsu 1989).

4.6　Equations of motion and spin propagation

The post-Newtonian equation of motion for black hole 1 now follows from matching Q_{00}^4 at $O(|\mathbf{w} - \mathbf{r}_1|)$ near the black-hole worldline. This computation uses eqn (4.5.6) together with eqn (4.5.10) to find the contribution of internal perturbations to the $|\mathbf{w} - \mathbf{r}_1|$ part of Q_{00}^4. It also uses the information which we derived previously on the matching quantities $A, B^k, B^{kl}, D^{ik}, D^{ikl}, D^{iklm}, F_{12}^{ik}$ of eqn (4.2.10). We do not give the details, but merely the final result

$$\begin{aligned}
\ddot{r}_{12}^i = {} & \mu_2 r_{22}^i S^{-3} - 3\mu_2 r_{22}^k S^k S^i S^{-5} - \mu_2 r_{12}^i S^{-3} + 3\mu_2 r_{12}^k S^k S^i S^{-5} \\
& + [3\mu_1\mu_2 + 2(\mu_2)^2] S^i S^{-4} \\
& - [2(\mu_1)^2 + 7\mu_1\mu_2 + \frac{1}{2}(\mu_2)^2]\mu_2(\mu_1)^{-2}\dot{r}_2^k \dot{r}_2^k S^i S^{-3} \\
& + \frac{1}{2}[3(\mu_1)^2 + 5\mu_1\mu_2 + 7(\mu_2)^2]\mu_2(\mu_1)^{-2}\dot{r}_2^k S^k \dot{r}_2^i S^{-3} \\
& + [3(\mu_1)^2 + \frac{15}{2}\mu_1\mu_2]\mu_2(\mu_1)^{-2}(\dot{r}_2^k S^k)^2 S^i S^{-5}.
\end{aligned} \tag{4.6.1}$$

The first four terms in eqn (4.6.1) are just the Newtonian terms which we would expect for small perturbations $\epsilon^2 r_{12}^i$, $\epsilon^2 r_{22}^i$ in the orbits. The remaining terms are relativistic in origin. Of course, the post-Newtonian equation of motion for the second black hole is found by interchanging the particle labels 1 and 2 in eqn (4.6.1) (so that $S^i \to -S^i$ also).

Next we turn to the post-Newtonian equation of spin propagation. We first justify our interpretation suggested in Section 4.2 of $F_{12}^{ij}(\epsilon u)$ as describing the lowest-order precession of the spin of the first black hole, where $F_{12}^{ij}(\epsilon U)$ is the antisymmetric matrix appearing in the transformation eqn (4.2.10). The Kerr spin parameters first appear in the slow-motion scheme at the $1\frac{1}{2}$ post-Newtonian order, in the terms Q_{0i}^4 and Q_{00}^5. Metric terms which represent the variation of the Kerr parameters over long time-scales first appear at the $2\frac{1}{2}$ post-Newtonian order, in Q_{0i}^6 and Q_{00}^7. These describe changes taking place over a post-Newtonian time-scale of $O(\epsilon^{-3})$. We pointed out in the Introduction that secular fractional changes of $O(1)$ in the masses and scalar angular momenta of the black holes only take place over a much longer time-scale of $O(\epsilon^{-10})$. Thus, the only significant effect on the Kerr spin parameters over a post-Newtonian time-scale will be a slow precession. This will appear in Q_{0i}^6 through the slow rotation of the Kerr parameters in

$$Q_{0i}^4 = -2(\mu_1)^2 \epsilon_{ikl} \chi_1^k (w^l - r_1^l) \, |\mathbf{w} - \mathbf{r}_1|^{-3}$$
$$- 2(\mu_2)^2 \epsilon_{ikl} \chi_2^k (w^l - r_2^l) \, |\mathbf{w} - \mathbf{r}_2|^{-3} . \tag{4.6.2}$$

By examining the matching conditions for the P_{ab}^n and the transformation to (u, w^i) coordinates, one finds that Q_{0i}^6 contains the term

$$2(\mu_1)^2 F_{12}^{ik} \epsilon_{klm} \chi_1^l (w^m - r_1^m) \, |\mathbf{w} - \mathbf{r}_1|^{-3} ,$$

together with a similar term for the second black hole. Other terms involving the spin and the transformation terms $D^{ik}, D^{ikl}, D^{iklm}$ are also present in Q_{0i}^6, but these will remain bounded over a post-Newtonian time-scale and are not of the form of precession terms. The expression

$$2(\mu_1)^2 F_{12}^{ik} \epsilon_{klm} \chi_1^l (w^m - r_1^m) \, |\mathbf{w} - \mathbf{r}_1|^{-3}$$

is the only term in Q_{0i}^6 which describes $O(1)$ post-Newtonian variations in the normalized Kerr parameter of the first black hole, and is just as if χ_1^i were replaced by the rotated parameter $(\delta_{ik} + \epsilon^2 F_{12}^{ki})\chi_1^k$ in Q_{0i}^4; a similar argument holds for the second black hole.

Equation (4.2.26) determining $\dot{F}_{12}^{ij}$ from the matching conditions now gives the post-Newtonian spin precession of the first black hole. Written in vectorial form, the equation for the propagation of the spin $\mathbf{\Sigma}$ of the first black hole is

$$\frac{d\mathbf{\Sigma}}{du} = \epsilon^2 (\mu_2)^2 (\mu_1 + \mu_2)^{-2} (2\mu_1 + \frac{3}{2}\mu_2) \, |\mathbf{r}_1|^{-3} (\mathbf{r}_1 \times \dot{\mathbf{r}}_1) \times \mathbf{\Sigma}$$
$$+ O(\epsilon^4 \, |\mathbf{\Sigma}|). \tag{4.6.3}$$

It might appear that we have calculated this spin precession without really using the field equations. However, our expression for $\dot{F}_{12}^{ij}$ arose from imposing the condition that the terms in P_{0i}^3 of eqn (4.2.8) due to internal perturbations are $O(r^2)$ as $r \to 0$. This follows from the condition that the QS internal perturbations are $O(\epsilon^6)$, which in turn depends on arguments of the type used in Chapter 3, involving the Teukolsky equation, to show that any lower-order internal perturbations are trivial. Thus the expression for $\dot{F}_{12}^{ij}$ does ultimately rely on the field equations near the black hole, together with a boundary condition of regularity at the future event horizon. The same comment can be made of all our other derivations of equations of motion and spin propagation.

We now proceed to find the $1 - \frac{1}{2}$ post-Newtonian equation of spin propagation by matching for Q_{0i}^4. We have already dealt with the matching up to $O(|\mathbf{w} - \mathbf{r}_1|^0)$ in Q_{0i}^4; this led to the expression for G^k in eqn (4.4.7). Equations (4.5.6) and (4.5.9) describing the dominant internal perturbations, together with eqns (4.5.10) and (4.5.11), allow us to treat the matching for Q^4_{0i} up to $O(|\mathbf{w} - \mathbf{r}_1|)$. By examining the $|\mathbf{w} - \mathbf{r}_1|$ part of Q^4_{0i} and using the previous expression for G^k, we find

$$
\begin{aligned}
\dot{F}^{ik}_{13} = & - 2(\mu_2)^2 \epsilon_{imk} \chi^m_2 S^{-3} - 2\mu_1 \mu_2 \epsilon_{imk} \chi^m_1 S^{-3} \\
& + 3(\mu_2)^2 \epsilon_{iml} \chi^m_2 S^k S^l S^{-5} \\
& + 3\mu_1 \mu_2 \epsilon_{iml} \chi^m_1 S^k S^l S^{-5} \\
& - 3(\mu_2)^2 \epsilon_{klm} \chi^m_2 S^i S^l S^{-5} \\
& - 3\mu_1 \mu_2 \epsilon_{klm} \chi^m_1 S^i S^l S^{-5},
\end{aligned}
\tag{4.6.4}
$$

$$
\begin{aligned}
G^{kl} = & - \frac{3}{2} (\mu_2)^2 \epsilon_{kmn} \chi^m_2 S^n S^l S^{-5} \\
& - \frac{3}{2} (\mu_2)^2 \epsilon_{lmn} \chi^m_2 S^n S^k S^{-5} \\
& + \frac{1}{2} \mu_1 \mu_2 \epsilon_{lmn} \chi^n_1 S^k S^m S^{-5} \\
& + \frac{1}{2} \mu_1 \mu_2 \epsilon_{kmn} \chi^n_1 S^l S^m S^{-5},
\end{aligned}
\tag{4.6.5}
$$

where G^{kl} is a matching term appearing in eqn (4.2.10).

The interpretation of F^{ik}_{13} as describing $1 - \frac{1}{2}$ post-Newtonian spin precession is analogous to the post-Newtonian interpretation of F^{ik}_{12}; the evolution of the spin of the first black hole over $1\frac{1}{2}$ post-Newtonian time-scales is determined by the rotated parameter $(\delta_{ik} + \epsilon^2 F^{kl}_{12} + \epsilon^3 F^{kl}_{13}) \chi^k_1$. In vectorial form, the spin propagation equation for black hole 1, allowing for the $1\frac{1}{2}$ post-Newtonian effects of eqn (4.6.4), is

$$
\begin{aligned}
\frac{d\mathbf{\Sigma}}{du} = \mathbf{\Sigma} \times \Big\{ & - \epsilon^3 (\mu_2)^2 (\mu_1 + \mu_2)^{-2} (2\mu_1 + \frac{3}{2}\mu^2) (\mathbf{r}_1 \times \dot{\mathbf{r}}_1) |\mathbf{r}_1|^{-3} \\
& + \epsilon^4 (\mu_2)^2 \chi_2 |\mathbf{r}_1 - \mathbf{r}_2|^{-3} \\
& - 3\epsilon^4 (\mu_2)^2 (\mathbf{r}_1 - \mathbf{r}_2) [\chi_2 \cdot (\mathbf{r}_1 - \mathbf{r}_2)] |\mathbf{r}_1 - \mathbf{r}_2|^{-5} \\
& + \epsilon^4 \mu_1 \mu_2 \chi_1 |\mathbf{r}_1 - \mathbf{r}_2|^{-3} \\
& - 3\epsilon^4 \mu_1 \mu_2 (\mathbf{r}_1 - \mathbf{r}_2) [\chi_1 \cdot (\mathbf{r}_1 - \mathbf{r}_2)] |\mathbf{r}_1 - \mathbf{r}_2|^{-5} \Big\} \\
& + O(\epsilon^5 |\mathbf{\Sigma}|).
\end{aligned}
\tag{4.6.6}
$$

As we pointed out previously, one way to extract gauge-independent information from these equations is to look for secular effects, such as the EIH perihelion shift. For example, the post-Newtonian spin precession leads to $O(1)$ changes in the orientation of the intrinsic spins, when the orbit is bound, over time-scales of $O(\epsilon^{-3})$. These secular changes can be computed with the help of some expressions given by Barker and O'Connell (1970) for averages such as $< S^{-3} >$, taken over Newtonian time-scales. This leads to the spin precession on post-Newtonian time-scales given by

$$\frac{d\Sigma_1}{d(\epsilon^{-3}u)} = \frac{G^{3/2}\mu_2(4\mu_1 + 3\mu_2)}{2c^2a^{5/2}(1 - e^2)(\mu_1 + \mu_2)^{1/2}}\mathbf{n} \times \Sigma_1,$$

where $\mathbf{n}$ is the unit vector in the direction of the orbital angular momentum, and we have again written the G and c dependence explicitly. Here a is the semi-major axis of the ellipse. Thus the spins precess at a steady rate around the orbital-angular-momentum vector. When we look over longer time-scales, the $1\frac{1}{2}$ and second post-Newtonian equations of motion can only produce small extra perihelion precessions for a bound system, which will be swamped by the post-Newtonian precession. This follows from the conservation of energy and angular momentum, since gravitational radiation only acts to change these 'conserved quantities' over a $2\frac{1}{2}$ post-Newtonian time-scale, and since the angular momentum has to be dominated by the orbital contribution, with our scaling of small parameters. The $1\frac{1}{2}$ post-Newtonian spin precession for a bound system can also be calculated with the help of Barker and O'Connell (1970). In order to find $O(1)$ changes in the spin, one must first average the $1\frac{1}{2}$ post-Newtonian effects in the spin precession equation (4.6.6) over several Newtonian orbital time-scales, regarding χ_1 and χ_2 as fixed. The result must then be averaged over several post-Newtonian time scales, allowing for the post-Newtonian spin precession about $\mathbf{n}$. The outcome is that (except in one special case) the active spin contributions in eqn (4.6.6) make little difference to the evolution of χ_1 and χ_2 over $1\frac{1}{2}$ post-Newtonian time-scales, since the projections of the spin along $\mathbf{n}$ and into the orbital plane remain esentially of constant length. The only exception occurs when $\mu_1 = \mu_2$, since then the post-Newtonian spin precession rates for the two black holes are identical, and there is some resonance behaviour which forces $\mathbf{n} \cdot \Sigma_1$ to change with time.

Next we find the $1\frac{1}{2}$ post-Newtonian equation of motion by matching the $|\mathbf{w} - \mathbf{r}_1|$ parts of Q_{00}^5. Here we use eqns (4.5.6) and (4.5.9) together with eqns (4.5.14) and (4.5.15) to find the explicit $O(\epsilon^7)$ QS metric perturbations which contribute to the $|\mathbf{w} - \mathbf{r}_1|$ part of Q_{00}^5 through the matching. We also need the expression for Q_{00}^5 in eqn (4.4.6) and the matching quantities $G^k, G^{kl}, \dot{F}_{13}^{ik}$. The resulting equation of motion is

$$\begin{aligned}
\ddot{r}_{13}^i = {}& \mu_2 r_{23}^i S^{-3} - 3\mu_2 r_{23}^k S^k S^i S^{-5} - \mu_2 r_{13}^i S^{-3} \\
& + 3\mu_2 r_{13}^k S^K S^i S^{-5} - 4\mu_2(\mu_1 + \mu_2)\epsilon_{ikl}\dot{r}_1^k\chi_2^l S^{-3} \\
& + 3\mu_1(\mu_1 + \mu_2)S^i\epsilon_{klm}\dot{r}_1^k\chi_1^l S^m S^{-5} \\
& + 6\mu_2(\mu_1 + \mu_2)S^i\epsilon_{klm}\dot{r}_1^k\chi_2^l S^m S^{-5} \\
& - 3\mu_1(\mu_1 + \mu_2)\epsilon_{ikl}\dot{r}_1^k S^l\chi_1^m S^m S^{-5} \\
& - 6\mu_2(\mu_1 + \mu_2)\epsilon_{ikl}\chi_2^k S^l\dot{r}_1^m S^m S^{-5}.
\end{aligned} \tag{4.6.7}$$

The first four terms arise as we might expect from the Newtonian force. The remaining terms describe various types of spin–orbit interaction.

To find the dominant second post-Newtonian part of the spin–spin force, we examine the matching for Q_{00}^6. Since $\ddot{r}_{14}^i$ appears in Q_{00}^6 in the form $2\dot{r}_{14}^k(w^k - r_1^k)$

when we match out the internal scheme, it is sufficient to look for terms in Q_{00}^6 which are proportional to both black-hole spins and have radial dependence $|\mathbf{w} - \mathbf{r}_1|$. Now when one computes Q_{00}^6 in the course of the slow-motion perturbation of flat space, no such terms depending on $\chi_1^k \chi_2^l$ appear. To see this, we first remark that there are no such terms in Q_{ij}^4, for Q_{ij}^4 satisfies the field equation

$$Q_{ij,kk}^4 + Q_{kk,ij}^4 - Q_{ik,jk}^4 - Q_{jk,ik}^4 = S_{ij}^4, \qquad (4.6.8)$$

where S_{ij}^4 is a complicated expression involving only Q_{ab}^n for $n < 4$, and hence contains no spin terms. Since there are no products $\chi_1^k \chi_2^l$ in the boundary conditions for Q_{ij}^4 near the two black holes found from the matching, it is possible to choose Q_{ij}^4 such that it involves no products $\chi_1^k \chi_2^l$. In fact, the author has computed an explicit form for Q_{ij}^4, which satisfies eqn (4.6.8) and the matching conditions near the holes; the only spin terms in this Q_{ij}^4 involve $\chi_1^k \chi_1^l$ or $\chi_2^k \chi_2^l$ through Kerr far-field terms of the type given in P_{ij}^4 of eqn (4.2.8?). Next consider Q_{0i}^5, which satisfies a field equation of the form

$$Q_{0i,kk}^5 - Q_{0k,ik}^5 = \epsilon^{-1} Q_{ik,0k}^4 - \epsilon^{-1} Q_{kk,0i}^4 + S_{0i}^5. \qquad (4.6.9)$$

Again S_{0i}^5 is a complicated quantity depending only on the Q_{ab}^n for $n < 4$, so that the right-hand side of eqn (4.6.9) contains no terms involving $\chi_1^k \chi_2^l$. It can be shown that the boundary conditions on Q_{0i}^5 near the black holes arising from the matching involve no $\chi_1^k \chi_2^l$ terms, and hence that Q_{0i}^5 may be chosen to involve no such product terms. Now Q_{00}^6 satisfies the field equation

$$Q_{00,kk}^6 = 2\epsilon^{-1} Q_{0k,0k}^5 + S_{00}^6, \qquad (4.6.10)$$

where S_{00}^6 depends only on the Q_{ab}^n for $n < 5$ and contains no spin product terms $\chi_1^k \chi_1^l$. The most singular parts of Q_{00}^6 near the black holes, found from the matching, involve no spin product terms. Hence the full expression for Q_{00}^6 involves no spin product terms $\chi_1^k \chi_2^l$.

We see that spin product terms involving $\chi_1^k \chi_2^l$ in Q_{00}^6 to compensate for $2\ddot{r}_{14}^k (w^k - r_1^k)$ can only arise themselves from matching out the internal scheme for black hole 1. To find such terms produced in Q_{00}^6 by the matching we need to understand the $O(\epsilon^8)$ QS metric perturbations, since these produce terms in P_{00}^6 which behave as r and hence terms in Q_{00}^6 which behave as $|\mathbf{w} - \mathbf{r}_1|$. We recall that the $O(\epsilon^8)$ QS metric perturbations are $O(R^3)$ as $R \to \infty$, consisting of $L = 2$ and $L = 3$ stationary modes. The $L = 3$ modes are determined by the r^3 parts of P_{ab}^3, and arise ultimately from derivatives of the curvature due to the Schwarzschild far field of black hole 2; they can only produce $|\mathbf{w} - \mathbf{r}_1|$ terms in Q_{00}^6 which are proportional to products $\chi_1^k \chi_1^l$. The $L = 2$ $O(\epsilon^8)$ QS modes are determined by the r^3 parts of P_{ab}^2 and the r^2 parts of P_{ab}^4. These $L = 2$ modes can only produce $\chi_1^k \chi_2^l$ terms in Q_{00}^6 with $|\mathbf{w} - \mathbf{r}_1|$ radial dependence if the corresponding electric and magnetic tensors E_{ij}^{18}, H_{ij}^{18} contain parts $^{\text{spin}}E_{ij}^{18}, ^{\text{spin}}H_{ij}^{18}$ which involve χ_2^k; such spin-dependent tensors can only arise from the r^2 terms in P^4_{ab} which are proportional to χ^k_2. We find that the only contribution to $^{\text{spin}}E^{18}_{ij}$, $^{\text{spin}}H^{18}_{ij}$ is (through the matching) from the term $-2(\mu_2)^2\epsilon_{ikl}\chi^k_2(w^l - r^l_2) \mid \mathbf{w} - \mathbf{r}_2 \mid^{-3}$ in Q^4_{0i}. This leads to

$$^{\text{spin}}E^{18}_{ij} = 0, \tag{4.6.11}$$

$$^{\text{spin}}H^{18}_{ij} = (\mu_2)^2(-3S^i\chi^j_2S^{-5} - 3S^j\chi^i_2S^{-5}$$
$$- 3\delta_{ij}S^k\chi^k_2S^{-5} + 15S^iS^k\chi^k_2S^{-7}). \tag{4.6.12}$$

When we use eqn (4.5.9) to find the explicit QS metric perturbations for a magnetic $L = 2$ stationary mode, and then match out again, we find that $^{\text{spin}}H^{18}_{ij}$ contributes $-2\mu_1 \, ^{\text{spin}}H^{18}_{kl}\chi^k_1(w^l - r^l_1)$ to the $\mid \mathbf{w} - \mathbf{r}_1 \mid$ part of Q^6_{00}. It may be checked that no other terms involving products $\chi^k_1\chi^l_2$ are contributed to the $\mid \mathbf{w} - \mathbf{r}_1 \mid$ part of Q^6_{00} when we match out the internal scheme for black hole 1. Hence the dominant second post-Newtonian part of the spin–spin force corresponds to the acceleration

$$\text{spin} - \text{spin} \ddot{r}^i_{14} = \mu_1 \, ^{\text{spin}}H^{18}_{ik}\chi^k_1. \tag{4.6.13}$$

It should be remarked that equations similar to those of this section, intended to describe interactions between ordinary weak-field macroscopic spinning bodies, have been derived by Barker and O'Connell (1975, 1976, 1987), Chan and O'Connell (1977), Börner et al. (1975), and Cho and Hari Dass (1976). These equations agree with those of this section, although their régime of validity is quite different. A very thorough treatment of post-Newtonian effects in the motion of a system of weakly self-gravitating bodies is given by Damour et al. (1991, 1992, 1993) and Will (1993). See also Ehlers and Rudolph (1977), and Breuer and Rudolph (1982).

This work can be extended to strongly self-gravitating bodies, allowing also for deviations from Einstein gravity, with a view to studying the binary pulsars PSR1913 +16 and PSR1534+12 (Damour and Taylor 1992; Will 1993). These binary pulsars give such accurate timing measurements that they allow a number of tests of the validity of general relativity, including tests of the strong-field régime.

4.7   Comments and conclusions

We have now fulfilled the aims set out in the introductory Section 4.1, by finding the dominant equations of motion and spin propagation describing the dynamics of two rotating black holes of comparable masses, which move at non-relativistic speeds in each other's far fields. As a by-product of our calculation, we have given descriptions of the gravitational field which together are accurate both in the combined far field of the black holes and also in the regions near the black holes where the curvature is large.

The equations of motion and spin propagation found in Section 4.6 agree with the expressions discussed in Section 4.1 which were found by previous authors and should apply to various special cases of the two-black-hole interaction considered here. To verify this, let us first compare our work with that of EIH (Einstein *et al.* 1938; Einstein and Infeld 1948). The EIH equations of motion refer to the interaction of non-rotating bodies, so that we only need to check our post-Newtonian equation of motion (4.6.1) against EIH's. Written in our language, the EIH post-Newtonian equations of motion describe the post-Newtonian shifts $\epsilon^{2'} r_{12}^i$ and $\epsilon^{2'} r_{22}^i$ of two bodies from their Newtonian positions r_1^i, r_2^i. The equation for the first body is

$$
\begin{aligned}
{}^{'}\ddot{r}_{12}^i ={}& \mu_2{}^{'}r_{22}^i S^{-3} - 3\mu_2{}^{'}r_{22}^k S^k S^i S^{-5} - \mu_2{}^{'}r_{12}^i S^{-3} \\
&+ 3\mu_2{}^{''}r_{12}^k S^k S^i S^{-5} + [5\mu_1\mu_2 + 4(\mu_2)^2]S^i S^{-4} \\
&- [2(\mu_1)^2 + 4\mu_1\mu_2 + (\mu_2)^2]\mu_2(\mu_1)^{-2}\dot{r}_2^k \dot{r}_2^k S^i S^{-3} \\
&+ [3(\mu_1)^2 + 7\mu_1\mu_2 + 4(\mu_2)^2]\mu_2(\mu_1)^{-2}\dot{r}_2^k S^k \dot{r}_2^i S^{-3} \\
&+ \frac{3}{2}\mu_2(\dot{r}_2^k S^k)^2 S^i S^{-5}.
\end{aligned} \tag{4.7.1}
$$

But this is equivalent to our eqn (4.6.1) if we define

$$
\begin{aligned}
{}^{'}r_{12}^i &= r_{12}^i - \frac{1}{2}\dot{r}_1^k r_1^k \dot{r}_1^i + \frac{1}{2}\dot{r}_1^i \dot{r}_1^k r_1^i - \mu_2 r_1^i \mid \mathbf{r}_2 \mid^{-1}, \\
{}^{'}r_{22}^i &= r_{22}^i - \frac{1}{2}\dot{r}_2^k r_2^k \dot{r}_2^i + \frac{1}{2}\dot{r}_2^i \dot{r}_2^k r_2^i - \mu_1 r_2^i \mid \mathbf{r}_2 \mid^{-1}.
\end{aligned} \tag{4.7.2}
$$

Now the functions $({}^{'}r_{12}^i - r_{12}^i)$ and $({}^{'}r_{22}^i - r_{22}^i)$ remain bounded over post-Newtonian time scales, whether or not the Newtonian orbit is bound. Hence we agree with EIH on all changes of $O(1)$ in the Newtonian orbit parameters which take place over a post-Newtonian time-scale. The only difference between our equations and EIH's lies in the terms $\epsilon^2({}^{'}r_{12}^i - r_{12}^i)$ and $\epsilon^2({}^{'}r_{22}^i - r_{22}^i)$ which remain of $O(\epsilon^2)$. This difference corresponds to the freedom of choosing reference points for the black holes anywhere inside regions of the dimension $O(\epsilon^2)$ on which the holes are smeared out, and is unimportant.

As mentioned in Section 4.1, the post-Newtonian perihelion precession of a bound system of two black holes is $6\pi(M_1 + M_2)Gp^{-1}c^{-2}$ radians per revolution, where M_1 and M_2 are the particle masses and p is the semi-latus rectum of the relative orbit. The exact solution of the post-Newtonian equations of motion gives $r_{12}{}^i(\epsilon u)$ and $r_{22}{}^i(\epsilon u)$—see Damour (1987). This gives more detailed information about the space–time than does the perihelion shift; to observe such post-Newtonian effects, one would have to devise a test or measurement which is gauge invariant at post-Newtonian order.

We can further compare our equations with eqns (4.1.2)–(4.1.6), which refer to the interaction of two rotating bodies with very disparate masses. We let $\mu_1/\mu_2 \to 0$ in our combined post-Newtonian and $1\frac{1}{2}$ post-Newtonian spin-propagation equation(4.6.6), and immediately recover the effects of eqns (4.1.5) and (4.1.6). When we take this limit for our $1\frac{1}{2}$ post-Newtonian equation of motion (4.6.7), we recover the two types of spin–orbit force given by eqns (4.1.2) and (4.1.3), on using the vector identity

$$\mathbf{r}[\mathbf{v}\cdot(\boldsymbol{\Sigma}\times\mathbf{r})] - (\mathbf{v}\cdot\mathbf{r})(\boldsymbol{\Sigma}\times\mathbf{r}) = \mathbf{v}\times[r^2\boldsymbol{\Sigma} - (\mathbf{r}\cdot\boldsymbol{\Sigma})\mathbf{r}]. \tag{4.7.3}$$

Here the effects of eqn (4.1.2) arise from the terms in eqn (4.6.7) proportional to $(\mu_2)^2\chi_2^k$, and those of eqn (4.1.3) arise from the terms proportional to $\mu_1\mu_2\chi_1^k$. Finally, our second post-Newtonian spin–spin equation (4.6.13) agrees with eqn (4.1.4).

Thus, by our use of the matching method, we have found a generalization of EIH's equations which allows for the rotation of the black holes. Similarly, our equations are more general versions of well-known expressions for the interaction of rotating bodies, since we have allowed the black holes to have comparable masses in addition to spin. In common with the EIH method, our treatment uses only the vacuum Einstein field equations; we have avoided the need for making dubious assumptions about the matter inside the black holes.

The assumptions which underlie our approximation scheme would be expected to hold in a wide class of interactions between pairs of black holes. They should only break down in two situations. One possibility is that the black holes approach within a few Schwarzschild radii of each other. In this case it seems very unlikely that a perturbation method can be made to work in the spheric case; instead one is forced back to the problem of solving the full non-linear Einstein equations. The second possibility is that the black holes move in each other's far fields, but that their relative velocity is not small compared to the speed of light (Thorne and Kovács 1975; Kovács and Thorne 1977, 1978). In this case perturbation theory is still useful, and a combination of our matching method with a fast-motion scheme might well succeed in describing the system adequately.

Similar methods (Manton 1977) have also been used for the force between 't Hooft–Polyakov monopoles ('t Hooft 1974; Polyakov 1974) in a flat-space theory with interacting Yang–Mills and scalar fields. The interaction of slowly moving monopoles is described by geodesic motion on the configuration space of static solutions. Further work based on this approach is described by Atiyah and Hitchin (1985). One can then use the monopole methods in general relativity (Gibbons and Ruback 1986) to study the special case of the motion of a system of extreme $(Q = M)$ Reissner–Nordström black holes. Here the multi-extreme-black-hole static solutions are given analytically by the Papapetrou–Majumdar metrics (Majumdar 1947; Papapetrou 1947). This approach will, of course, not work for a generic system of black holes, where such static multi-black-hole solutions do not exist.

5

GRAVITATIONAL RADIATION FROM HIGH-SPEED BLACK-HOLE ENCOUNTERS

5.1 Introduction

Let us turn now to the gravitational radiation emitted in black-hole encounters, which becomes more and more significant as one increases the approach velocity. A review which covers the subject of gravitational wave generation is given by Thorne (1987). As described in Chapter 4, one can study the interaction of two black holes using perturbation theory in various different régimes. First, one can take, as in Chapter 4, the case of a low-velocity quasi-Newtonian interaction, where the black holes' paths are either approximately bound or unbound Newtonian orbits. Second, one can take the case of a distant encounter of two black holes at relativistic speed (but not too close to the speed of light), with deflection by a small angle (Thorne and Kovács 1975; Kovács and Thorne 1977; 1978; D'Eath 1979b). If the relativistic encounter is close, then the usual perturbation methods will break down, as strong-field effects will appear. Third, one can take an encounter of two black holes with relative velocity close to the speed of light. Here, instead of taking (for example) the masses as small parameters, one uses γ^{-1} as a small parameter, where γ is a typical Lorentz factor (measured in a centre-of-mass frame) of an incoming black hole. This leads to a new type of perturbation theory, which is valid even for head-on collisions (Chapters 6, 7, and 8), giving the gravitational radiation emitted near the forward and backward directions (D'Eath 1978; D'Eath and Payne 1992a–c). The results on gravitational radiation for all three régimes merge smoothly together. The first two régimes will be summarized briefly in Section 5.2. The third régime is the topic of the rest of this chapter, and is related to the work on collisions at the speed of light described in the following Chapters 6, 7 and 8.

Although there have been many calculations of gravitational wave generation within the context of general relativity, little is known about truly strong-field radiation which originates in highly non-linear interaction processes. In particular, no realistic exact solutions of the field equations have yet been found which are asymptotically flat and contain gravitational radiation. Radiation calculations are usually carried out by perturbation methods, with the result that most analytic expressions for waves emitted from local systems are valid only in the limit that the strength of the radiation tends to zero. For example, the radiative field of a small body of mass m moving in a background space–time scales like m as $m \to 0$, while in a post-Newtonian treatment of a bound system of masses moving with low velocities of order ϵ (Chapter 4), the radiative part of the metric scales down rapidly as ϵ^5. This can be expressed more physically by saying that most perturbation approaches are only useful for calculating radiation fields when the power involved is much less than $c^5/G \simeq 3.6 \times 10^{59}$ erg/sec, the characteristic power associated with fully non-linear gravitational interactions (Misner et al. 1973). General reviews of the generation of gravitational radiation are given in Thorne (1977, 1980a).

It is, then, important to try to extend the range of perturbation theory as far as possible, in order to obtain more information about the two types of processes which can lead to fully non-linear radiation, namely non-spherical stellar collapse and collisions between strongly relativistic bodies with comparable masses or energies. We shall see in this chapter that perturbation methods can, as above, be used to make analytic statements about strong-field gravitational radiation produced by collisions of black holes at velocities close to the speed of light, when one uses γ^{-1} as a small parameter. Related methods are used for collisions at the speed of light.

There are two reasons for considering the black-hole collision problem in the limit that the relative velocity tends to c, rather than a situation in which the two black holes approach each other with only moderate velocities at early times. First, there is the possibility that the form of the space–time metric produced by the collision simplifies in the high-speed limit. The metrics due to collisions at precisely the speed of light will be special boundary cases among the family of all black-hole collision metrics, and may well be more tractable analytically. Any attempts to find exact solutions of the vacuum Einstein equations, describing black-hole collisions, should then start with the simplest of these limiting space–times, in which the impact parameter is zero and the geometry axisymmetric (although it may be excessively optimistic to think of ever finding an exact solution in this case!).

Second, some geometrical simplifications also take place for the high-speed collision. The gravitational field of an incoming black hole with mass M_1 and energy $\mu = M_1\gamma_1$ becomes concentrated owing to special-relativistic effects in a shock wave close to a null plane in the surrounding nearly Minkowskian space–time, when γ_1 is large; the shock will be accurately approximated by a certain type of impulsive plane-fronted wave (Section 5.3; see also Ehlers and Kundt 1962; Aichelburg and Sexl 1971), with strength proportional to μ, provided that we ignore the detailed structure near the plane. Then suppose that we work in a centre-of-mass frame and examine the collision of two black holes with comparable masses M_1 and M_2 moving in opposite directions with high velocities, such that the incoming energies $M_1\gamma_1$ and $M_2\gamma_2$ are both equal to μ. In this case we can divide the space–time up into four basic regions (see Fig. 5.1), if we again ignore the detailed structure of the gravitational field near the shocks. Ahead of both incoming shocks the space-time will be nearly flat (region I). Behind shock 1 due to the black hole M_1, there will be another nearly flat region II, in which the geometry has not yet been disturbed by information propagating from the collision. Shock 1 will be effectively a null surface in the subset of Minkowski space formed by the union of regions I and II, and can be given, say, by the

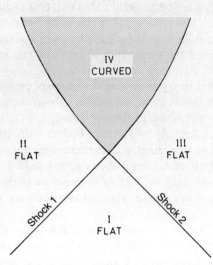

FIG. 5.1. Schematic space–time diagram for a collision of two black holes approaching at the speed of light. Before the collision, the space–time curvature due to black hole 1 resides on a null hyperplane, denoted by shock 1, which separates two regions I and II of flat space–time. Similarly, there is a null shock 2 produced by black hole 2, which separates region I from another flat region III. At the collision, the null generators of each shock acquire shear and begin to converge; thereafter the shock surfaces are curved. Space–time curvature develops in the region IV to the future of the collision.

relation $z - t = 0$ in Minkowskian coordinates. There will be a similar nearly flat region III behind shock 2, which will be given by the relation $z + t = 0$ in Minkowskian coordinates defined on the union of regions I and III, and which has the same strength as shock 1. Regions II and III will also be bounded by shocks which have emanated from the collision. These shocks will lie on curved surfaces when viewed from the Minkowskian regions to their immediate past, and are found by following the null geodesic generators of the incoming plane shocks through the spacelike two-surface on which the incoming shocks meet (the collision two-surface). In fact, the curved shocks will constitute the boundary of the future of the collision two-surface, and form an achronal surface (Hawking and Ellis 1973)—containing no two points with timelike separation—which is almost everywhere null. The generators of shock 1, say, will have acquired some shear at the collision with shock 2, and will thereafter be both shearing and converging. See eqns (3.5.12) and (3.5.27) for the evolution equations for shear σ and divergence ρ. Equation (3.5.12) shows how the non-trivial Weyl curvature delta-function Ψ_0 generates a non-trivial shear σ as shock 1 passes through the collision with shock 2. Then eqn (3.5.27) shows how the non-zero $\sigma\bar{\sigma}$ shear term generates a non-zero divergence ρ. Alternatively, one can compute σ and ρ directly for the curved shock 1 surface from eqn (5.6.3). Behind both curved shocks 1 and 2 there will be a strongly curved interaction zone (region IV) in which the gravitational field behaves in a highly non-linear manner. One might conjecture that the union of curved shocks 1 and 2 might be a characteristic initial surface for the solution of the vacuum Einstein equations in region IV (provided that a certain caustic region is treated carefully).

This geometrical framework has been exploited by Penrose (1974), who has found apparent horizons (compact spacelike two-surfaces such that the convergence of the outgoing normal family of null geodesics is zero) (Hawking and Ellis 1973) on the union of the incoming plane null surfaces, for a range of values of the quantity (impact parameter)$/\mu$. This information can then be used to bound the amount of radiation arriving at future null infinity due to the collision, if one makes the cosmic censorship hypothesis (Chapter 2; see also Hawking and Ellis 1973), which claims that any singularities formed during the evolution of reasonable initial data will be hidden behind an event horizon. Under this assumption, the areas of the apparent horizons found by Penrose provide lower bounds for the areas of black holes left as final products of the collision (see Subsection 2.5.3). Energy conservation then limits the strength of the radiation; for example, Penrose has shown in this way that the energy radiated in the axisymmetric speed-of-light collision must be less than $100(1 - (1/\sqrt{2})) = 29\%$ of the original 2μ, under the Cosmic Censorship Hypothesis. Conversely, if too much energy is radiated away in a collision, then the hypothesis must be false.

The collision at precisely the speed of light has various mildly undesirable features, caused by the infinite Lorentz contraction of the incoming states, which pushes all the curvature into δ functions on the null planes. This implies that the space–time is, strictly speaking, not asymptotically flat in the past, and also that the curvature falls off only as r^{-2} instead of the usual r^{-3} at large spacelike distances. Further, we shall see that the space–time is not quite asymptotically flat at future null infinity, owing to some (fairly harmless) singularities in the radiation field. The speed-of-light collision is studied in Chapters 6, 7, and 8. Here, to avoid these difficulties, it will be instructive in this chapter to keep γ_1 and γ_2 large but finite, and to consider limiting properties of the space–times as $\gamma_1, \gamma_2 \to \infty$ with M_1, M_2 varying so that $M_1\gamma_1 = M_2\gamma_2 = \mu$ is fixed; then the singular features of the $\gamma = \infty$ collision will be smeared out. We emphasize that the energy μ is to be regarded as a quantity or $O(1)$ in perturbation theory. More precisely, we shall consider one-parameter families of space–times containing colliding black holes of comparable mass, where we also fix the ratio M_1/M_2 as $\gamma_1, \gamma_2 \to \infty$. To give a complete prescription for a one-parameter family, the scaling of the impact parameter as $\gamma \to \infty$ must further be specified. Then we shall examine the behaviour of the gravitational field in the various qualitatively different regions of space–time, described by different asymptotic expansions.

The various asymptotic expansions are needed to allow for the different length and time-scales characteristic of the gravitational field. For example, the incoming plane shocks 1, 2 and the curved shocks 1, 2 have detailed structure on length scales of $O(\gamma^{-1})$ measured through the shock, due to the special-relativistic contraction, while the structure only varies over scales of $O(1)$ in space–time directions parallel to the shock. More complicated behaviour, with rapid variations in two or three space-time directions, occurs near the collision two-surface where the shocks meet, and in two caustic regions on the curved shocks, where one shock has been almost infinitely focused by the other. Another type of behaviour occurs in the interaction zone (region IV) where the gravitational field will be slightly perturbed away from the exact $\gamma = \infty$ metric, since γ_1, γ_2 are not quite infinite, but where there is no rapid variation in preferred directions. A separate perturbation expansion must be set up for each qualitatively different region, adapted to these characteristic length and time-scales. One expects that the expansions holding in adjacent regions of space–time, such as the curved shock 1 and the interaction region IV, match smoothly onto each other; thus boundary conditions on one asymptotic expansion are provided by all the adjacent expansions. By proceeding in this way we shall be able to obtain some analytic information about radiation in the strong-field region IV, without knowledge of the $\gamma = \infty$ interaction metric. This will be found by matching from expansions which describe the evolution of the curved shocks at late times out in the far field; we can, in fact, describe the radiation heading out at angles of $O(\gamma^{-1})$ from the initial collision direction, while remaining within the far-field curved shocks.

We should also mention another approach to radiation calculations for high-speed encounters due to Curtis (1975, 1978a, b), following a suggestion of Penrose. In this approach, one considers only the collision of shock waves at precisely the speed of light. Curtis has calculated the results of scattering a linearized plane-fronted wave against a fully non–linear plane-fronted wave and of scattering two weak plane-fronted waves off each other. But note that the statements in perturbation theory resulting from these calculations can be rephrased as statements about behaviour in certain regions of the fully non-linear space–time, produced by the collision of two shocks with equal strengths μ; for one can always arrange that two incoming null particles have equal energies, by performing a Lorentz boost. Then a transformation $g_{ab} \rightarrow N^2 g_{ab}$ with a constant conformal factor N^2 will scale up the energies by a factor N, without disturbing the vacuum Einstein equations. In particular, this implies that Curtis' calculation of the scattering of a weak shock on a strong shock should relate to the radiation calculations of this book, since the radiation which he finds can be reinterpreted as radiation close to the direction of motion in a centre-of-mass non-linear collision. This speed-of-light approach will be taken in Chapters 6, 7, and 8.

Although we shall always describe our perturbation schemes as representing collisions or encounters between high-speed non-rotating black holes, most of the statements will be valid for interactions between any pair of bodies with comparable masses M_1, M_2 at large γ_1, γ_2, such that $M_1\gamma_1 \gg R_1, M_2\gamma_2 \gg R_2$, where R_1, R_2 are typical proper radii of the bodies. This occurs because the metric is generated predominantly by the collision of the boosted Coulomb fields of the bodies, out at distances of order μ or further from their centres. We only choose to refer to interactions between bodies which are initially Schwarzschild black holes because of the need to specify all the space–times in the one-parameter families with $\gamma_1, \gamma_2 \to \infty$, used in the perturbation theory; the choice of incoming Schwarzschild states fixes a one-parameter family uniquely, once we have decided on the scaling of the impact parameter. One could instead treat a suitable family of Kerr geometries. The results would only differ by higher-order terms. As Barrow and Carr (1978) have pointed out, this work may be of relevance to the formation of primordial black holes in the anisotropic universe, which might provide the 'energy' for ultra-relativistic collisions. When one studies black-hole formation with large shear in the universe, one expects that the energy in outgoing gravitational waves will be of the same order as the energy in the initial shearing motion.

In Section 5.2, we shall survey the results on gravitational radiation for weak-field encounters. In Section 5.3, we shall turn to high-speed encounters, studying the geometry of an isolated Schwarzschild black hole which moves at high speed, considering both the detailed shock structure for large γ and the impulsive $\gamma = \infty$ shock. In Sections 5.4, 5.5, and 5.6, we restrict attention to the axisymmetric collision with zero impact parameter. In Section 5.4, we show how to follow the evolution of the detailed shock structure through the collision region and in the region in which the shock surface is curved, at distances of $O(1)$ from the axis; the curved shock evolution is governed by geometrical optics, with the shock strength obeying an ordinary differential equation along the shock generators. The strength of each curved shock appears to grow indefinitely as we approach the caustic region on the axis, mentioned previously. In fact, some qualitatively new behaviour takes over at distances of $O(\gamma^{-1})$ from the caustic region, in which the gravitational field becomes wavelike and a singularity can be averted. This self-interaction of the shock is of interest in its own right, and is described in Section 5.5. The calculation of radiation at angles of $O(\gamma^{-1})$ from the axis is carried out in Section 5.6. It requires the description of curved shock evolution at large distances of $O(\gamma)$ away from the axis, at late times of $O(\gamma^2)$ after the collision; the shock evolution is qualitatively different from that described in Section 5.4 at distances of $O(1)$ away from the axis. In Section 5.6, we find the dominant radiation field analytically, close to the axis, and present numerical results. Similar methods can also be used for collisions and encounters with non-zero impact parameter. In particular, we describe in Section 5.7 the régime in which the impact parameter scales linearly with γ, and hence is of order γ^2 (mass), since we regard the energy μ as a quantity of $O(1)$. The black holes then only deflect each other slightly, and the radiation generated is strongly beamed near the direction of motion. With this scaling of the impact parameter, the radiation in the beam with half-angle of $O(\gamma^{-1})$ still has the strength characteristic of fully non-linear interactions in general relativity. These calculations describe encounters which are much closer, and which produce much more radiation, than those amenable to the usual flat-space perturbation methods for large γ; only when we let the impact parameter become much larger than $\gamma^2 \times$ (mass) do we find a radiation field proportional to $M_1 M_2$, as in the standard 'fast-motion' approaches (Thorne and Kovács 1975; Kovács and Thorne 1977, 1978). In Section 5.8, we summarize the various results.

5.2   Gravitational radiation from distant encounters

We begin by outlining the language used by Thorne (1977) in describing gravitational radiation. Complete information about the wave field in an asymptotically flat space–time is carried by the transverse-traceless part of the spatial metric, h_{jk}^{TT}. In the asymptotic wave region, h_{jk}^{TT} obeys

$$\partial_k h_{jk}^{TT} = 0, \qquad h_{kk}^{TT} = 0, \tag{5.2.1a}$$

and in the wave region, the condition $\partial_k h_{jk}^{TT} = 0$ reduces to $n_k h_{jk}^{TT} = 0$, where n_k gives the radially outward direction in which the radiation propagates. The massless spin-2 field is resolved into its two polarization states by projection onto two orthogonal basis tensors:

$$h_{jk}^{TT}(t, r, \theta, \phi) = A_+ e_{jk}^+ + A_\times e_{jk}^\times. \tag{5.2.1b}$$

Here, polar coordinates r, θ, ϕ are used in the asymptotic region, and the basis tensors are chosen to be

$$\mathbf{e}^+ = \mathbf{e}_\theta \otimes \mathbf{e}_\theta - \mathbf{e}_\phi \otimes \mathbf{e}_\phi, \qquad \mathbf{e}^\times = \mathbf{e}_\theta \otimes \mathbf{e}_\phi + \mathbf{e}_\phi \otimes \mathbf{e}_\theta, \tag{5.2.2}$$

where $\mathbf{e}_\theta$ and $\mathbf{e}_\phi$ are unit vectors in the θ and ϕ directions. Thorne (1977) has shown that A_+ and $A_\times$ are frame-independent scalar fields: if a boost is applied to h_{jk}^{TT}, which is then projected onto new spatial transverse-traceless basis tensors constructed from the boosted $e_{jk}^+, e_{jk}^\times$, one finds that the same A_+ and $A_\times$ result. In a given frame, the rate of energy output per unit solid angle in gravitational waves is then

$$\frac{dE}{d\Omega dt} = \frac{r^2}{16\pi}\left[\left(\frac{\partial A_+}{\partial t}\right)^2 + \left(\frac{\partial A_\times}{\partial t}\right)^2\right], \tag{5.2.3}$$

using geometrical units with $c = G = 1$. This description of gravitational waves can straightforwardly be related to the description in terms of the Bondi news function (Bondi *et al.* 1962) adopted in Section 5.6 and in Chapters 6, 7, and 8.

The perturbation methods developed for distant encounters fall essentially into three classes. The domains of validity depend on the encounter velocity and impact parameter, although there is some overlap, and the predictions for different régimes mesh smoothly together. The basic reference which summarizes information on the first two approaches, namely post-Newtonian and second-linearized methods, is the paper by Kovács and Thorne (1978). The third approach involving plane-fronted waves will be treated in the remainder of this chapter.

(a) Post-Newtonian methods

Suppose that the two bodies concerned have masses m_A, m_B, and that they approach from infinity with relative velocity v and impact parameter b. Suppose further that typical radii r_A, r_B of the bodies are much smaller than the minimum separation between A and B, so that they interact as point particles. As described in Chapter 4, because of matching properties, the gravitational radiation emitted in the distant interactions of two black holes is, at leading order, the same as the gravitational radiation emitted in the distant interaction of two material bodies. If the conditions

$$v \ll 1, \qquad (m_A + m_B) \ll bv \qquad (5.2.4)$$

hold, then it is valid to treat the gravitational field around the bodies as a slow-motion perturbation of flat space, where time derivatives are much smaller than spatial derivatives, and the field equations become elliptic. The bodies move on nearly Newtonian orbits, and the radiation is given at leading order by the quadropole-moment expression

$$h_{jk}^{TT} = (2/r)\ddot{I}_{jk}^{TT}(t - r), \qquad (5.2.5)$$

derived in Sections 36.9–10 of Misner et $al.$ (1973). The tracefree Newtonian quadrupole moment I_{ij}^T is defined in eqn (4.1.10). One then forms the transverse-traceless quantity I_{ij}^{TT} by projecting I_{ij}^T orthogonally to n_j, the unit outward radial vector.

If, moreover, the condition

$$(m_A + m_B) \ll bv^2 \qquad (5.2.6)$$

holds, then the gravitational deflection produced by the encounter is small, i.e. the encounter is 'distant'. This gives the simplest low-velocity case in which the radiation can be computed, and allows a straightforward comparison with the second-linearized weak-field relativistic encounters of Subsection 5.2.2. Of course, one could also study the most general quasi-Newtonian hyperbolic encounter, where the scattering is through a large angle, and compute radiation formulae. In describing the radiation, use a frame in which A starts at rest at the origin, while B moves in the x–z plane, with initial velocity in the positive z-direction. Arrange that $t = 0$ at closest approach, and measure retarded time by defining

$$T = (t - r)v/b, \qquad (5.2.7)$$

corresponding to the typical time-scale b/v of the radiation. Define

$$\ell(T) = (l + T^2)^{\frac{1}{2}}, \qquad (5.2.8)$$

$$\mathcal{A}_+ = (br/4m_A m_B)A_+, \qquad \mathcal{A}_\times = (br/4m_A m_B)A_\times. \qquad (5.2.9)$$

Then (Kovács and Thorne 1978) the quadrupole radiation has

$$\mathcal{A}_+ = \frac{1}{2}(\ell^{-3} + \ell^{-1})\sin^2\theta + \frac{1}{2}\ell^{-3}(\sin^2\phi - \cos^2\theta\cos^2\phi)$$
$$+ (T\ell^{-3} + T\ell^{-1} + 1)\cos\theta\sin\theta\cos\phi, \qquad (5.2.10)$$
$$\mathcal{A}_\times = \ell^{-3}\cos\theta\cos\phi\sin\phi - (T\ell^{-3} + T\ell^{-1} + 1)\sin\theta\sin\phi.$$

Note that the radiation patterns given by A_+ and A_x are quadratic in the masses, proportional to $m_A m_B$. This is a feature in common with the weak-field radiation from a distant encounter at relativistic speeds (Subsection 5.2.2; see also Kovács and Thorne 1978). The total energy radiated (Ruffini and Wheeler 1971; Hansen 1972) is

$$\Delta E = (37\pi/15)m_A^2 m_B^2 v/b^3. \qquad (5.2.11)$$

These calculations can be refined to include higher-order terms in the post-Newtonian expansion (Epstein and Wagoner 1975; Wagoner and Will 1976). Such improvements applied to the distant encounter problem (Turner and Will 1978) yield results which remain accurate well into the régime covered by second-linearized theory, for speeds $v \lesssim 0.3$.

(b) Second-linearized methods

We next describe the method of Thorne and Kovács, used in analysing the bremsstrahlung problem. This again treats the geometry produced by the two interacting bodies as a perturbation of flat space–time:

$$\mathcal{G}^{\mu\nu} = \eta^{\mu\nu} - \bar{h}^{\mu\nu}, \qquad |\bar{h}^{\mu\nu}| \ll 1, \qquad (5.2.12)$$

where

$$\mathcal{G}^{\mu\nu} = (-g)^{\frac{1}{2}} g^{\mu\nu}, \qquad g = \det(g_{\mu\nu}), \qquad (5.2.13)$$

and $\eta^{\mu\nu}$ is the Minkowski metric. The gauge condition

$$\bar{h}^{\mu\nu}{}_{,\nu} = 0 \qquad (5.2.14)$$

is imposed, and the field equations are conveniently written in the form (Section 20.3 of Misner et al. 1973)

$$H^{\mu\alpha\nu\beta}{}_{,\alpha\beta} = 16\pi(-g)(T^{\mu\nu} + t^{\mu\nu}_{L-L}), \qquad (5.2.15)$$

$$H^{\mu\alpha\nu\beta} = \mathcal{G}^{\mu\nu}\mathcal{G}^{\alpha\beta} - \mathcal{G}^{\alpha\nu}\mathcal{G}^{\mu\beta}, \qquad (5.2.16)$$

where $T^{\mu\nu}$ is the energy–momentum tensor and $t^{\mu\nu}_{L-L}$ is the pseudo-tensor of Landau and Lifschitz (1962).

In the bremsstrahlung calculation, one builds up a sequence of approximations to $\bar{h}^{\mu\nu}$, in powers of the masses m_A, m_B. The first-order term $_1\bar{h}^{\mu\nu}$, is the sum of the linearized far-fields of A and B, where, for example, the contribution from A can be found by taking a 'Coulomb' m_A/r linearized gravitational field in the gauge (5.2.14), and giving it a Lorentz boost up to A's velocity. There is no restriction on the relative velocity v, provided that the encounter is distant; time derivatives are treated on a par with spatial derivatives, and the field equations are regarded as hyperbolic. In particular, the second approximation $_2\bar{h}^{\mu\nu}$ satisfies

$$\eta^{\alpha\beta} {}_2\bar{h}^{\mu\nu}{}_{,\alpha\beta} = {}_2T^{\mu\nu}, \qquad (5.2.17)$$

where $_2T^{\mu\nu}$ is a source containing matter terms plus a contribution quadratic in $_1\bar{h}^{\mu\nu}$ and its derivatives. The resulting terms in $_2\bar{h}^{\mu\nu}$ proportional to $(m_A)^2$ and

$(m_B)^2$ arise from the self-fields of A and B, and are non-radiative. The radiation field is contained in the part of $_2\bar{h}^{\mu\nu}$ produced by the interaction, proportional to $m_A m_B$. Thorne and Kovács (1975) show how to calculate this field using Green's functions in the space–time, which is weakly curved by the perturbation $_1\bar{h}^{\mu\nu}$. The details of the application to a distant encounter are given by Kovács and Thorne (1977). The original bremsstrahlung calculation of Peters (1970) gives an alternative derivation of the same radiation field. Peters assumes $m_B \ll m_A$, and considers perturbations about the Schwarzschild far-field of A; however, the $m_A m_B$ part of the metric which he calculates is actually valid whatever the ratio m_B/m_A.

The second-linearized radiation calculation is accurate for encounters sufficiently distant that the condition

$$(m_A + m_B) \ll bv^2(1 - v^2)^{\frac{1}{2}} \tag{5.2.18}$$

holds. At low velocities, the condition is just eqn (5.2.6), and the method reproduces the results of Section 5.2(a). The restriction (5.2.18) at high speeds will be explained in Section 5.7. Such encounters produce only small orbital deflections.

As before, let B approach with speed v in the positive z-direction in the rest-frame S of A, with the orbits lying in the x–z plane and $x = +b$ initially for B. Define

$$\gamma = (1 - v^2)^{-\frac{1}{2}}. \tag{5.2.19}$$

It is also convenient to consider the centre-of-velocity frame $\tilde{S}$, in which A and B move initially with velocities $-\tilde{v}e_{-z}, +\tilde{v}e_{-z}$, where

$$\tilde{v} = \gamma v/(\gamma + 1), \qquad \tilde{\gamma} \equiv (1 - \tilde{v}^2)^{-\frac{1}{2}} = [(\gamma + 1)/2]^{\frac{1}{2}}. \tag{5.2.20}$$

The spatial origin of $\tilde{S}$ is chosen to lie midway between the two trajectories. Now the quadratic radiation field depends only on the product $m_A m_B$, and hence can be found by considering an equal-mass encounter, which has complete forward–backward symmetry in frame $\tilde{S}$. This leads (Kovács and Thorne 1978) to the symmetry relations

$$A_+(\tilde{t}, \tilde{r}, \tilde{\theta}, \tilde{\phi}) = A_+(\tilde{t}, \tilde{r}, \pi - \tilde{\theta}, \tilde{\phi} + \pi),$$
$$A_x(\tilde{t}, \tilde{r}, \tilde{\theta}, \tilde{\phi}) = -A_x(\tilde{t}, \tilde{r}, \pi - \tilde{\theta}, \tilde{\phi} + \pi). \tag{5.2.21}$$

The wave-forms for arbitrary v are given in Kovács and Thorne (1978), but are too long to be written down here. Provided that v is only moderately relativistic, they describe radiation with the same general time-scale b/v and amplitude $A_{+,x} \sim m_A m_B/br$ as in the low-velocity limit. However, when v is close to 1, Kovács and Thorne distinguish two characteristic time-scales. First, in frame S, A sees B's gravitational field Lorentz-contracted, and undergoes accelerations over a time-scale $b/(v\gamma)$. The appropriate retarded-time quantity for measuring the resulting radiation is then

$$T_A = v\gamma(t - r)/b. \qquad (5.2.22)$$

Second, B accelerates in its initial rest frame, with time-scale $b/(v\gamma)$ in that frame, and corresponding wave generation. The relevant time-scale in frame S is found by Lorentz transformation, and hence depends on the angular coordinates (θ, ϕ) of the distant observer. The natural retarded-time parameter with which to view this radiation in frame S is

$$T_B = \frac{T_A + \beta b\gamma}{\gamma(1 - \alpha v)}, \qquad (5.2.23)$$

where $\alpha = \cos\theta$ and $\beta = \sin\theta \cos\phi$.

At high speeds, most of the radiated energy is thrown forward into a narrow cone with a half-angle $\sim \gamma^{-1}$ in S, and its angular structure is best viewed by using the variable

$$\psi = \theta\gamma, \qquad (5.2.24)$$

which is of order unity in the beam. The basic time-scale in this 'forward region' is $b/(v\gamma)$; Kovács and Thorne (1978) are led to describe the leading radiation field using the time variable

$$\hat{T} = T_A + \psi \cos\phi. \qquad (5.2.25)$$

With the definitions

$$c = \cos\phi, \qquad s = \sin\phi, \qquad (5.2.26)$$

$$\ell_A = [1 + (\hat{T} - \psi c)^2]^{\frac{1}{2}}, S^2 = 1 - \frac{4\psi c}{(1 + \psi^2)}\hat{T} + \frac{4\psi^2}{(1 + \psi^2)^2}\hat{T}^2, \qquad (5.2.27, 5.2.28)$$

the leading 'forward region' field is given by

$$\begin{aligned}
\mathcal{A}_+ = &\frac{4\gamma^2}{(1 + \psi^2)^2 S^2}\left\{\psi c S^2 + \left(\frac{1 + \psi^4}{1 + \psi^2}\right)|\hat{T}|\left(2c^2 - 1 - \frac{2\psi c}{1 + \psi^2}\hat{T}\right)\right. \\
&\frac{1}{\ell_A}\left[\frac{1}{2} < (1 - 2c^2) - (1 + 2c^2)\psi^2 - 2c^2\psi^4 > \right. \\
&+ \psi c < 1 + 2c^2 + 2(1 + c^2)\psi^2 > \hat{T} \\
&\left.\left. + < (1 - 2c^2) - (1 + 4c^2)\psi^2 > \hat{T}^2 + 2c\psi\hat{T}^3\right]\right\},
\end{aligned}$$

$$\begin{aligned}
\mathcal{A}_\times = &\frac{4\gamma^2 s}{(1 + \psi^2)^2 S^2}\left\{-\psi(1 + \psi^2)S^2 + 2(1 - \psi^2)|\hat{T}|\left(-c + \frac{\psi}{1 + \psi^2}\hat{T}\right)\right. \\
&\left. + \frac{1}{\ell_A}\left[c(1 + \psi^2) - 2 < (1 + c^2) + c^2\psi^2 > \psi\hat{T} + 2c(1 + 2\psi^2)\hat{T}^2 - 2\psi\hat{T}^3\right]\right\}.
\end{aligned}$$
$$(5.2.29)$$

There is an apparent discontinuity in the time derivatives at $\hat{T} = 0$, produced by the $|\hat{T}|$ terms, but the wave-forms are in fact smooth when viewed on the shorter time-scale $\Delta T_B \sim 1[\Delta\hat{T} \sim (1 + \psi^2)/\gamma]$ near $\hat{T} = 0$.

The wave-forms (5.2.29) are accurate in the forward region:

$$\theta \ll \gamma^{-\frac{1}{2}}.$$

(5.2.30)

In the centre-of-velocity frame $\tilde{S}$, they describe a beam at angles $\tilde{\theta} \ll \pi/2$. By the symmetry properties (5.2.21), there is a corresponding beam in $\tilde{S}$ in the backward direction, where $(\pi - \tilde{\theta}) \ll \pi/2$. Hence, by making Lorentz transformations and using eqns (5.2.21) and (5.2.29), one can find the wave-forms in the 'backward region' in S:

$$\theta \gg \gamma^{-\frac{1}{2}}.$$

(5.2.31)

Overlapping with both forward and backward regions is the 'intermediate region':

$$\gamma^{-1} \ll \theta \ll \pi/2,$$

(5.2.32)

(i.e. $\tilde{\gamma}^{-1} \ll \tilde{\theta}, \tilde{\gamma}^{-1} \ll \pi - \tilde{\theta}$), where the radiation changes smoothly in character as θ varies from forward to backward. The typical intermediate-region radiation time-scale is $b\theta$, and the wave-forms there are very simple (Kovács and Thorne, eqns (4.14a–c)), being found most easily using plane-fronted wave methods. It can be verified that the wave-forms in the intermediate region (i.e. at all angles except very close to the axes) are given by the collision of two weak impulsive plane-frontal shocks, travelling at the speed of light. The radiation is found by a second-linearized calculation (a simplified version of the treatment of Kovács and Thorne 1978) as carried out by Curtis (1975, 1978a). Thus the large-γ radiation behaviour meshes with the $\gamma = \infty$ behaviour.

The characteristic power/solid angle emitted into the forward region at high speeds is

$$\frac{dE}{d\Omega dt} \sim \frac{\gamma^6 (m_A m_B)^2}{b^4}$$

(5.2.33)

for $\theta \leq \gamma^{-1}$. Since $\Delta t \sim \gamma^{-1} b$ and $\Delta \Omega \sim \gamma^{-2}$ in the forward region, and since nearly all the energy is beamed forward in S, one can estimate the total energy ΔE radiated in S. Numerical results provide an estimate of the relevant coefficient (Peters 1970; Kovács and Thorne 1978):

$$\Delta E = (20.0 \pm 0.3)(m_A m_B)^2 \gamma^3 b^{-3}.$$

(5.2.34)

This estimate holds whenever γ is large, provided that $\gamma \ll b/(m_A + m_B)$ (eqn (5.2.18)).

5.3 The incoming states

Suppose we take the Schwarzschild metric with mass M in isotropic coordinates,

$$ds^2 = -\left(\frac{1 - M/2\bar{r}}{1 + M/2\bar{r}}\right)^2 dt^2$$

$$+ \left(1 + \frac{M}{2\bar{r}}\right)^4 [d\bar{r}^2 + \bar{r}^2(d\theta^2 + \sin^2\theta\, d\phi^2)], \tag{5.3.1}$$

and make a coordinate transformation such that the black hole moves with a velocity $\beta = (1 - \gamma^{-2})^{\frac{1}{2}}$ close to the speed of light, which is 1 in our units. We write the energy $M\gamma = \mu$. Then we consider the one-parameter family of such space–times found by letting $\gamma \to \infty$ with μ fixed. It has been shown by Aichelburg and Sexl (1971) that the metric approaches a limit, which can be put in the form

$$ds^2 = du\, dv + dx^2 + dy^2 - 4\mu \log(x^2 + y^2)\delta(u)du^2, \tag{5.3.2}$$

where $u = z + t, v = z - t$ and the motion is in the negative z direction.

It is clear from this form that the limiting metric is a special case of an axisymmetric plane-fronted wave (Ehlers and Kundt 1962) lying between two regions of Minkowski space. The detailed structure of the shock and the black-hole property have been lost by taking the limit $\gamma \to \infty$; the only length-scale in the metric is now μ rather than M. Moreover, the infinite Lorentz boosting has changed the algebraic type of the Weyl tensor from type D for the Schwarzschild solution to type N for the limiting metric (5.3.2) (Penrose and Rindler 1986). To calculate the curvature, it is best to make a discontinuous coordinate transformation which puts the metric in a C^0 form, eliminating the δ function in eqn (5.3.2). A suitable transformation is

$$\begin{aligned}
x &= \hat{x} + 4\mu\hat{u}\theta(\hat{u})\hat{x}/(\hat{x}^2 + \hat{y}^2), \\
y &= \hat{y} + 4\mu\hat{u}\theta(\hat{u})\hat{y}/(\hat{x}^2 + \hat{y}^2), \\
v &= \hat{v} + 4\mu\theta(\hat{u})\log(\hat{x}^2 + \hat{y}^2) \\
&\quad - 16\mu^2\hat{u}\theta(\hat{u})/(\hat{x}^2 + \hat{y}^2), \\
u &= \hat{u},
\end{aligned} \tag{5.3.3}$$

where $\theta(\hat{u})$ is the Heaviside function. If we then write

$$\hat{x} = \hat{\rho}\cos\phi, \qquad \hat{y} = \hat{\rho}\sin\phi, \tag{5.3.4}$$

the metric becomes

$$ds^2 = d\hat{u}\, d\hat{v} + [1 - 4\mu\hat{u}\theta(\hat{u})/\hat{\rho}^2]^2 d\hat{\rho}^2 + \hat{\rho}^2[1 + 4\mu\hat{u}\theta(\hat{u})/\hat{\rho}^2]^2 d\phi^2. \tag{5.3.5}$$

In these coordinates the metric only has a discontinuity in its first derivative at the shock surface $\{\hat{u} = 0\}$, but there is the disadvantage that the Minkowskian form in $\{\hat{u} > 0\}$ behind the shock has been lost. The only non-zero components of the Riemann tensor are

$$R_{\hat{x}\hat{u}\hat{x}\hat{u}} = 4\mu(\hat{x}^2 - \hat{y}^2)\delta(\hat{u})/\hat{\rho}^4,$$

$$R_{\hat{x}\hat{u}\hat{y}\hat{u}} = 8\mu\hat{x}\hat{y}\delta(\hat{u})/\hat{\rho}^4, \qquad (5.3.6)$$

$$R_{\hat{y}\hat{u}\hat{y}\hat{u}} = 4\mu(\hat{y}^2 - \hat{x}^2)\delta(\hat{u})/\hat{\rho}^4,$$

together with the components found from these by the use of symmetry operations. With our conventions,

$$Z^a_{;dc} - Z^a_{;cd} = R^a_{bcd}Z^b \qquad (5.3.7)$$

for any vector field Z_a, where a semi-colon denotes covariant differentiation. The $\delta(\hat{u})$ behaviour of the Riemann tensor implies that the shock is impulsive in the terminology of Penrose (1972b).

Returning to the metric form, we see from eqn (5.3.3) that the two pieces of Minkowski space on either side of the shock have been joined with a warp. For example, any geodesic crossing the shock will be a continuous curve in $(\hat{u}, \hat{v}, \hat{x}, \hat{y})$ coordinates, so that the value of v along the curve will change discontinuously by $8\mu\log\hat{\rho}$ at the shock, while x and y do not jump. Geodesics crossing the shock will also be deflected, when viewed in (u, v, x, y) coordinates. Such a geodesic will have a continuous tangent vector in $(\hat{u}, \hat{v}, \hat{x}, \hat{y})$ coordinates; its deflection will be found from the differential coordinate transformation at the shock. This transformation matrix will define a Lorentz transformation, since the metric form (5.3.5) is Minkowskian just behind the shock. The Lorentz transformation is

$$dt = (1 + 8\mu^2\hat{\rho}^{-2})d\hat{t} - 4\mu\hat{x}\hat{\rho}^{-2}d\hat{x} - 4\mu\hat{y}\hat{\rho}^{-2}d\hat{y} + 8\mu^2\hat{\rho}^{-2}d\hat{z},$$

$$dx = -4\mu\hat{x}\hat{\rho}^{-2}d\hat{t} + d\hat{x} - 4\mu\hat{x}\hat{\rho}^{-2}d\hat{z},$$

$$dy = -4\mu\hat{y}\hat{\rho}^{-2}d\hat{t} + d\hat{y} - 4\mu\hat{y}\hat{\rho}^{-2}d\hat{z}, \qquad (5.3.8)$$

$$dz = -8\mu^2\hat{\rho}^{-2}d\hat{t} + 4\mu\hat{x}\hat{\rho}^{-2}d\hat{x} + 4\mu\hat{y}\hat{\rho}^{-2}d\hat{y} + (1 - 8\mu^2\hat{\rho}^{-2})d\hat{z},$$

where

$$\hat{u} = \hat{z} + \hat{t}, \qquad \hat{u} = \hat{z} - \hat{t}. \qquad (5.3.9)$$

We shall need this transformation in order to find the effect of shock 2 on shock 1.

Next, we consider the detailed shock structure which is present when γ is large but not infinite. We now wish to describe a Schwarzschild black hole moving in the positive z direction, at speed β close to 1, in the region of thickness $O(\gamma^{-1})$ close to the plane $\{z = \beta t\}$. For later convenience in studying the interaction problem, we have chosen to describe the detailed structure of a shock which moves in the opposite direction to the impulsive shock just discussed. Thus, we shall make statements about asymptotic behaviour of the one-parameter family of Schwarzschild space–times with $M = \mu/\gamma$, which are valid as t, x, y, z, and $\gamma(z - \beta t)$ tend to constants, while $\gamma \to \infty$, with μ fixed.

Starting from the isotropic form (5.3.1) of the Schwarzschild metric, we make a Lorentz boost

$$\bar{t} = \gamma(t' - \beta z'), \qquad \bar{x} = x', \qquad \bar{y} = y', \qquad \bar{z} = \gamma(z' - \beta t'), \qquad (5.3.10)$$

where

$$\bar{x} = \bar{r}\sin\theta\cos\phi, \qquad \bar{y} = \bar{r}\sin\theta\sin\phi, \qquad \bar{z} = \bar{r}\cos\theta, \qquad (5.3.11)$$

to arrive at an asymptotic expansion for the metric in the shock:

$$\begin{aligned}
g_{t't'} &= (4\mu/\bar{r})\gamma - (1 + \mu^2/2\bar{r}^2) \\
&\quad + (2\mu^3/\bar{r}^3 - 2\mu/\bar{r})\gamma^{-1} + O(\gamma^{-2}), \\
g_{t'x'} &= g_{t'y'} = 0, \\
g_{t'z'} &= -(4\mu\bar{r})\gamma + (\mu^2/2\bar{r}^2) \\
&\quad + (2\mu/\bar{r} - 2\mu^3/\bar{r}^3)\gamma^{-1} + O(\gamma^{-2}), \\
g_{x'x'} &= 1 + (2\mu/\bar{r})\gamma^{-1} + O(\gamma^{-2}), \\
g_{x'y'} &= g_{x'z'} = 0, \\
g_{y'y'} &= 1 + (2\mu/\bar{r})\gamma^{-1} + O(\gamma^{-2}), \\
g_{y'z'} &= 0, \\
g_{z'z'} &= (4\mu/\bar{r})\gamma + (1 - \mu^2/2\bar{r}^2) \\
&\quad + (2\mu^3/\bar{r}^3 - 2\mu/\bar{r})\gamma^{-1} + O(\gamma^{-2}).
\end{aligned} \qquad (5.3.12)$$

This metric can be treated further to remove all the γ^1 and γ^0 parts, except for the Minkowskian $\mathrm{diag}(-1, 1, 1, 1)$. First, we remove the γ^1 parts by using a coordinate transformation

$$\begin{aligned}
t' &= t'' - 2\mu\log[r'' + \gamma(z'' - \beta t'')] \\
&\quad + 4\mu^2\gamma^{-1}[r'' - \gamma(z'' - \beta t'')](\rho'')^{-2}, \\
x' &= x'' - 2\mu\gamma^{-1}[r'' - \gamma(z'' - \beta t'')]x''(\rho'')^{-2}, \\
y' &= y'' - 2\mu\gamma^{-1}[r'' - \gamma(z'' - \beta t'')]y''(\rho'')^{-2} \\
&\quad + 4\mu^2\beta\gamma^{-1}[r'' - \gamma(z'' - \beta t'')](\rho'')^{-2},
\end{aligned} \qquad (5.3.13)$$

where

$$\begin{aligned}
(r'')^2 &= (x'')^2 + (y'')^2 + \gamma^2(z'' - \beta t'')^2, \\
(\rho'')^2 &= (x'')^2 + (y'')^2.
\end{aligned} \qquad (5.3.14)$$

This leads to a metric

$$g_{t''t''} = -1 + \frac{15}{2}\mu^2(r'')^{-2} - 8\mu^2\gamma(z'' - \beta t'')(r'')^{-3}$$
$$+ O(\gamma^{-1}),$$

$$g_{t''x''} = O(\gamma^{-1}),$$

$$g_{t''y''} = O(\gamma^{-1}),$$

$$g_{t''z''} = -\frac{15}{2}\mu^2(r'')^{-2} + 8\mu^2\gamma(z'' - \beta t'')(r'')^{-3}$$
$$+ O(\gamma^{-1}),$$

$$g_{x''x''} = 1 + \gamma^{-1}(y''^2 - x''^2)f(\rho'', \gamma(z'' - \beta t'')) \qquad (5.3.15)$$
$$+ O(\gamma^{-2}),$$

$$g_{x''y''} = -2\gamma^{-1}x''y''f(\rho'', \gamma(z'' - \beta t'')) + O(\gamma^{-2}),$$

$$g_{x''z''} = O(\gamma^{-1}),$$

$$g_{y''y''} = 1 + \gamma^{-1}(x''^2 - y''^2)f(\rho'', \gamma(z'' - \beta t''))$$
$$+ O(\gamma^{-2}).$$

$$g_{y''z''} = O(\gamma^{-1}),$$

$$g_{z''z''} = 1 + \frac{15}{2}\mu^2(r'')^{-2} - 8\mu^2\gamma(z'' - \beta t'')(r'')^{-3} + O(\gamma^{-1}),$$

where the function f is defined by

$$f(p,q) = 4\mu[q - (p^2 + q^2)^{\frac{1}{2}}]p^{-4} + 2\mu(p^2 + q^2)^{-1/2}p^{-2}. \qquad (5.3.16)$$

Next, the geometry can be put in the form of a Minkowski metric plus corrections of $O(\gamma^{-1})$ using a coordinate transformation close to the identity:

$$t'' = t''' - \frac{15}{4}\mu^2\gamma^{-1}(\rho''')^{-1}\arctan[\gamma(z''' - \beta t''')/\rho'''] - 4\mu^2\gamma^{-1}/r''',$$

$$x'' = x''',$$
$$\qquad\qquad (5.3.17)$$
$$y'' = y''',$$

$$z'' = z''' - \frac{15}{2}\mu^2\gamma^{-1}(\rho''')^{-1}\arctan[\gamma(z''' - \beta t''')/\rho'''] - 4\mu^2\gamma^{-1}/r''',$$

which removes the extra γ^0 terms in $g_{t''t''}, g_{t''z''}$ and $g_{z''z''}$. Here

$$(r''')^2 = (x''')^2 + (y''')^2 + \gamma^2(z''' - \beta t''')^2,$$
$$\qquad\qquad (5.3.18)$$
$$(\rho''')^2 = (x''')^2 + (y''')^2.$$

Finally, a coordinate transformation can be made which removes all γ^{-1} corrections to the Minkowski metric η_{ab}, except for those in the transverse components $g_{\hat{x}\hat{x}}, g_{\hat{x}\hat{y}}$, and $g_{\hat{y}\hat{y}}$. We arrive at the asymptotic expansion

$$g_{ab}(\hat{t}, \hat{x}, \hat{y}, \hat{z}, \gamma) = \eta_{ab} + \gamma^{-1}h_{ab}^{(1)}(\hat{x}, \hat{y}, \gamma(\hat{z} - \beta\hat{t}))$$

$$+ \gamma^{-2} h_{ab}^{(2)}(\hat{x}, \hat{y}, \gamma(\hat{z} - \beta\hat{t})) + \cdots, \qquad (5.3.19)$$

where the only non-zero components of $h_{ab}^{(1)}$ are

$$
\begin{aligned}
h_{\hat{x}\hat{x}}^{(1)} &= (\hat{y}^2 - \hat{x}^2) f(\hat{\rho}, \gamma(\hat{z} - \beta\hat{t})), \\
h_{\hat{x}\hat{y}}^{(1)} &= -2\hat{x}\hat{y} f(\hat{\rho}, \gamma(\hat{z} - \beta\hat{t})), \qquad\qquad (5.3.20) \\
h_{\hat{y}\hat{y}}^{(1)} &= (\hat{x}^2 - \hat{y}^2) f(\hat{\rho}, \gamma(\hat{z} - \beta\hat{t})),
\end{aligned}
$$

and the function f is defined in eqn (5.3.16). This is the most useful form for the detailed shock structure in the region in which $\hat{t}, \hat{\rho}, \hat{z}$ and $\gamma(\hat{z} - \beta\hat{t})$ are of $O(1)$.

We remark that $h_{ab}^{(1)}$ tends to zero as we move well ahead of the shock, with $\gamma(\hat{z} - \beta\hat{t}) \to +\infty$. But as we let $\gamma(\hat{z} - \beta\hat{t}) \to -\infty$, moving well behind the shock, $h_{ab}^{(1)}$ grows linearly with $\gamma(\hat{z} - \beta\hat{t})$. This behaviour has been designed for matching onto the impulsive shock geometry in the C^0 form (5.3.5), which should give an accurate description of the metric, once we let $|\gamma(\hat{z} - \beta\hat{t})|$ become so large that $|\hat{z} - \hat{t}|$ is of $O(1)$. For this purpose, we must replace $\hat{u}$ by $-\hat{v}$ and $\hat{v}$ by $-\hat{u}$ in eqn (5.3.5), while keeping $\hat{u} = \hat{z} + \hat{t}$, $\hat{v} = \hat{z} - \hat{t}$, so that the impulsive $\gamma = \infty$ shock also moves in the positive $\hat{z}$ direction. Then the linear behaviour of the combination

$$\gamma^{-1}(h_{\hat{x}\hat{x}}^{(1)} + i h_{\hat{x}\hat{y}}^{(1)}) \sim 8\mu\gamma^{-1}(\hat{y} - i\hat{x})^2 \gamma(\hat{z} - \beta\hat{t})\theta(\gamma(\beta\hat{t} - \hat{z}))(\hat{\rho})^{-4}$$

when $|\gamma(\hat{z} - \beta\hat{t})|$ is large matches precisely onto the $\hat{v}\theta(-\hat{v})$ terms near the shock in the impulsive metric (5.3.5). One might wish to extend this desirable property, so that $g_{ab}(\hat{t}, \hat{x}, \hat{y}, \hat{x}, \gamma)$ in the shock region matches out to the complete impulsive $\gamma = \infty$ metric (5.3.5) with $O(\gamma^{-1})$ corrections, in the region in which $\hat{t}, \hat{\rho}, \hat{z}$, and $(\hat{z} - \hat{t})$ are of $O(1)$. This could presumably be done by using some of the coordinate freedom in choosing $h_{ab}^{(2)}$ to arrange for the correct $\gamma^{-2}[\gamma(\hat{z} - \beta\hat{t})]^2\theta(\gamma(\beta\hat{t} - \hat{z}))$ behaviour in $\gamma^{-2} h_{ab}^{(2)}$, and by choosing $h_{ab}^{(n)}$ for $n > 2$ to increase less rapidly than $[\gamma(\hat{z} - \beta\hat{t})]^n$ as $|\gamma(\hat{z} - \beta\hat{t})| \to \infty$.

The leading curvature terms in the shock region were found originally by Pirani (1959). These are $O(\gamma)$, and reside in the components $R_{\hat{x}\hat{v}\hat{x}\hat{v}}, R_{\hat{x}\hat{v}\hat{y}\hat{v}}$, and $R_{\hat{y}\hat{v}\hat{y}\hat{v}}$, together with the components found from these by symmetry operations. We have

$$
\begin{aligned}
R_{\hat{x}\hat{v}\hat{x}\hat{v}} &= 3\gamma\mu(\hat{y}^2 - \hat{x}^2)[\hat{x}^2 + \hat{y}^2 + \gamma^2(\hat{z} - \beta\hat{t})^2]^{-\frac{5}{2}} + O(1), \\
R_{\hat{x}\hat{v}\hat{y}\hat{v}} &= -6\gamma\mu\hat{x}\hat{y}[\hat{x}^2 + \hat{y}^2 + \gamma^2(\hat{z} - \beta\hat{t})^2]^{-\frac{5}{2}} + O(1), \qquad (5.3.21) \\
R_{\hat{y}\hat{v}\hat{y}\hat{v}} &= 3\gamma\mu(\hat{x}^2 - \hat{y}^2)[\hat{x}^2 + \hat{y}^2 + \gamma^2(\hat{z} - \beta\hat{t})^2]^{-\frac{5}{2}} + O(1)
\end{aligned}
$$

in the shock region, found from the second derivatives $h_{ab,\hat{v}\hat{v}}^{(1)}$. These leading $O(\gamma)$ curvature terms are of Petrov type N (Penrose and Rindler 1984) when taken by themselves, and their net effect on integration through the $O(\gamma^{-1})$

shock thickness is to produce the δ-function impulsive shock strength given in eqn (5.3.6), after allowing for the reversed direction of motion. But note that the curvature (5.3.21) in the smeared-out shock at finite γ still has the $(\hat{r})^{-3}$ behaviour at large spatial distances, characteristic of asymptotically flat space–times. It is only when we take the limit $\gamma \to \infty$ that we reach a plane-fronted metric with $(\hat{r})^{-2}$ falloff in the curvature, due to the loss of a spatial dimension.

5.4   Evolution of the curved shocks by geometrical optics

We now turn to the interaction problem, in which black hole 1 moves initially in the positive z direction and black hole 2 moves in the negative z direction, both with energy μ at large $\gamma_1, \gamma_2 \to \infty$; in other words, the impact parameter is to be of the general size μ. We start by describing the geometry of the shocks in the speed-of-light collision, and then consider the evolution of the detailed shock structure, after a collision at finite γ_1, γ_2 in the special axisymmetric case with zero impact parameter. The shocks provide boundary data for the interaction region of the space–time, region IV of Fig. 5.1, and, hence, it is appropriate to study them before studying other regions.

The situation before the collision of the impulsive shocks at $\gamma = \infty$ is depicted in Fig. 5.2. Using the C^0 form (5.3.5) of the metric for a single impulsive shock, we can write the precollision metric as

$$
\begin{aligned}
ds^2 =&\, d\hat{u}\, d\hat{v} + [1 - 4\mu\hat{u}\theta(\hat{u})(\hat{\rho})^{-2}]^2 d\hat{\rho}^2 \\
&+ [8\mu\hat{v}\theta(-\hat{v})(\hat{\rho}_1)^{-2} + 16\mu^2(\hat{v})^2\theta(-\hat{v})(\hat{\rho}_1)^{-4}]d\hat{\rho}_1^2 \\
&+ \hat{\rho}^2[1 + 4\mu\hat{u}\theta(\hat{u})(\hat{\rho})^{-2}]^2 d\phi^2 \\
&+ \hat{\rho}_1^2[-8\mu\hat{v}\theta(-\hat{v})(\hat{\rho}_1)^{-2} + 16\mu^2(\hat{v})^2\theta(-\hat{v})(\hat{\rho}_1)^{-4}]d\phi_1{}^2, \qquad (5.4.1)
\end{aligned}
$$

where

$$
\begin{aligned}
\hat{u} &= \hat{z} + \hat{t}, \qquad \hat{v} = \hat{z} - \hat{t}, \\
\hat{\rho}_1 \cos\phi_1 &= \hat{\rho}\cos\phi + a, \qquad \hat{\rho}_1 \sin\phi_1 = \hat{\rho}\sin\phi.
\end{aligned} \qquad (5.4.2)
$$

This form of the metric will be valid in region I ($\hat{u} < 0, \hat{v} > 0$) between the two incoming shocks, in a neighbourhood of shock 1 extending into region II ($\hat{u} < 0, \hat{v} > 0$), and in a neighbourhood of shock 2 extending into region III ($\hat{u} > 0, \hat{v} > 0$). When $\hat{v} > 0$, it reduces to eqn (5.3.5), and when $\hat{u} < 0$ it reduces to the metric (5.3.5) with the shock moving in the opposite direction, so that $\hat{u} \to -\hat{v}$ and $\hat{v} \to -\hat{u}$. Here a is the impact parameter. Note that the notation b will be reserved for the scaling impact parameter $= b\gamma$ (eqn (5.7.3)) in Section 5.7.

To find the continuation of (say) shock 1 after its collision with shock 2, we follow the null geodesic generators through the spacelike collision two-surface $\{\hat{u} = \hat{v} = 0\}$. As explained in Section 5.3, a geodesic passing through shock 2 can be described in (u, v, x, y) coordinates as suffering a jump of $8\mu \log \rho \delta$ and a deflection produced by the Lorentz transformation (5.3.8) at the shock, while the geodesic is a C^1 curve when viewed in $(\hat{u}, \hat{v}, \hat{x}, \hat{y})$ coordinates. Since the incoming

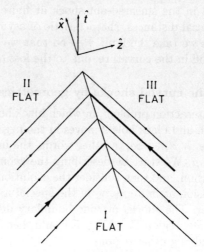

FIG. 5.2. The two null shocks are shown approaching before the collision; one spatial dimension has been suppressed. The incoming shock 1 lies on the surface $\hat{z} = \hat{t}$ separating the flat regions I and II, while shock 2 lies on the surface $\hat{z} = -\hat{t}$ between the flat regions I and III. Some null generators have been drawn on each null surface; the heavy lines show the null worldlines at the centre of the shocks—these are all that remain of the black holes after boosting to the speed of light. In the collision drawn here, the central worldlines do not intersect, and the spacetime is non-axisymmetric.

generators of shock 1 have $d\hat{z}/d\hat{t} = 1, d\hat{x}/d\hat{t} = 0, d\hat{y}/d\hat{t} = 0$, the curved shock 1 can be given the parametric representation

$$t = -2\mu \log(\xi^2 + \eta^2) + [1 + 16\mu^2/(\xi^2 + \eta^2)]\lambda,$$
$$x = \xi - 8\mu\xi\lambda/(\xi^2 + \eta^2),$$
$$y = \eta - 8\mu\eta\lambda/(\xi^2 + \eta^2),$$
$$z = 2\mu \log(\xi^2 + \eta^2) + [1 - 16\mu^2/(\xi^2 + \eta^2)]\lambda. \tag{5.4.3}$$

Here $\lambda \geq 0$ is an affine parameter along each of the null geodesic generators, and (ξ, η) give the values of (x, y) on the geodesic at the collision. Curved shock 2 can be described in a similar way. The geometry of the curved shocks was first found by Penrose (1974).

Figure 5.3 shows how region III is bounded by part of the plane $\{z + t = 0\}$ and by the achronal curved shock 1. This achronal set contains a one-dimensional spacelike caustic region on the axis $\rho = (x^2 + y^2)^{\frac{1}{2}} = 0$, given by

$$z - t = 4\mu \log[4\mu(z + t)] - 4\mu, \tag{5.4.4}$$

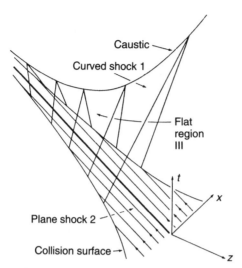

FIG. 5.3. In an axisymmetric collision at the speed of light, the region III of flat
space–time is bounded by the plane shock 2 and the curved shock 1. One spatial
dimension has been suppressed, and the coordinates are Minkowskian for region
III. Some null generators of shock 2 have been drawn, including the heavy line at
the centre of the shock. Each null generator of shock 2 eventually meets shock 1
on the collision surface, which is two-dimensional, spacelike and non-compact. In
the diagram, the collision surface appears as a pair of lines. The null generators of
the curved shock 1 emerge through the collision surface, and come to a focus on
the caustic region at the top of the diagram. The caustic region is a spacelike line
which becomes asymptotically null at both ends, and lies on the axis of symmetry.

where the generators of curved shock 1 have been focused by shock 2 and intersect
each other. At large values of ρ the geodesics have been deflected towards the
axis by a small angle of approximately $8\mu/\rho$. Correspondingly, the curved shock
surface has the logarithmic shape

$$z = t + 8\mu \log \rho + o(1) \tag{5.4.5}$$

as $\rho \to \infty$ with t fixed. As we consider geodesics which start with smaller and
smaller ρ values, the deflection becomes larger; for a geodesic starting at $\rho =
4\mu$, dz/dt is zero, and for smaller ρ values dz/dt is negative. The caustic region
becomes asymptotically null, with $dz/dt \to +1$, in the region in which geodesics
starting from large ρ values hit the axis. But the other end of the caustic region,
due to geodesics from small ρ values hitting the axis, is asymptotic to the surface
$\{z + t = 0\}$. One expects that a space–time singularity will be produced by the
infinite focusing at the caustic region in the speed-of-light collision. This would
lead to a loss of predictability of the geometry to the future of the caustic region;

for example, there may be a Cauchy horizon (Hawking and Ellis 1973) on the null surface, formed by continuing the generators of curved shock 1 through the caustic region. That is, it might (at the speed of light) become impossible to evolve the space–time to the future of the caustic. However, when the collision is at finite γ_1, γ_2, we shall see that the metric can remain quite smooth near the caustic region and beyond.

An alternative view of curved shock 1 is provided in Figs 5.4(a), 5.4(b), and 5.4(c), which give three successive snapshots of the shock at different times t.

Next, we wish to describe the detailed behaviour of the curved shocks, produced by an axisymmetric collision at speeds less than 1. A similar but more complicated description could also be given for the shock dynamics, when the impact parameter a is non-zero. For simplicity, we shall further assume, in this and in the following sections, that the two incoming black holes have equal masses M and, hence, the same factor $\gamma = \mu/M$. The extension to the case of unequal but comparable masses M_1, M_2, given by the scaling $\gamma_1 = \gamma c_1, \gamma_2 = \gamma c_2$ along a family with $\gamma \to \infty$, is always straightforward, and will be mentioned whenever appropriate.

At early times before the collision, the detailed structure of shock 1 will be described by the expansion (5.3.19) and (5.3.20), together with corrections tending to zero rapidly as $\hat{t} \to -\infty$, due to the field of black hole 2, which can be felt slightly by black hole 1 before the collision when γ is finite. Similarly, the detailed structure of shock 2 will be represented by eqns (5.3.19) and (5.3.20), with β replaced by $-\beta$, apart from terms giving early warning of the arrival of black hole 1. At finite γ, the collision no longer takes place on a spacelike two-surface. Instead, the collision will be smeared out over a region at separations of $O(\gamma^{-1})$ away from the surface $\{\hat{z} = \hat{t} = 0\}$. This leads us to set up an asymptotic expansion in $(\hat{t}, \hat{x}, \hat{y}, \hat{z})$ coordinates,

$$g_{ab}(\hat{t}, \hat{x}, \hat{y}, \hat{z}, \gamma) = \eta_{ab} + \gamma^{-1} k_{ab}^{(1)}(\gamma\hat{t}, \hat{x}, \hat{y}, \gamma\hat{z})$$
$$+ \gamma^{-2} k_{ab}^{(2)}(\gamma\hat{t}, \hat{x}, \hat{y}, \gamma\hat{z}) + \cdots, \qquad (5.4.6)$$

which is valid as $\gamma\hat{t}, \hat{x}, \hat{y}, \gamma\hat{z} \to$ constants, with $\gamma \to \infty$. The vacuum field equations in this region are of the form

$$\mathcal{L}[k_{ab}^{(n)}] = \mathcal{F}^{(n)}[k_{ab}^{(1)}, \ldots, k_{ab}^{(n-1)}], \qquad (5.4.7)$$

where $\mathcal{L}$ is a linear operator, involving only second derivatives with respect to the variables $(\gamma\hat{t})$ and $(\gamma\hat{z})$, while $\mathcal{F}^{(n)}$ is a functional of the lower-order perturbations $k_{ab}^{(1)}, \ldots, k_{ab}^{(n-1)}$, involving their derivatives. In particular, $k_{ab}^{(1)}$ satisfies a linear homogeneous field equation. The boundary conditions on this expansion are that it should reduce to eqns (5.3.19) and (5.3.20) for the incoming shock 1, when we let $(\gamma\hat{t}) \to -\infty$ and $(\gamma\hat{z}) \to -\infty$, in such a way that $\gamma(\hat{z} - \beta\hat{t})$ remains finite, and also a corresponding condition for matching onto the expansion for the detailed structure of the incoming shock 2.

Because of the linearity of the lowest-order field equation, we see that

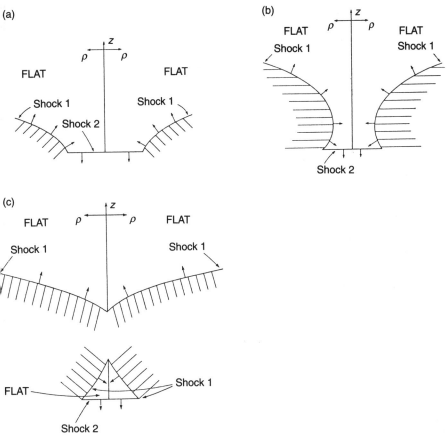

FIG. 5.4. The curved shock is seen at three successive times t as it moves at the
speed of light into the flat region III. The diagrams show the instantaneous shock
front in a neighbourhood of the symmetry axis; z measures distance along the axis,
and ρ measures distance away from the axis in the Minkowskian region III. Also
shown is the plane front of shock 2, which moves down the axis at the speed of
light. The arrows on the shocks give their direction of motion. At the earliest time
(a) all of shock 1 moves in the positive z direction; the null generators which have
been deflected by the largest angle are those which were closest to the axis at the
collision with shock 2—further out from the axis the shock front has a logarithmic
shape. Later (b) shock 2 has moved down to reveal a part of shock 1 which moves in
the negative z direction, due to the very strong deflection produced near the axis.
Still later (c) shock 1 has reached the axis; one part of shock 1 moves upward, and
the other downward. Shock 1 meets the axis at the caustic region, which travels
along the axis at more than the speed of light.

$$k_{ab}^{(1)}(\gamma\hat{t}, \hat{x}, \hat{y}, \gamma\hat{z}) = h_{ab}^{(1)}(\hat{x}, \hat{y}, \gamma(\hat{z} - \hat{t})) + h_{ab}^{(1)}(\hat{x}, \hat{y}, \gamma(\hat{z} + \hat{t})),(5.4.8)$$

apart from possible gauge terms, where the function $h_{ab}^{(1)}$ is defined by eqns
(5.3.20) and (5.3.16). The removal of the coefficient β in the arguments of $h_{ab}^{(1)}$ is
permissible since $\beta = 1 - \frac{1}{2}\gamma^{-2} + O(\gamma^{-4})$. The linearity of eqn (5.4.8) is deceptive:
it does not mean that the shocks have failed to interact. In fact, shock 1 (say)
emerges through shock 2 into a region in which the coordinates $(\hat{t}, \hat{x}, \hat{y}, \hat{z})$ are no
Minkowskian. To understand the interaction of the shocks more clearly, we should
transform to coordinates (t, x, y, z), which show that the region behind shock 2
is almost flat before the arrival of curved shock 1. Thus, we should follow shock 1
through the collision to points at which $\hat{x}, \hat{y}, \gamma(\hat{z} - \hat{t})$ are of $O(1)$ but $\gamma(\hat{z} + \hat{t}) \gg 1$.
Then, we are well past the detailed structure of shock 2, and into the region
where shock 2 can be accurately represented by eqn (5.3.5). If we then make the
transformation (5.3.3), so that the metric due to shock 2 becomes manifestly
flat, then the leading corrections to flat space will be found by transforming
$\gamma^{-1}h_{ab}^{(1)}(\hat{x}, \hat{y}, \gamma(\hat{z} - \hat{t}))$. Using the inverse of the Lorentz transformation (5.3.8)
we find that $h_{ab}^{(1)}(\hat{x}, \hat{y}, \gamma(\hat{z} - \hat{t}))$ is given in (t, x, y, z) coordinates just behind the
shock 2 by

$$\begin{aligned}
&h_{tt}^{(1)} = -16\mu^2 f, \quad\quad h_{tx}^{(1)} = -4\mu\xi f, \\
&h_{ty}^{(1)} = -4\mu\eta f, \quad\quad h_{tz}^{(1)} = -16\mu^2 f, \\
&h_{xx}^{(1)} = (\eta^2 - \xi^2)f, \; h_{xy}^{(1)} = -2\xi\eta f, \quad\quad h_{xz}^{(1)} = -4\mu\xi f, \\
&h_{yy}^{(1)} = (\xi^2 - \eta^2)f, \; h_{yz}^{(1)} = -4\mu\eta f, \quad\quad h_{zz}^{(1)} = -16\mu^2 f.
\end{aligned} \quad (5.4.9)$$

Here we follow the notation of eqn (5.4.3), and write (ξ, η) for the values of (x, y)
on curved shock 1 just after the collision; thus, the thin curved shock is centred
around $t = -2\mu\log(\xi^2 + \eta^2)$, $z = 2\mu\log(\xi^2 + \eta^2)$, due to the discontinuous
jump produced by shock 2. Also f in eqn (5.4.9) refers to $f((\xi^2 + \eta^2)^{\frac{1}{2}}, \gamma(\hat{z} - \hat{t}))$
with the functional form of f defined by eqn (5.3.16).

The initial data for curved shock 1 just behind shock 2 can be rewritten in a
more satisfactory form without mention of the variables $\hat{z}$ and $\hat{t}$. Let us represent
the curved shock surface (5.4.3) locally by the equation $z = j(t, \rho)$. Thus we are
at present working near the start of curved shock 1, where

$$j(-2\mu\log(\xi^2 + \eta^2), (\xi^2 + \eta^2)^{\frac{1}{2}}) = 2\mu\log(\xi^2 + \eta^2). \quad (5.4.10)$$

Using the inverse of the Lorentz transformation (5.3.8), we find that in the
small region occupied by curved shock 1, $\gamma(\hat{z} - \hat{t})$ can effectively be replaced by
$[1 - 16\mu^2/(\xi^2 + \eta^2)]\gamma(z - j(t, \rho))$ just behind shock 2. Hence, in the initial data
(5.4.9), we can regard f as

$$f((\xi^2 + \eta^2)^{\frac{1}{2}}, [1 - 16\mu^2/(\xi^2 + \eta^2)]\gamma(z - j(t, \rho))).$$

We have now set up the initial data for the curved shock 1, showing how the detailed structure of shock 1 has been rearranged near the surface $\{z = j(t, \rho)\}$ during the collision with shock 2.

One expects that curved shock 1 will be concentrated at distances of $O(\gamma^{-1})$ away from the surface $\{z = j(t, \rho)\}$ during its evolution, just as the incoming shock was spread over a thickness of $O(\gamma^{-1})$ near the plane $\{z = t\}$. This leads one to the asymptotic expansion

$$g_{ab}(t, x, y, z, \gamma) = \eta_{ab} + \gamma^{-1} l_{ab}^{(1)}(t, x, y, \gamma(z - j(t, \rho)))$$
$$+ \gamma^{-2} l_{ab}^{(2)}(t, x, y, \gamma(z - j(t, \rho))) = \cdots \quad (5.4.11)$$

for the detailed structure of curved shock 1. The expansion should be valid as t, x, y and $\gamma(z - j(t, \rho))$ tend to constants, while $\gamma \to \infty$. The initial data just described provide a possible choice for $l_{ab}^{(1)}$ on the initial surface, unique up to the addition of gauge terms.

In order to calculate the evolution of the detailed structure for curved shock 1, we make use of the symmetries of the space–times. Apart from the rotational Killing vector field $\partial/\partial\phi$, the space–times also possess a discrete isometry $\phi \to -\phi$, inherited from the corresponding property of the incoming Schwarzschild solutions. Under these symmetry conditions we can write the components $l_{ab}^{(1)}$, giving the dominant shock structure, as

$$l_{tt}^{(1)} = A, \qquad l_{tx}^{(1)} = xB, \qquad l_{ty}^{(1)} = yB, \qquad l_{tz}^{(1)} = C,$$
$$l_{xx}^{(1)} = D + (y^2 - x^2)E, \qquad l_{xy}^{(1)} = -2xyE, \qquad l_{xz}^{(1)} = xF, \qquad (5.4.12)$$
$$l_{yy}^{(1)} = D + (x^2 - y^2)E, \qquad l_{yz}^{(1)} = yF, \qquad l_{zz}^{(1)} = G,$$

where A, B, C, D, E, F, G are functions

$$A = A(t, \rho, \gamma(z - j(t, \rho))),$$

etc. The metric can be further simplified by choice of a convenient gauge. Gauge transformations respecting the symmetries of the space–time involve three free functions, and we can use these to remove three of the seven functions $A, \ldots, G$ in eqns (5.4.12). We choose

$$B = D = F = 0. \quad (5.4.13)$$

In describing the field equations, we write $j_t = \partial j/\partial t, j_\rho = \partial j/\partial\rho$. Since $z = j(t, \rho)\}$ is a null surface,

$$(j_t)^2 = 1 + (j_\rho)^2. \quad (5.4.14)$$

We also label the variables $t, \rho, \gamma(z - j(t, \rho))$ as x^1, x^2, x^3, respectively. Then three of the field equations give

$$A_{33} = \rho^2 E_{33}, \qquad C_{33} = 0, \qquad A_{33} + G_{33} = 0. \quad (5.4.15)$$

If we require that the field $l_{ab}^{(1)}$ tends to zero well ahead of curved shock 1, when $\gamma(z - j(t, \rho)) \to \infty$, then we can integrate these equations to find

$$A = \rho^2 E, \qquad C = 0, \qquad A + G = 0. \tag{5.4.16}$$

Now $l_{ab}^{(1)}$ is described simply in terms of the single function E. There is one remaining field equation for E, found by combining the $R_{xx} = 0$ and $R_{xy} = 0$ equations. This evolution equation for the curved shock gives

$$2j_\rho E_2 - 2j_t E_1 + (5\rho^{-1}j_\rho + j_{\rho\rho} - j_{tt})E = 0. \tag{5.4.17}$$

Using the parametric equation (5.4.30) for the surface $\{z = j(t, \rho)\}$, we can write (5.4.17) as a first-order ordinary differential equation along the null generators

$$\frac{dE}{d\lambda} = [20\mu/(\xi^2 + \eta^2 - 8\mu\lambda) - 4\mu/(\xi^2 + \eta^2 + 8\mu\lambda)]E. \tag{5.4.18}$$

This has the general solution

$$E = K(\rho, \gamma(z - j(t, \rho)))[1 - 8\mu\lambda/(\xi^2 + \eta^2)]^{-\frac{5}{2}}[1 + 8\mu\lambda/(\xi^2 + \eta^2)]^{-\frac{1}{2}} \tag{5.4.19}$$

for an arbitrary function K. We see that the curved shock evolution is dominated by 'geometrical optics' behaviour at ρ values of $O(1)$, where each part of the shock evolves separately along its own null generator, and transverse derivatives across the generators have little effect.

A gauge transformation can be found which puts our initial data (5.4.9) for the detailed structure of curved shock 1 into the form (5.4.12) with eqns (5.4.13) and (5.4.16) holding. This leads to the initial condition

$$E \mid_{\lambda=0} = K(\rho, \gamma(z - j(t, \rho)))$$
$$= f((\xi^2 + \eta^2)^{\frac{1}{2}}, [1 - 16\mu^2/(\xi^2 + \eta^2)]\gamma(z - j(t, \rho))) \tag{5.4.20}$$

with f defined by eqn (5.3.16). Together, eqns (5.4.19) and (5.4.20) describe the evolution of $l_{ab}^{(1)}$, giving the dominant structure of curved shock 1, beyond the collision region.

Note that $\rho^2 E$ tends to infinity as $\lambda \to (\xi^2 + \eta^2)/8\mu$, where shock 1 reaches the caustic region. This behaviour does not necessarily imply that a space–time singularity is reached, although it does indicate that the present asymptotic expansion (5.4.11) has broken down and must be matched onto another expansion which allows for some qualitatively different behaviour in the caustic region, as described in the following section.

The information presented here can be used to provide characteristic initial data for the geometry in the interaction region IV. This geometry will also be described by an asymptotic expansion as $\gamma \to \infty$, where the leading term is just the metric produced by the collision at exactly the speed of light. The interaction region for the $\gamma = \infty$ problem is bounded by curved shocks 1 and 2, which are again impulsive, with δ functions in the curvature which can be found from the detailed structure just discussed by integrating across the shock.

Only a small modification is needed for the case of unequal masses. The detailed structure of curved shock 1 (say) will be represented by an expansion like eqn (5.4.11), except that γ should be replaced by γ_1, so that the leading perturbation of flat space is

$$(\gamma_1)^{-1} l_{ab}^{(1)}(t, x, y, \gamma_1(z - j(t, \rho))).$$

The functional form of $l_{ab}^{(1)}$ is unchanged.

5.5 The caustic region

In this section we shall consider the behaviour of the gravitational field produced by the axisymmetric collision close to the spacelike caustic region $\{\rho = 0, z = j(t, 0)\}$ on the axis. This is because one will need subsequently to understand the space–time geometry to the future of the curved shocks and of the caustics. It is clearly necessary to have a separate asymptotic expansion describing the caustic region. Following the remarks in Section 5.4, one should be able to continue through the caustic even in the singular speed-of-light collision, by taking the limit $\gamma \to \infty$ of the calculations below. We showed in Section 5.4 that the leading metric perturbations $\gamma^{-1} l_{ab}^{(1)}$ in curved shock 1 grow like $\gamma^{-1} \rho^2 E$ as $\rho \to 0$, where $\rho^2 E$ is proportional to $\rho^2 [1 - 8\mu\lambda/(\xi^2 + \eta^2)]^{-\frac{5}{2}}$, which grows like $\rho^{-\frac{1}{2}}$ as the caustic region is approached. This singular behaviour implies that the asymptotic expansion (5.4.11) is breaking down, and suggests that the gravitational field possesses some qualitatively different features in the caustic region. To find a qualitative change, note that in the curved shock expansion we neglected ρ derivatives along the shock by comparison with derivatives across the shock, by using the variables ρ and $\gamma(z - j(t, \rho))$. But as we approach the caustic region, ρ derivatives should gain in importance until they may be able to balance derivatives across surfaces of constant $[z - j(t, \rho)]$. To investigate this possibility, we examine the detailed structure of the region at distances of $O(\gamma^{-1})$ away from the line $\{\rho = 0, z = j(t, 0)\}$, allowing for rapid variations in two independent directions. We shall see how the shock can interact strongly with itself in this region, so as to avoid forming a singularity.

The asymptotic expansion for the caustic region should be valid as $t, \gamma\rho, \gamma(z - j(t, 0)) \to$ constants, while $\gamma \to \infty$. If this expansion is to match onto eqn (5.4.11) for curved shock 1, when $(\gamma\rho) \to \infty$, we shall need half-integral powers of γ to allow for the $\gamma^{-1} \rho^{-\frac{1}{2}} (\gamma\rho)^{-\frac{1}{2}}$ growth of $\gamma^{-1} l_{ab}^{(1)}$ as $\rho \to 0$ in the intermediate matching region in which $\rho \ll 1$ but $\gamma\rho \gg 1$. Thus we consider the expansion

$$g_{ab}(t, x, y, z, \gamma) = \eta_{ab} + \gamma^{-\frac{1}{2}} m_{ab}^{(1)}(t, \gamma\rho, \gamma(z - j(t, 0)))$$
$$+ \gamma^{-1} m_{ab}^{(2)}(t, \gamma\rho, \gamma(z - j(t, 0)))$$

$$+ \gamma^{-\frac{3}{2}} m_{ab}^{(3)}(t, \gamma\rho, \gamma(z - j(t, 0))) + \cdots . \qquad (5.5.1)$$

Fixing attention on the leading perturbations $\gamma^{-\frac{1}{2}} m_{ab}^{(1)}$, we can exploit the symmetries of the space–time and make gauge transformations so that $m_{ab}^{(1)}$ is described by four functions A, C, E, G of $t, \gamma\rho$ and $\gamma(z - j(t, 0))$:

$$m_{tt}^{(1)} = A, \qquad m_{tx}^{(1)} = m_{ty}^{(1)} = 0, \qquad m_{tz}^{(1)} = C,$$

$$m_{xx}^{(1)} = \gamma^2(\gamma^2 - x^2)E, \qquad m_{xy}^{(1)} = -2\gamma^2 xy E, \qquad m_{xz}^{(1)} = 0, \qquad (5.5.2)$$

$$m_{yy}^{(1)} = \gamma^2(x^2 - y^2)E, \qquad m_{yz}^{(1)} = 0, \qquad m_{zz}^{(1)} = G.$$

Because of the rapid variations in two independent directions implied by the presence of the arguments $\gamma\rho$ and $\gamma(z - j(t, 0))$, the gravitational field resembles locally a cylindrical Einstein–Rosen wave (Einstein and Rosen 1937). The spacelike nature of the caustic region implies that $[j_t(t, 0)]^2 > 1$ and hence that the field equations in the caustic region are actually wavelike rather than elliptic. Note also that the curvature has reached a size of $\gamma^{-\frac{1}{2}}(\gamma)^2 = O(\gamma^{\frac{3}{2}})$ owing to second derivatives of $\gamma^{-\frac{1}{2}} m_{ab}^{(1)}$. This shows that the gravitational field is stronger here than in any other region which we have considered, owing to the focusing effect. The curvature components are only of $O(\gamma)$ in the plane and curved shocks, and only of $O(1)$ in the interaction zone just behind the curved shocks.

Boundary conditions on the field in the caustic region are provided by matching out in timelike and null directions to the past of the caustic region. This leads us into the nearly flat region ahead of curved shock 1 and into the detailed structure of curved shock 1. In particular, the field is quite smooth on the axis in the matching region. Hence we should impose a condition of regularity at the axis $\{\gamma\rho = 0\}$ on the solution $m_{ab}^{(1)}$ of the linear field equations for A, C, E, G.

In describing the field equations for the caustic region, let us write

$$X^1 = t, \qquad X^2 = \gamma\rho, \qquad X^3 = \gamma(z - j(t, 0)). \qquad (5.5.3)$$

Then two of the field equations for $m_{ab}^{(1)}$ give

$$C_{22} + C_2/X^2 = 0,$$
$$(A_{22} + G_{22}) + (A_2 + G_2)/X^2 = 0. \qquad (5.5.4)$$

Since A, C, and G must be regular on the axis, the solutions are of the form

$$C = L(X^1 + X^3), \qquad (A + G) = N(X^1, X^3) \qquad (5.5.5)$$

for some functions L, N. But now we can use a gauge transformation by a vector field $\gamma^{-\frac{3}{2}} \xi^a$, such that only $\xi^t = \xi^t(X^1, X^3)$ and $\xi^z = \xi^z(X^1, X^3)$ are non-zero, in order to ensure

$$C = 0, \qquad (A + G) = 0, \qquad (5.5.6)$$

while preserving the form (5.5.2).

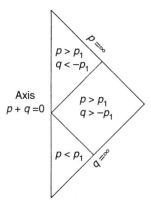

FIG. 5.5. The domain of the coordinates p, q, used in describing the caustic region. The domain is bounded by the axis $p + q = 0$ and the two 'null infinities' $q = \infty$, $p = \infty$; waves are fed into the caustic region at $q = \infty$ by shock 1, and then interact with themselves before emerging in another shock behind the caustic region, at $p = \infty$. Also shown are the three domains appearing in eqn (5.5.19).

The remaining equations give

$$A_{22} - (j_\rho)^2 A_{33} + A_2/X^2 = 0, \tag{5.5.7a}$$
$$A_{23} - (X^2)^2 E_{23} - 4X^2 E_3 = 0, \tag{5.5.7b}$$
$$A_{22} - A_2/X^2 - (X^2)^2 (j_\rho)^2 E_{33} = 0, \tag{5.5.7c}$$
$$2A_2/X^2 - (X^2)^2 E_{22} - (X^2)^2 (j_\rho)^2 E_{33} - 7X^2 E_2 - 8E = 0, \tag{5.5.7d}$$

where $j_\rho = j_\rho(t, 0)$. These equations can be shown to be all compatible. As for exact Einstein–Rosen waves, one first solves a cylindrical wave equation (5.5.7a), and then finds the other metric functions by quadrature.

We solve eqn (5.5.7a) by use of a Green function. First, make the change of variables

$$p = j_\rho X^2 - X^3, \qquad q = j_\rho X^2 + X^3, \tag{5.5.8}$$

to put eqn (5.5.7a) in the standard Euler–Darboux form

$$2(p + q)A_{pq} + A_p + A_q = 0. \tag{5.5.9}$$

To be specific, consider the part of the caustic region which moves up the z axis (see Fig. 5.4(c)). Then $j_\rho > 0$, and the domain of eqn (5.5.9) is $\{p + q \geq 0\}$. The Green function must be adapted to the boundary conditions on A, found via matching. Apart from regularity of A on the axis $\{p + q = 0\}$, we require that $A \to 0$ rapidly in 'timelike' directions, with $p \to -\infty, q \to +\infty$ (see Fig. 5.5), for this corresponds to matching out into the nearly flat region of space–time ahead of the caustic region and of curved shock 1. The input for the cylindrical wave equation is found by matching out in 'null' directions with $p \to$ const., $q \to +\infty$, since this takes us essentially along the generators of curved shock 1, which is feeding waves into the caustic region. By matching in from curved shock 1, we find

$$A(p,q) \sim q^{-\frac{1}{2}} h(p) \tag{5.5.10}$$

as $p \to$ const., $q \to \infty$, where

$$
\begin{aligned}
h(p) = 2^{\frac{5}{2}} \mu^{3/2} [1 - 16\mu^2/(\xi^2 + \eta^2)]^{-\frac{1}{2}} \times \\
\times \Big(\{\xi^2 + \eta^2 + [1 - 16\mu^2/(\xi^2 + \eta^2)]^2 p^2\}^{-\frac{1}{2}} \\
- 2\{\xi^2 + \eta^2 + [1 - 16\mu^2/(\xi^2 + \eta^2)]^2 p^2\}^{\frac{1}{2}}/(\xi^2 + \eta^2) \\
- 2[1 - 16\mu^2/(\xi^2 + \eta^2)]p/(\xi^2 + \eta^2) \Big).
\end{aligned}
\tag{5.5.11}
$$

Here we have used the expressions (5.4.19) and (5.4.20) for the function E in curved shock 1, and let $\lambda \to (\xi^2 + \eta^2)/8\mu$ so that $\rho \to 0$.

Accordingly, we look for a Green function $H(p, q; p_1)$ satisfying (in a distributional sense)

$$2(p + q)H_{pq} + H_p + H_q = 0, \tag{5.5.12}$$

which is regular on the axis and has the asymptotic behaviour

$$H(p, q; p_1) \sim q^{1 - 1/2}\theta(p - p_1) + o(q^{-\frac{1}{2}}) \tag{5.5.13}$$

as $q \to \infty$ with p, p_1 fixed. This will allow us to build up a solution in the form

$$A(p, q) = \int_{-\infty}^{\infty} dp_1 H(p, q; p_1)h'(p_1). \tag{5.5.14}$$

Such a Green function can be found by starting from well-known properties of the Euler–Darboux equation (Mackie 1965) as used, for example, by Szekeres (1972) in studying space–times which contain colliding plane gravitational waves. The Riemann–Green function used by Szekeres is adapted to a characteristic initial-value problem with data given on a pair of intersecting lines $\{p =$ const.$\}$, $\{q =$ const.$\}$, and must then be modified to cope with our problem in which initial data are given only by the incoming field at $q = \infty$.

One begins with the function

$$J(p, q; p_1) = -(p + q)^{-\frac{1}{2}} P_{-\frac{1}{2}}(1 + 2(p_1 - p)/(p + q))\theta(p_1 - p), \tag{5.5.15}$$

where $P_{-\frac{1}{2}}$ is the Legendre function of degree $-\frac{1}{2}$. This satisfies the homogeneous equation distributionally and has asymptotic behaviour

$$J(p, q; p_1) \sim -q^{-\frac{1}{2}}\theta(p_1 - p) \tag{5.5.16}$$

as $q \to \infty$ with p, p_1 fixed. It also has singularities on the axis, given by

$$J(p, q; p_1) \sim \pi^{-1}(p_1 - p)^{-\frac{1}{2}}\log(p + q)\theta(p_1 - p) \qquad (5.5.17)$$

as $(p + q) \to 0$. To eliminate the singularities on the axis, we add a solution of the cylindrical wave equation defined by the usual retarded integral for the flat-space wave equation in two spatial dimensions, with a source along the axis which compensates for the coefficient $\pi^{-1}(p_1 - p)^{-\frac{1}{2}}\theta(p_1 - p)$ of $\log(p+q)$ in eqn (5.5.17). This leads to the function

$$H(p, q; p_1) = J(p, q; p_1)$$
$$+ \frac{1}{\pi} \int_{-\infty}^{\min(2p_1, -2q)} \frac{d\zeta}{(p_1 - \frac{1}{2}\zeta)^{\frac{1}{2}}[(p - q - \zeta)^2 - (p + q)^2]^{\frac{1}{2}}}. \qquad (5.5.18)$$

The integral in eqn (5.5.18) is elliptic. By standard methods it can be expressed in terms of Legendre functions, which can then be simplified with the help of quadratic transformation formulae for the hypergeometric function (Abramovitz and Stegun 1972). The simplest form for the resulting Green function in the three different regions shown in Fig. 5.5 is

$$H(p, q; p_1) = 0, \qquad p < p_1$$
$$= (p + q)^{-\frac{1}{2}} P_{-\frac{1}{2}}\left(1 + 2(p_1 - p)/(p + q)\right), \qquad p > p_1, \quad q > -p_1$$
$$= |p_1 + q|^{-\frac{1}{2}} P_{-\frac{1}{2}}\left(1 - 2(p + q)/(p_1 + q)\right), \qquad p > p_1, \quad q < -p_1.$$
$$(5.5.19)$$

We see that $J(p, q; p_1)$ has been precisely cancelled in the region $\{p < p_1\}$ by the integral in eqn (5.5.18). The Green function $H(p, q; p_1)$ of eqn (5.5.19) now satisfies all our requirements, since it is regular on the axis and obeys the differential equation (5.5.12) with the asymptotic behaviour (5.5.13). Note that the 'retarded' property that $H(p, q; p_1) = 0$ for $p < p_1$ is to be expected with our boundary conditions, by the usual energy conservation arguments for the wave equation, since there is no radiation input for H at $q = \infty$ when $p < p_1$.

We now define $A(p, q)$ by eqn (5.5.14), and note that the integral converges since $h'(p_1)$ is $O((p_1)^{-2})$ as $p_1 \to -\infty$. It can be checked that $A(p, q)$ satisfies the Euler–Darboux equation (5.5.9) and the asymptotic condition (5.5.10), while being regular on the axis; uniqueness follows again by standard 'energy' arguments.

Next we find E by integrating eqn (5.5.7c). Since E should tend to zero rapidly as $X^3 = \gamma(z - j(t, 0)) \to \infty$, when we match out into the nearly flat region ahead of the caustic region and of curved shock 1, we must have

$$E(X^1, X^2, X^3) =$$

$$(j_\rho)^{-2} \int_{X^3}^{\infty} d\xi' \int_{\xi'}^{\infty} d\xi'' [(X^2)^{-2} A_{22}(X^1, X^2, \xi'') - (X^2)^{-3} A_2(X^1, X^2, \xi'')].$$

$$(5.5.20)$$

This expression for E also satisfies eqns (5.5.7b) and (5.5.7d), so that eqns (5.5.7a–d) are consistent. We have now found completely the metric perturbations $\gamma^{-\frac{1}{2}} m_{ab}^{(1)}$ which give the dominant structure of the caustic region.

As a result of its wavelike self-interaction near the caustic region, the gravitational field has avoided producing a singularity, although the interaction in this region has managed to alter the profile of the curved shock. Beyond the caustic region, the curved shock continues travelling in almost a null direction, and lies within the interaction zone, so that we can no longer describe it analytically (see Fig. 5.6). This null direction corresponds in (p, q) coordinates to fixing q and letting $p \to \infty$. Thus, to find the modified profile, we should examine the 'output' from our cylindrical wave equation, given the 'input' specified by $h(p)$. The behaviour of $A(p, q)$ as $p \to \infty$ with q fixed is not very revealing since A grows as const. $\times p^{\frac{1}{2}}$ in this direction, because of the linear growth of $h(p)$ as $p \to \infty$. This is a gauge effect, produced since we are moving well behind the front of the incoming curved shock 1.

To see clearly the modification in the shock profile which takes place in the caustic region, we should examine instead the gauge-invariant leading curvature terms at $O(\gamma^{\frac{3}{2}})$. These curvature terms are essentially of five types (depending on the number of indices from the set $\{x, y\}$), which can all be written just in terms of A by using eqns (5.5.7a–d). Examples of each type are

$$R_{tztz} = \frac{1}{2} \gamma^{\frac{3}{2}} (j_\rho)^2 A_{33} + O(\gamma),$$

$$R_{tztx} = -\frac{1}{2} \gamma^{\frac{3}{2}} x\rho^{-1} A_{23} + O(\gamma),$$

$$R_{xzyz} = \frac{1}{2} \gamma^{\frac{3}{2}} [2 + (j_\rho)^2] xy\rho^{-2} (j_\rho)^{-2} (A_{22} - A_2/\gamma\rho)$$
$$+ O(\gamma),$$

$$R_{xyyz} = \frac{1}{2} \gamma^{\frac{3}{2}} x\rho^{-1} A_{23} + O(\gamma),$$

$$R_{xyxy} = -\frac{1}{2} \gamma^{\frac{3}{2}} (j_\rho)^2 A_{33} + O(\gamma).$$

$$(5.5.21)$$

Each of the leading curvature components satisfies a flat-space wave equation in two spatial dimensions with respect to the variables $\gamma x, \gamma y$, and $\gamma(z - j(t, 0))$, since A satisfies such an equation (eqn (5.5.7a)).

Consider, for example, the evolution of A_{33} in the caustic region, which then gives the components R_{tztz} and R_{xyxy}. The incoming data for A_{33} are

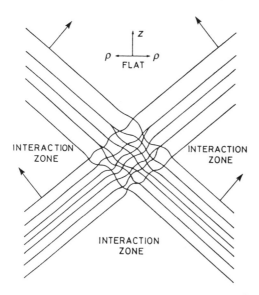

FIG. 5.6. A schematic diagram showing the caustic region produced as curved shock
1 crosses over itself near the symmetry axis, while moving upwards into the flat
space–time region III. The distance along the axis is given by z, while ρ measures
the distance away from the axis in this representation of the instantaneous shock
profile. The incoming shock 1 is shown in the top half of the diagram, separating the
flat region III from the interaction zone IV, where the space–time is strongly curved.
As it moves in the arrowed direction, it undergoes a wavelike self-interaction in the
centre of the diagram. The shock re-forms and continues to move in the arrowed
direction after its self-interaction, at the bottom of the diagram; however, the shock
profile has been altered in the caustic region.

$$A_{33} \sim q^{-\frac{1}{2}} h''(p), \tag{5.5.22}$$

as $q \to \infty$ with p fixed. Then

$$A_{33}(p,q) = \int_{-\infty}^{\infty} dp_1 \, H(p,q;p_1) h'''(p_1). \tag{5.5.23}$$

We find the asymptotic behaviour of $A_{33}(p,q)$ as $p \to \infty$ with q fixed from
the corresponding behaviour of $H(p,q;p_1)$ in eqn (5.5.19). For this, we need the
asymptotic expansion

$$P_{-\frac{1}{2}}(1+2\omega) = \pi^{-1}[4\log 2 - \log(1+\omega)] + O((1+\omega)\log(1+\omega)) \tag{5.5.24}$$

as $\omega \to -1$. Hence the 'output' for A_{33} is

$$A_{33}(p, q) \sim -\pi^{-1} p^{-\frac{1}{2}} \int_{-\infty}^{\infty} d\zeta \log |\zeta| \, h'''(\zeta - q), \qquad (5.5.25)$$

as $p \to \infty$ with q fixed, where the coefficient of $p^{-\frac{1}{2}} \log p$ vanishes, since

$$\int_{-\infty}^{\infty} d\zeta \, h'''(\zeta) = 0. \qquad (5.5.26)$$

The integral in eqn (5.5.25) can be evaluated, giving

$$A_{\zeta\zeta} \sim p^{-\frac{1}{2}} m(q) \qquad (5.5.27)$$

as $p \to \infty$ with q fixed, where

$$m(q) = \frac{K_1}{\pi q} \left\{ \frac{-2q^2(2q^2 + 5d^2)}{(q^2 + d^2)^2} \right.$$
$$\left. + \frac{3d^4 q}{(q^2 + d^2)^{\frac{5}{2}}} \log \left[\frac{(q^2 + d^2)^{\frac{1}{2}} - q}{(q^2 + d^2)^{\frac{1}{2}} + q} \right] \right\}, \qquad (5.5.28)$$
$$K_1 = 2^{\frac{5}{2}} \mu^{\frac{3}{2}} [1 - 16\mu^2/(\xi^2 + \eta^2)]^{\frac{1}{2}} (\xi^2 + \eta^2)^{-1},$$
$$d^2 = (\xi^2 + \eta^2)[1 - 16\mu^2/(\xi^2 + \eta^2)]^{-2}.$$

Notice how the shock has been spread out during this self-interaction, since $m(q)$ decreases only as q^{-1} for large $|q|$, while the incoming profile $h''(p)$ decreases like p^{-3} at large $|p|$.

The other curvature terms behave in a similar way, where the field emerging behind the caustic region is given by

$$A_{23} \sim j_\rho p^{-1/2} m(q),$$
$$(A_{22} - A_2/\gamma\rho) \sim (j_\rho)^2 p^{-\frac{1}{2}} m(q) \qquad (5.5.29)$$

as $p \to \infty$.

We have now seen in some detail how the shock structure is modified in a region with linear dimensions of $O(\gamma^{-1})$ around the caustic region, where the curvature reaches the large size of $O(\gamma^{\frac{3}{2}})$, before the shock re-emerges with altered profile, moving into the interaction zone. Had we instead examined the speed-of-light collision at $\gamma = \infty$, we would probably have encountered difficulties with a space–time singularity at the caustic region, possibly accompanied by a Cauchy horizon along the continuation of the shock beyond the caustic region. The considerations of this section show that any such singularity must be of a fairly mild kind, since the shock has very similar structures both before and after it reaches the caustic region, and there should be a natural extension of the $\gamma = \infty$ space-time through the impulsive shock continuation.

In the case of unequal masses, the statements in this section still hold when we replace γ by γ_1, if we are dealing (say) with the caustic region in curved shock 1.

5.6   Wave generation in the far-field curved shocks

So far, by considering the evolution of the curved shocks at distances of $O(1)$ away from the axis in the head-on collision, we have not seen any behaviour which resembles that of radiation propagating out near future null infinity. The curved shock structure analysed in Sections 5.4 and 5.5 may well eventually lead to a short burst of high-frequency radiation, which would carry a power of $O(1)$ in typical frequencies of $O(\gamma)$, giving a small total radiated energy of $O(\gamma^{-1})$ in the short time of $O(\gamma^{-1})$ that such radiation would last. But any calculation of this type of radiation would require us to follow the curved shocks beyond their caustic regions and into the interaction zone, where we are ignorant of the metric; our description of the curved shocks at ρ of $O(1)$ has not taken us out to the region in which they are radiative.

However, the collision problem with γ large but finite has the desirable property that we can locate a region of each curved shock, where perturbation methods can be used to view the shock evolution all the way out to future null infinity. For shock 1 (say) this region is to be found at t values of $O(\gamma^2)$, ρ values of $O(\gamma)$, and $[z - j(t, p)]$ values of $O(1)$; thus we can perform radiation calculations by examining the curved shock near the surface $\{z = j(t, \rho)\}$ at large distances from the axis and at late times after the collision. We shall call this region the far-field curved shock 1.

There are various reasons for expecting behaviour in the far-field curved shocks which is qualitatively different from that described in Section 5.3 for the curved shocks closer to the axis. First note that γ^2 is indeed a characteristic time for the far-field curved shock evolution, since the null geodesic generators of eqn (5.4.3) have been deflected inward at the collision by an angle of $O(\gamma^{-1})$, approximately $8\mu/\rho$ in this region, and hence take a time of $O(\gamma^2)$ to reach the axis. This is further suggested by eqn (5.4.19) giving the detailed evolution of shock 1 at ρ values of $O(1)$, where the evolution time-scale grows like $(\xi^2 + \eta^2)$ as $(\xi^2 + \eta^2)$ becomes large. Note also that we consider $[z - j(t, \rho)]$ values of $O(1)$ in this region, rather than $[z - j(t, \rho)]$ values of $O(\gamma^{-1})$, in order to see the shock structure properly. For the incoming shock depends on $\hat{z}$ through quantities such as $[\hat{x}^2 + \hat{y}^2 + \gamma^2(\hat{z} - \beta\hat{t})^2]^{\frac{1}{2}}$, so that we have to alter $\hat{z}$ by amounts of $O(1)$ to move through the shock when $(\hat{x}^2 + \hat{y}^2)$ is of $O(\gamma^2)$; a similar statement should hold for the far-field shock after collision.

We can now see that the far-field curved shock structure must allow for the fact that the incoming velocity $\beta = 1 - \frac{1}{2}\gamma^{-2} + \ldots$ is not quite unity. For the curved shock would also progress with a velocity differing from 1 by an amount of $O(\gamma^{-2})$, and the cumulative effects of this velocity difference should alter the location of the shock in relation to $\{z = j(t, \rho)\}$ by an amount of $O(1)$ over the time-scale of $O(\gamma^2)$. This effect should be visible in coordinates $\gamma^{-2}t, \gamma^{-1}\rho, [z - j(t, \rho)]$, and must be reflected in the governing partial differential equations.

We also see that the region in which the shock generators focus near the axis at late times must be different from the caustic region described in Section

5.5. For eqn (5.4.3) shows that the vertical velocity of the caustic region in $\{z = j(t, \rho)\}$ is approximately $1 + 32\mu^2/(\xi^2 + \eta^2)$, only slightly above unity, when $(\xi^2 + \eta^2)$ is large. Hence, when $(\xi^2 + \eta^2)$ is of $O(\gamma^2)$, the excess velocity $32\mu^2/(\xi^2 + \eta^2)$ may be outweighed by the fact that the shock does not quite travel at the speed of light. Hence, we should expect new qualitative features in the behaviour of the far-field curved shocks near the axis, without the 'spacelike' properties of a caustic region.

The far-field curved shock structure allows us to see gravitational radiation, provided that we restrict attention to directions at angles of $O(\gamma^{-1})$ away from the z axis. Consider a null geodesic travelling at an angle θ of $O(\gamma^{-1})$. This geodesic will have reached ρ values of $O(\gamma)$ after a time of $O(\gamma^2)$, and $[z - j(t, \rho)]$ can remain of $O(1)$ along the geodesic over this period of time, since the z velocity will only differ from 1 by $O(\gamma^{-2})$. Thus we can observe the null geodesic for most of its way out to future null infinity, while remaining within far-field curved shock 1, which can be described by perturbation theory. In this way we can compute strong-field gravitational radiation using only perturbation methods.

As suggested by these remarks, the qualitative difference between the far-field curved shocks and the behaviour described in Section 5.4 lies in the wavelike evolution of the far field, as compared to the goemetrical optics of the shocks closer in. In the far-field shocks, there is enough time for transverse derivatives in the x and y directions to come into play, whereas only evolution along the rays is important while the shock is travelling from (ξ, η) values of $O(1)$ to the caustic region. The radiative behaviour of the far-field curved shocks is produced when the wavelike incoming field (5.3.19) and (5.3.20) with (5.3.16) is distorted by the collision, so that the 'waves' cease to represent a Schwarzschild field, and start to propagate in all directions. This idea of 'virtual quanta' of gravitational radiation being made real by a collision was exploited by Matzner and Nutku (1974) in a related calculation of gravitational waves generated by a high-speed encounter, in which a small mass passes a large mass with large impact parameter. There one regards the small mass B as being at rest before being struck by a large mass A with sufficiently small impact parameter that shock A can be treated as an exact plane wave, not to be confused with a plane-fronted wave, which scatters as a weak perturbation off B's Schwarzschild far field. This method will not give the dominant part of the radiation, to be calculated below, since it ignores the logarithmic transverse variations of the shocks (Kovács and Thorne 1978). Rather, it complements other methods by describing the high-frequency domain.

In analysing the far-field curved shock 1, we prefer to use the variable $(z - t - 8\mu \log \gamma)$ instead of $[z - j(t, \rho)]$. This simplifies the form of the field equations and is permissible since eqn (5.4.3) shows that

$$z - j(t, \rho) = (z - t - 8\mu \log \gamma)$$
$$- 8\mu \log[\frac{1}{2}\gamma^{-1}\rho + (\frac{1}{4}\gamma^{-2}\rho^2 + 8\mu\gamma^{-2}t)^{\frac{1}{2}}]$$

$$+ 32\mu^2\gamma^{-2}t[\frac{1}{2}\gamma^{-1}\rho + (\frac{1}{4}\gamma^{-2}\rho^2 + 8\mu\gamma^{-2}t)^{\frac{1}{2}}]^{-2} + O(1) \ (5.6.1)$$

as $\gamma \to \infty$, when regarded as a function of γ with $\gamma^{-2}t, \gamma^{-1}\rho$, and $(z-t-8\mu\log\gamma)$ fixed. Thus any function of $\gamma^{-2}t, \gamma^{-1}x, \gamma^{-1}y, [z - j(t,\rho)]$ can be rewritten as a function of $\gamma^{-2}t, \gamma^{-1}x, \gamma^{-1}y, (z - t - 8\mu\log\gamma)$. Then we consider the asymptotic expansion for the metric in far-field curved shock 1:

$$g_{ab}(t,x,y,z,\gamma) = \eta_{ab} + \gamma^{-2}g_{ab}^{(2)}(\gamma^{-2}t, \gamma^{-1}x, \gamma^{-1}y, z - t - 8\mu\log\gamma)$$
$$+ \gamma^{-3}g_{ab}^{(3)}(\gamma^{-2}t, \gamma^{-1}x, \gamma^{-1}y, z - t - 8\mu\log\gamma) + \cdots \ (5.6.2)$$

in the limit $\gamma \to \infty$ with $\gamma^{-2}t, \gamma^{-1}x, \gamma^{-1}y, (z - t - 8\mu\log\gamma)$ fixed. We should also explain the omission of any term $\gamma^{-1}g_{ab}^{(1)}$ in eqn (5.6.2). This arises from the form of the incoming shock at distances of $O(\gamma)$ from the axis:

$$g_{ab}(\hat{t}, \hat{x}, \hat{y}, \hat{z}, \hat{\gamma}) = \eta_{ab} + \gamma^{-2}j_{ab}^{(2)}(\gamma^{-1}\hat{x}, \gamma^{-1}\hat{y}, \hat{z} - \beta\hat{t})$$
$$+ \gamma^{-3}j_{ab}^{(3)}(\gamma^{-1}\hat{x}, \gamma^{-1}\hat{y}, \hat{z} - \beta\hat{t}) + \cdots, \ (5.6.3)$$

where the only non-zero components of $j_{ab}^{(2)}$ are

$$j_{\hat{x}\hat{x}}^{(2)} = \gamma^{-2}(\hat{y}^2 - \hat{x}^2)f(\gamma^{-1}\hat{\rho}, \hat{z} - \beta\hat{t}),$$
$$j_{\hat{x}\hat{y}}^{(2)} = -2\gamma^{-2}\hat{x}\hat{y}f(\gamma^{-1}\hat{\rho}, \hat{z} - \beta\hat{t}), \ (5.6.4)$$
$$j_{\hat{y}\hat{y}}^{(2)} = \gamma^{-2}(\hat{x}^2 - \hat{y}^2)f(\gamma^{-1}\hat{\rho}, \hat{z} - \beta\hat{t}),$$

and the function f is defined in eqn (5.3.16). The incoming far-field shock 1 will be distorted by the collision with shock 2, but the perturbation order γ^{-2} of the leading term will not be altered.

To find initial data for the far-field curved shock 1, we proceed as in Section 5.4. There is a collision region where the two far-field shocks first meet, with

$$g_{ab}(\hat{t}, \hat{x}, \hat{y}, \hat{z}, \gamma) = \eta_{ab} + \gamma^{-2}n_{ab}^{(2)}(\hat{t}, \gamma^{-1}\hat{x}, \gamma^{-1}\hat{y}, \hat{z})$$
$$+ \gamma^{-3}n_{ab}^{(3)}(\hat{t}, \gamma^{-1}\hat{x}, \gamma^{-1}\hat{y}, \hat{z}) + \cdots, \ (5.6.5)$$

which is valid as $\gamma \to \infty$ and $\hat{t}, \gamma^{-1}\hat{x}, \gamma^{-1}\hat{y}, \hat{z} \to$ constants. Again there is the deceptive linearity of the term

$$n_{ab}^{(2)}(\hat{t}, \gamma^{-1}\hat{x}, \gamma^{-1}\hat{y}, \hat{z}) = j_{ab}^{(2)}(\gamma^{-1}\hat{x}, \gamma^{-1}\hat{y}, \hat{z} - \hat{t})$$
$$+ j_{ab}^{(2)}(\gamma^{-1}\hat{x}, \gamma^{-}\hat{y}, \hat{z} + \hat{t}), \ (5.6.6)$$

where the shocks have in fact interacted since the coordinates $\hat{t}, \hat{x}, \hat{y}, \hat{z}$ are not Minkowskian behind shock 2. Using the coordinate transformation (5.3.3) just behind shock 2, we arrive at initial data for the far-field curved shock in the form

(5.6.2), at $\gamma^{-2}t = 0$. The effect of the collision is only to give a logarithmic delay to the incoming shock 1; there is no extra contribution to $g_{ab}^{(2)}|_{\text{initial}}$ from the differential coordinate transformation behind shock 2 (the Lorentz tranformation (5.3.8)), since this only differs from the identity by corrections of $O(\gamma^{-1})$ when $\hat{\rho}$ is of $O(\gamma)$. We find that the initial data

$$g_{ab}^{(1)}|_{\text{initial}} = g_{ab}^{(2)}(\gamma^{-2}t = 0, \gamma^{-1}x, \gamma^{-1}y, z - t - 8\mu\log\gamma)$$

only have non-zero components:

$$
\begin{aligned}
g_{xx}^{(2)}|_{\text{initial}} &= \gamma^{-2}(\gamma^2 - x^2)f(\gamma^{-1}\rho, z - t - 8\mu\log\gamma \\
&\quad - 8\mu\log(\gamma^{-1}\rho)), \\
g_{xy}^{(2)}|_{\text{initial}} &= -2\gamma^{-2}xy f(\gamma^{-1}\rho, z - t - 8\mu\log\gamma \\
&\quad - 8\mu\log(\gamma^{-1}\rho)), \\
g_{yy}^{(2)}|_{\text{initial}} &= \gamma^{-2}(x^2 - y^2)f(\gamma^{-1}\rho, z - t - 8\mu\log\gamma \\
&\quad - 8\mu\log(\gamma^{-1}\rho)).
\end{aligned}
\tag{5.6.7}
$$

One might expect this to act as characteristic initial data, since it refers to the state of the far-field curved shock 1 just behind the almost-null surface of shock 2.

Next consider the field equations for the far-field curved shock 1. To simplify the notation, we write

$$
\begin{aligned}
U &= \gamma^{-2}t, \qquad X = \gamma^{-1}x, \qquad Y = \gamma^{-1}y, \\
V &= z - t - 8\mu\log\gamma, \qquad P = (X^2 + Y^2)^{\frac{1}{2}}.
\end{aligned}
\tag{5.6.8}
$$

This anticipates our finding that 'slow time' U and 'distance through the shock' V both behave as null variables. For solving the field equations, we again choose a special gauge; in the present context, we use the symmetries of the problem to fix the gauge at all orders in perturbation theory:

$$
\begin{aligned}
g_{tt} &= -1 + H^1(U, P, V, \gamma), \qquad g_{tx} = g_{ty} = 0, \\
g_{tz} &= H^2(U, P, V, \gamma), \\
g_{xx} &= 1 + (Y^2 - X^2)H^3(U, P, V, \gamma), \\
g_{xy} &= -2XYH^3(U, P, V, \gamma), \\
g_{xz} &= 0, \qquad g_{yy} = 1 + (X^2 - Y^2)H^3(U, P, V, \gamma), \\
g_{yz} &= 0, \qquad g_{zz} = 1 + H^4(U, P, V, \gamma),
\end{aligned}
\tag{5.6.9}
$$

for some functions H^1, H^2, H^3, H^4. In particular, we have

$$g_{tt}^{(2)} = A(U,P,V), \qquad g_{tx}^{(2)} = g_{ty}^{(2)} = 0, \qquad g_{tz}^{(2)} = C(U,P,V),$$
$$g_{xx}^{(2)} = (Y^2 - X^2)E(U,P,V), \qquad g_{xy}^{(2)} = -2XYE(U,P,V), \qquad g_{xz}^{(2)} = 0,$$
$$g_{yy}^{(2)} = (X^2 - Y^2)E(U,P,V), \qquad g_{yz}^{(2)} = 0, \qquad g_{zz}^{(2)} = G(U,P,V),$$
$$\tag{5.6.10}$$

for some functions A, C, E, G.

We consider first the (tt) component of the field equations, giving

$$A_{VV} + 2C_{VV} + G_{VV} = 0. \tag{5.6.11}$$

Since we require all perturbations to die out as we let $V \to +\infty$ and move well ahead of the curved shock, we find

$$A + 2C + G = 0. \tag{5.6.12}$$

The remaining field equations for $g_{ab}^{(2)}$ are

$$C_{PP} + C_P/P = -2P^4 E E_{VV} - P^4 E_V E_V, \tag{5.6.13a}$$
$$A_P + C_P = P^2 E_P + 4PE, \tag{5.6.13b}$$
$$2E_{UV} + E_{PP} + 5E_P/P = 0. \tag{5.6.13c}$$

It is the last of these equations which determines the wavelike evolution of curved shock 1 at the leading order; eqn (5.6.13) shows that each of the 'transverse' components $g_{xx}^{(2)}, g_{xy}^{(2)}$, and $g_{yy}^{(2)}$ obeys a flat-space wave equation with respect to the variables (U, X, Y, V). Having found E, one can use eqns (5.6.12) and (5.6.13a,b) to calculate A, C, and G; the boundary conditions are that A, C, and G should tend to zero as $P \to \infty$ with U, V fixed, and that they should be regular at $P = 0$. The boundary condition at large P should be imposed because the gravitational field tends to flatness at large radii, and the condition at small P arises since we are considering a one-parameter family of space–times which should each be smooth on the axis in the far-field curved shock (we saw in Section 5.5 that the geometry is smooth around the axis in the shock region at early times, and there is no reason for this to change as the shock moves up the axis at later times).

Our initial data as presented in eqn (5.6.7) are not compatible with eqns (5.6.13a, b); that is, they are not part of a g_{ab} which satisfies the gauge conditions (5.6.9) to all orders in perturbation theory. This can be remedied by a gauge transformation which introduces non-zero components $g_{tt}^{(2)}$, $g_{tz}^{(2)}$, and $g_{zz}^{(2)}$, but does not alter $g_{xx}^{(2)}, g_{xy}^{(2)}$, and $g_{yy}^{(2)}$. Hence our characteristic initial data for eqn (5.6.13c) on the 'null surface' $\{U = 0\}$ are

$$
\begin{aligned}
E(U = 0, X, Y, V) &= f(P, V - 8\mu \log P) \\
&= 4\mu P^{-4}\{(V - 8\mu \log P) \\
&\quad - [P^2 + (V - 8\mu \log P)^2]^{\frac{1}{2}}\}
\end{aligned}
$$

$$+ 2\mu P^{-2}[P^2 + (V - 8\mu \log P)^2]^{-\frac{1}{2}}. \quad (5.6.14)$$

We must solve the equation

$$2\mathcal{E}_{UV} + \mathcal{E}_{XX} + \mathcal{E}_{YY} = 0 \quad (5.6.15)$$

for $\mathcal{E}(U, X, Y, V) = e^{2i\phi} P^2 E(U, P, V)$, subject to the initial condition (5.6.14) and a requirement that $\mathcal{E} \to 0$ rapidly when we let $V \to +\infty$ and move well ahead of the curved shock. Such characteristic initial-value problems are treated by Penrose (1980), and in our case the solution in $\{U > 0\}$ is

$$\mathcal{E}(U, X, Y, V) = \frac{-1}{2\pi U} \int_0^\infty \int_0^{2\pi} dP' d\phi' P' \frac{\partial}{\partial V'} \mathcal{E}(0, X', Y', V'), \quad (5.6.16)$$

where

$$\begin{aligned} X &= P \cos \phi, & Y &= P \sin \phi, \\ X' &= P' \cos \phi', & Y' &= P' \sin \phi', \end{aligned} \quad (5.6.17)$$

and V' is determined as a function of X', Y' by

$$V' = V + [P^2 - 2PP' \cos(\phi - \phi') + P'^2]/2U. \quad (5.6.18)$$

There is no difficulty with convergence in eqn (5.6.16), since

$$(\partial/\partial V')\mathcal{E}(0, X'.Y', V') \to 0$$

as $P' \to 0$, and

$$(\partial/\partial V')\mathcal{E}(0, X', Y', V')$$

is $O((P')^{-4})$ as $P' \to \infty$ with V' given by eqn (5.6.18).

The dominant radiation field at angles of $O(\gamma^{-1})$ close to the axis is carried by the transverse metric components g_{xx}, g_{xy}, and g_{yy}. To find the radiation, we should move out to large distances in null directions. Thus we should examine the behaviour of the gravitational field when $w \to \infty$ along the line

$$\begin{aligned} t &= \tau + \gamma^2 w - 8\mu \log \gamma, & x &= \gamma \psi w \cos \phi, \\ y &= \gamma w \sin \phi, & z &= (1 - \gamma^{-2} \psi^2)^{\frac{1}{2}} \gamma^2 w, \end{aligned} \quad (5.6.19)$$

which becomes asymptotically null, in a direction at a small angle θ of approximately $\psi\gamma^{-1}$ from the axis. In (U, X, Y, V) coordinates this line becomes

$$\begin{aligned} U &= w, & X &= \psi w \cos \phi, & Y &= \psi w \sin \phi, \\ V &= -\tau - \frac{1}{2}\psi^2 w, \end{aligned} \quad (5.6.20)$$

disregarding corrections of higher order in γ^{-1}. The quantity τ gives the retarded time at future null infinity. From eqn (5.6.16) we find

$$
\mathcal{E}(U, X, Y, V) = \frac{-\mu e^{2i\phi}}{\pi w} \int_0^\infty P' dP' \int_0^{2\pi} e^{2i\phi'} d\phi' \times
$$
$$
\times (\psi P' \cos\phi' + \tau + 8\mu \log P')
$$
$$
\times \{[P'^2 + (\psi P' \cos\phi' + \tau + 8\mu \log P')^2]^{-\frac{3}{2}}
$$
$$
+ 2(P')^{-2}[P'^2 + (\psi P' \cos\phi' + \tau + 8\mu \log P')^2]^{-\frac{1}{2}}\}
$$
$$
+ e^{2i\phi} \Lambda(\psi) w^{-1} + o(w^{-1}) \tag{5.6.21}
$$

as $w \to \infty$, where $\Lambda(\psi)$ is a function of ψ only and does not contribute to the radiation.

For a space–time invariant under the symmetries $\phi \to \phi + \mathrm{const.}$ and $\phi \to -\phi$, the gravitational radiation is characterized by a real-valued function $c_\tau(\tau, \theta)$ of retarded time and angle, known as the 'news function'(Bondi *et al.* 1962). In terms of Section 5.2, one has $A_\times = 0$ and $c_\tau = \frac{1}{2}(\partial A_+/\partial t)$. This determines the rate of mass loss of an isolated gravitating system with these symmetries, due to emission of gravitational waves, according to

$$
\frac{d(\mathrm{mass})}{d\tau} = -\frac{1}{2} \int_0^\pi d\theta \sin\theta [c_\tau(\tau, \theta)]^2. \tag{5.6.22}
$$

For our problem, the leading behaviour of the news function at angles θ of $O(\gamma^{-1})$ is given by

$$
c_\tau(\tau, \theta = \psi\gamma^{-1}) = -\frac{1}{2} \lim_{w\to\infty} \left[w \frac{\partial}{\partial\tau} \mathcal{E}(U, X, Y = 0, V) \right] + o(1) \tag{5.6.23}
$$

as $\gamma \to \infty$, when regarded as a function of γ with τ and ψ fixed. Hence we find the news function

$$
c_\tau(\tau, \theta = \psi\gamma^{-1}) = \frac{3\mu}{2\pi} \int_0^\infty dP \int_0^{2\pi} d\phi P^3 \cos 2\phi \times
$$
$$
\times [P^2 + (\psi P \cos\phi + \tau + 8\mu \log P)^2]^{-\frac{5}{2}} + o(1) \tag{5.6.24}
$$

as $\gamma \to \infty$ with τ, ψ fixed. Allowing for higher-order corrections, the news function at angles of $O(\gamma^{-1})$ should be represented by an asymptotic expansion such as

$$
c_\tau(\tau, \theta = \psi\gamma^{-1}) = \sum_{n=0}^\infty \gamma^{-n} Q_n(\tau, \psi), \tag{5.6.25}
$$

of which we have so far calculated only $Q_0(\tau, \psi)$. There will be a similar expansion for angles of $O(\gamma^{-1})$ away from the backward direction $\theta = \pi$.

Note that we have managed in this way to calculate truly strong-field gravitational radiation, with power/solid angle of the typical size c^5/G, characteristic of fully non-linear gravitational interactions, since there is no small scaling factor γ^{-n} in front of the leading term in eqn (5.6.25). Although we used perturbation

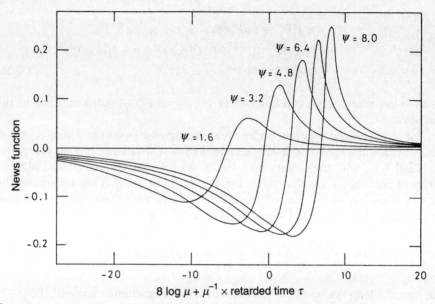

FIG. 5.7. The dominant news function, giving the amount of gravitational radiation
emitted in a high-speed axisymmetric collision, at angles $\theta = \psi\gamma^{-1}$ close to the
axis. Plotted horizontally is retarded time τ, measured in units of the energy μ
of one incoming black hole. The square of the news function gives $(4\pi)\times$ power
radiated/solid angle.

methods to analyse the small correction $\gamma^{-2}g^{(2)}_{ab}(\gamma^{-2}t, \gamma^{-1}x, \gamma^{-1}y, z-t-8\mu\log\gamma)$
to the flat-space metric in the far-field curved shock, the resulting radiation field
includes terms behaving like $\gamma^{-2}/\gamma^{-2}(x^2 + y^2 + z^2)^{\frac{1}{2}}$ at large radii, which have
lost their negative powers of γ. It has only been possible to calculate strong-field
radiation in perturbation theory by exploiting the very special properties of the
large-γ collision space–times. The underlying physical reason for the remarkable
effectiveness of perturbation methods in this case seems to be that the waves in
the far-field curved shock can escape out to future null infinity near the forward
direction without being caught and overtaken by information from the highly
non-linear interaction zone; these waves are assisted in their escape by the logarithmic delay across shock 2, which allows them to set off with large positive
values of z.

The integral in eqn (5.6.24) is apparently not tractable analytically—only
one variable can be integrated out. However, it can be computed numerically,
and the results of computing $Q_0(\tau, \psi)$ as a function of τ for five different values
of ψ are shown in Fig. 5.7. There is no need to recalculate $Q_0(\tau, \psi)$ for different
energies μ, since μ simply provides a time-scale for the radiation via

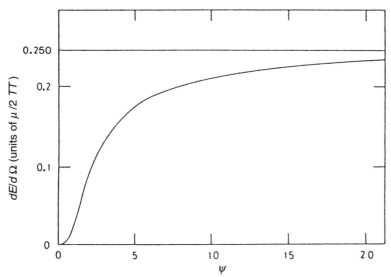

FIG. 5.8. The time-integrated power/solid angle in gravitational waves produced by a high-speed axisymmetric collision, at angles $\theta = \psi\gamma^{-1}$ close to the axis. On the vertical axis, energy/solid angle is measured in units of $\mu/2\pi$ in order to show the efficiency of the process in converting collision energy into gravitational waves; the units give the efficiency which the collision would have if the radiation were distributed isotropically with the news function seen at angle $\psi\gamma^{-1}$. At angles θ satisfying $\gamma^{-1} \ll \theta \ll 1, dE/d\Omega$ will be close to the limiting value $0.250\mu/2\pi$. However, $dE/d\Omega$ may be different for larger values of θ.

$$Q_0(\tau, \psi) \mid_{\text{energy 1}} = Q_0(\mu\tau - 8\mu\log\mu, \psi) \mid_{\text{energy } \mu}.$$

As ψ is increased from zero, $Q_0, (\tau, \psi)$ also changes slowly away from zero; this is to be expected for radiation in space–times which are axisymmetric, $\phi \to -\phi$ symmetric, and smooth near the axis (Bondi *et al.* 1962). For $\psi > 0, Q_0(\tau, \psi)$ always has two extrema and changes sign once, when regarded as a function of τ. As ψ is made larger, the amplitude grows and the pattern shifts to later and later retarded times, until the pattern stabilizes at a limiting shape, with the exception of a narrow spike at the second peak, which continues to grow as $\psi \to \infty$. The time-integrated power/solid angle in $Q_0(\tau, \psi)$ is also plotted in Fig. 5.8; this rises monotonically from zero to a limit as $\psi \to \infty$. It is given by $[\psi^2/(1 + \psi^2)]\mu/8\pi$. The limit is such that an energy 0.500μ would be given out as gravitational radiation, if radiation were emitted isotropically over the whole celestial sphere with the limiting power/solid angle. In that case 25% of the original incoming energy 2μ would have been converted into gravitational waves. This is somewhat below the 29% upper bound calculated by Penrose on the basis of the Cosmic Censorship Hypothesis, but still provides an estimate that the high-speed collision is an efficient generator of gravitational waves, removing a substantial fraction of the rest mass of the system. Indeed, such collisions may be the most efficient of all processes generating gravitational waves. An improved efficiency estimate which takes partial account of the angular distribution of the gravitational radiation is described in Chapter 8.

This limiting behaviour of the news function as $\psi \to \infty$ suggests that the radiation from the high-speed axisymmetric collision does not emerge in a beam near the axis, with angular width of $O(\gamma^{-1})0$, as is familiar from the theory of distant encounters at large γ. Rather, $Q_0(\tau, \psi)$ apparently represents some detailed structure in the radiation on angular scales of $O(\gamma^{-1})$ near the axis, which should fit onto a radiation pattern spread over the whole celestial sphere, on angular scales of $O(1)$.

The limiting radiation pattern as $\psi \to \infty$ can be described analytically. As $\psi \to \infty, Q_0(\tau, \psi)$ becomes centred around the retarded time $\tau = 8\mu \log \psi$ (this can be traced to the logarithmic delay across shock 2), and the dominant contribution to the integral comes from P values of order ψ^{-1}. We write

$$\tau = 8\mu \log (\psi/\mu) + \tau', \qquad P = \mu\psi^{-1}P', \tag{5.6.26}$$

and then consider

$$Q_0(\tau, \psi) = \frac{3\psi}{2\pi} \int_0^\infty dP' P'^3 \int_0^{2\pi} d\phi \cos 2\phi \times$$
$$\times [P'^2 + \psi^2(P' \cos \phi + 8 \log P' + \tau'/\mu)^2]^{-\frac{5}{2}}. \tag{5.6.27}$$

For a given value of P', the main contribution from the angular integration arises from those ϕ values which minimize $(P' \cos \phi + 8 \log P' + \tau'/\mu)^2$. If P' is such that $(P' \cos \phi + 8 \log P' + \tau'/\mu)$ crosses zero, then the main contribution to the angular integral is produced by a narrow range of ϕ values, with width of order ψ^{-1}; the contribution to the whole integral from other P' values is suppressed. The angular integration can be performed, and we find the limit

$$Q_0(\tau = 8\mu \log (\psi/\mu) + \tau', \psi)$$
$$\to \frac{4}{\pi} \int_D \frac{dP'}{P'^2} \left[\frac{2(8 \log P' + \tau'/\mu)^2}{P'^2} - 1 \right] \times$$
$$\times \left[1 - \frac{(8 \log P' + \tau'/\mu)^2}{P'^2} \right]^{-\frac{1}{2}} \tag{5.6.28}$$

as $\psi \to \infty$ with τ' fixed $(\tau'/\mu \neq 8 - 8 \log 8)$. Here the domain D consists of those values P' such that

$$(8 \log P' - P' + \tau'/\mu) < 0 < (8 \log P' + P' + \tau'/\mu).$$

When $\tau'/\mu < 8 - 8 \log 8$, D is one connected region, as in Fig. 5.9(a), but when

$\tau'/\mu > 8 - 8\log 8$, $\mathcal{D}$ is in two disjoint pieces, as in Fig. 5.9(b). A graph of the limiting news function in eqn (5.6.28) is shown in Fig. 5.10. The function has a logarithmic singularity at $\tau'/\mu = 8 - \log 8$, which is mild in the sense that it only carries a finite amount of energy when we integrate the power across the singularity. As $\psi \to \infty$, the functions $Q_0(\tau, \psi)$, regarded as functions of τ, can only approximate the singular limiting news function by developing higher and higher spikes near $\tau'/\mu = 8 - 8\log 8$. The energy spectrum for the news function (5.6.28) is shown in Fig. 5.11.

Let us consider further the radiation over angular scales of $O(1)$. This should be described by an asymptotic expansion of the form

$$c_{\tau'}(\tau', \theta) = \sum_{n=0}^{\infty} \gamma^{-n} S_n(\tau', \theta), \tag{5.6.29}$$

valid as $\gamma \to \infty$ with τ', θ fixed ($\theta \neq 0, \pi$), where the leading term $S_0(\tau', \theta)$ is precisely the news function for the collision at the speed of light. The functions $S_n(\tau', \theta)$ could only be calculated in their entirety by finding whole collision spacetimes, including their strong-field interaction zones. But we can obtain partial information on the $S_n(\tau', \theta)$ by using knowledge of the other expansion (5.6.25) for the one-parameter family of news functions, provided that the two expansions match in an intermediate region near the axis in which $\gamma^{-1} \ll \theta \ll 1$. Suppose, for example, that the one-parameter family of news functions is so well behaved in a neighbourhood of ($\gamma = \infty, \theta = 0$) that the news function can be expanded out in a double power series in γ^{-1} and θ (allowing both positive and negative powers of θ) valid in the intemediate region. This is consistent with our knowledge of $\gamma^0 Q_0(\tau, \psi)$, which provides the $\gamma^{-n}\theta^{-n}$ parts of the series via an expansion in negative powers of $\psi = \gamma\theta$ about $\psi = \infty$. In particular, the limiting news function of eqn (5.6.28) gives the $\gamma^0\theta^0$ term, and hence should be just $S_0(\tau', \theta = 0)$. Thus, if these ideas of matching are correct, we can actually make an exact statement about the radiation field in the fully non-linear space–time formed by the axisymmetric collision of two black holes at the speed of light, while using only perturbation methods.

Provided that we are correct in our identification of the limiting news function as $S_0(\tau', 0)$, we can make two remarks about the advantage of working with collisions at large but finite γ, rather than with impulsive wave collisions. First, note that the news function $S_0(\tau', \theta)$ will only be non-zero at $\theta = 0$ if the space–time is not smooth on the axis near future null infinity (Bondi *et al.* 1962). Second, the logarithmic singularity in $S_0(\tau', \theta = 0)$ again implies that the $\gamma = \infty$ space–time is, strictly speaking, not asymptotically flat; the impulsive shocks have focused each other so as to produce infinite radiative fields near future null infinity. Both remarks indicate that great care must be taken in any discussion which attempts to treat the $\gamma = \infty$ collision space–time as a typical isolated radiating system, and that it can be understood more easily as a limit of the better-behaved finite-γ solutions.

If it is valid to match the two expansions for the news function as outlined

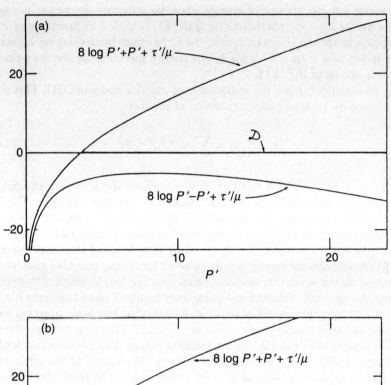

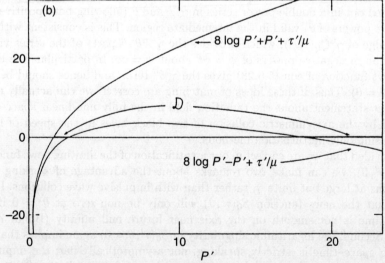

FIG. 5.9. The domain $\mathcal{D}$ of integration for eqn (5.6.28), consisting of those numbers $P' > 0$ such that $8 \log P' - P' + \tau'/\mu < 0 < 8 \log P' + P' + \tau'/\mu$. In (a), $\tau'/\mu = -14.0 < 8 - 8 \log 8$, so that $\mathcal{D}$ is one connected region. In (b), $\tau'/\mu = -5.0 > 8 - 8 \log 8$ and $\mathcal{D}$ is in two disjoint pieces.

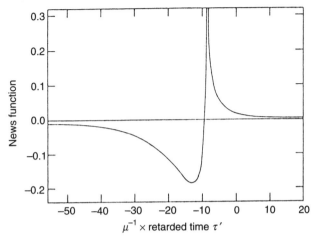

FIG. 5.10. The news function showing the gravitational radiation produced by a high-speed axisymmetric collision, at angles θ satisfying $\gamma^{-1} \ll \theta \ll 1$. This is found by taking the limit as $\psi \to \infty$ of the news function shown in Fig. 5.7. The peak has become infinitely high in this limit, with a logarithmic singularity; the time interval between trough and peak is approximately 4.5μ.

above, then further information about $S_0(\tau', \theta)$ can be extracted by perturbation calculations of higher order $Q_n(\tau, \psi)$. This will be found from the $\gamma^0 \theta^n = \gamma^{-n} \psi^n$ parts of the news function, for $n > 0$. Thus by calculating $Q_n(\tau, \psi)$ and matching out, one would find the θ^n part of $S_0(\tau', \theta)$. By this means, perturbation methods could be used to build up our knowledge of $S_0(\tau', \theta)$ near $\theta = 0$. Since we are using a centre-of-mass frame, $S_0(\tau', \theta)$ will be symmetrical about $\theta = \pi/2$, and (if sufficiently smooth) can be expanded as a convergent series

$$S_0(\tau', \theta) = \sum_{m=0}^{\infty} a_m(\tau')(\sin \theta)^m, \qquad (5.6.30)$$

of which we have so far found only $a_0(\tau')$. The perturbation problem has the unusual feature that we can continue to gather important information by proceeding to high orders, since $\gamma^{-n-1} Q_{n+1}(\tau, \psi)$ can lead to an $a_{n+1}(\tau')$ at least as large as $a_n(\tau')$ found from the lower-order $\gamma^{-n} Q_n(\tau, \psi)$. But it may well be that $S_0(\tau', \theta)$ is dominated by the first few multipoles, so that it can be reconstructed accurately by carrying on our perturbation treatment to find the next few $a_m(\tau')$. for $n \geq 2$, the calculation of $Q_n(\tau, \psi)$ will start to involve the solution of inhomogeneous flat-space wave equations with complicated source terms, so that technical difficulties could become prohibitive in an attempt to find the angular structure of the $\gamma = \infty$ news function by this method. Progress can be made on the $m = 2$ case—see Chapters 6, 7, and 8.

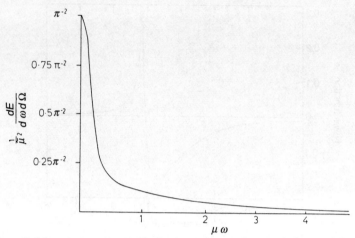

FIG. 5.11. The energy spectrum for the news function shown in Fig. 5.10.

It can in fact be shown that the angular expansion (5.6.30) for $S_0(\tau',\theta)$ involves only coefficients $a_m(\tau')$ for even m. This arises via matching, since the terms $\gamma^{-n}Q_n(\tau,\psi)$ in eqn (5.6.25) are actually zero when n is odd. One can see this most easily by first carrying out the perturbation calculations in a different frame, where black hole 1 (having the low mass $M\gamma^{-1}$), is initially at rest before being struck by a large shock 2 with energy $M\gamma^{-1}(2\gamma^2 - 1)$. In this frame one has a problem in which the far field of black hole 1 is a weak perturbation on the strong non-linear shock 2. This is essentially a situation with a small black hole in a background space–time (Chapter 3, and see D'Eath 1975a), apart from modifications due to the fact that the 'background space–time' provided by shock 2 itself depends on the small parameter γ^{-1}. The initial far field of black hole 1 is partially converted into radiation when shock 2 passes, and the relevant field equations to describe this process involve perturbation theory with respect to the metric of shock 2. The perturbation theory can be phrased in a way which only involves even powers of γ^{-1}. For example, the initial field of black hole 1 depends on the variable $(M\gamma^{-1})/r = M\gamma^{-2}/(\gamma^{-1}r)$, which scales as γ^{-1} in the region under consideration, where $\gamma^{-1}r$ is of $O(1)$. The even powers of γ^{-1} in the incoming field of shock 2 arise from the γ^2 in the ratio energy/mass $= 2\gamma^2 - 1$. Finally, by transforming back to the original frame, one finds that the asymptotic expansion (5.6.25) takes the form

$$c_\tau(\tau, \theta = \psi\gamma^{-1}) = \sum_{n=0}^{\infty} \gamma^{-2n}Q_{2n}(\tau, \psi). \qquad (5.6.31)$$

Hence, if it is valid to match the expansions for the news function in the manner suggested, eqn (5.6.30) should be replaced by

$$S_0(\tau',\theta) = \sum_{m=0}^{\infty} a_{2m}(\tau')(\sin\theta)^{2m}. \qquad (5.6.32)$$

In particular, a calculation (Chapters 6, 7, and 8) to the next non-trivial order in perturbation theory would yield $a_2(\tau')$ after matching out $Q_2(\tau,\psi)$; this calculation would require us to solve a flat-space wave equation in $\{U > 0\}$ with a source quadratic in the lowest-order perturbations of flat space.

When the black-hole masses are unequal, the only modification to the radiation field near the axis is that γ should be replaced by γ_1 near the forward direction ($\theta = 0$) and by γ_2 near the backward direction ($\theta = \pi/2$). For example, $c_\tau(\tau, \theta = \psi/\gamma_1) = Q_0(\tau,\psi) + O(\gamma^{-2})$. Since only the energy μ determines the radiation for the $\gamma = \infty$ collision on angular scales of $O(1)$, the leading term $S_0(\tau',\theta)$ in eqn (5.6.29) will be unchanged; however, the functional form of $S_n(\tau',\theta)$ for $n > 0$ may be different.

5.7   The non-axisymmetric case

The methods used to describe the curved shocks and wave generation in the previous three sections can also be applied to those space–times formed by non-axisymmetric collisions or encounters of black holes moving at nearly the speed of light. As in the axisymmetric problem, the space–times will again possess curved shock regions which can be analyzed using perturbation methods.

There is a wide range of possibilities available, since we now have the freedom of scaling the impact parameter while considering a one-parameter family of space–times labelled by the small parameter γ^{-1}. First consider the scaling (impact parameter) $\to$ constant as $\gamma \to \infty$, which would describe a situation in which the impact parameter is of a size comparable with the energy μ. Then the limiting $\gamma = \infty$ space–time will be quite different from the axisymmetric $\gamma = \infty$ space–time of the preceding section, and the radiation heading out at angles θ of $O(1)$ will also differ from the axisymmetric radiation described by eqn (5.6.29). The non-axisymmetric gravitational radiation will be characterized by a complex news function $c_{\tau'}(\tau',\theta,\phi)$ (Sachs 1962). In terms of Section 5.2, one has

$$c_{\tau'} = \frac{1}{2}\left(\frac{\partial A_+}{\partial t} + i\frac{\partial A_\times}{\partial t}\right).$$

The news function has an asymptotic expansion

$$c_{\tau'}(\tau',\theta,\phi) = \sum_{n=0}^{\infty} \gamma^{-n} S_n(\tau',\theta,\phi), \qquad (5.7.1)$$

where the leading term $S_0(\tau',\theta,\phi)$ gives the strong-field gravitational radiation produced in the interaction zone of the $\gamma = \infty$ space–time. There will also be some detailed structure in the radiation on angular scales of $O(\gamma^{-1})$ near the axis, given by an asymptotic expansion

$$c_\tau(\tau, \theta = \psi\gamma^{-1}, \phi) = \sum_{n=0}^{\infty} \gamma^{-n} Q_n(\tau, \psi, \phi), \qquad (5.7.2)$$

where the successive terms could be calculated by examining in greater and greater detail the structure of the far-field curved shock 1. However, by contrast with the radiation (5.7.1) on angular scales of $O(1)$, we find that the leading term $Q_0(\tau, \psi, \phi)$ is identical with the leading term $Q_0(\tau, \psi)$ of eqn (5.6.25) for the radiation close to the axis in the axisymmetric problem. This arises since the structure of the far-field curved shock 1 is again found by solving a characteristic initial-value problem behind shock 2, and the displacement of the initial data by distances of $O(1)$ in the x and y directions, produced by the non-zero impact parameter, does not alter the perturbation fields $g_{xx}^{(2)}$, $g_{xy}^{(2)}$, and $g_{yy}^{(2)}$. The non-zero impact parameter only produces higher-order effects in this region since the displacement is small by comparison with the $O(\gamma)$ length scales in the x and y directions, appropriate to the far-field shock structure. If matching of the two expansions for the news function is permissible, then we find that $S_0(\tau', \theta = 0, \phi)$ is just the limiting news function of eqn (5.6.28), regardless of the value of the impact parameter. Higher-order calculations in the curved shock and matching would then enable us to reconstruct more details of the angular structure of $S_0(\tau', \theta, \phi)$, in the same way as suggested for the axisymmetric problem in the preceding section.

The impact parameter can also be scaled up as $\gamma \to \infty$, so that we can describe a one-parameter family of space–times in which the black holes undergo a fairly distant encounter, rather than a collision or close encounter. The scaling which treats the impact parameter as $O(\gamma)$ is of particular interest, and will be the subject of the remainder of this section. Thus, we shall consider the case

$$\text{impact parameter} = b\gamma, \qquad (5.7.3)$$

where b is a positive constant. Recall that the mass M is scaled as $\mu\gamma^{-1}$, so that the impact parameter is now treated as $O(M\gamma^2)$. We have chosen this scaling because the interaction of the black holes will be sufficiently weak that the whole space–time can now be described using perturbation methods; but, as we shall see, the encounter is still sufficiently close that it generates strong radiation, which could not be calculated by the more usual perturbation techniques applied to the bremsstrahlung problem.

We can distinguish at least four different types of gravitational radiation generated by such an encounter; each type of radiation will need a separate asymptotic expansion for its description, and the expansions should match on the various overlap regions where one type of radiation merges into another. This situation is depicted in Fig. 5.12, which shows the spatial regions containing different types of radiation, at a time of $\frac{1}{4}b\gamma$ after the shocks have passed through each other. First, there will be radiation near the axis, generated by the collision of the far-field shocks at distances of $O(\gamma)$ away from the black holes; this radiation will have structure on angular scales of $O(\gamma^{-1})$ close to the forward and

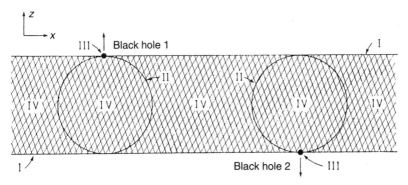

FIG. 5.12. A high-speed encounter with impact parameter $b\gamma$ leads to a variety of types of gravitational radiation, generated in different parts of the space–time. The various different regions are seen here in the plane of motion of the black holes, at a time $\frac{1}{4}b\gamma$ after the shocks have met. The dominant beamed part of the radiation is produced in the far-field shocks (I), which are just beginning to evolve after the collision. Weak radiation is generated at distances of $O(1)$ from each black hole as it passes through the other shock, and propagates out in all directions (II). High-frequency beamed radiation generated at distances of $O(\gamma^{-1})$ from each black hole is still close to the holes, in the regions (III). Region IV contains weak low-frequency radiation, left in the wake of the shocks.

backward directions. If we fix our attention on angles $\theta = \psi\gamma^{-1}$ of $O(\gamma^{-1})$ close to the forward direction, then the news function and the radiation time-scale are both of $O(1)$, when $\gamma \to \infty$. Thus the radiation near the axis again carries power/solid angle of the size c^5/G characteristic of fully non-linear interactions. But this intensity is now spread only over a solid angle of $O(\gamma^{-2})$, since for $b > 0$ the radiation is beamed, as might be expected in a fairly distant encounter at large γ. This type of radiation carries the bulk of the energy emitted to future null infinity, and will be discussed in more detail below.

Second, there will be radiation generated at distances of $O(1)$ away from black hole 2 when it passes through shock 1, and also at distances of $O(1)$ away from black hole 1 when it passes through shock 2; this travels outwards in the regions denoted by II in Fig. 5.12. The part of shock 1 through which black hole 2 passes can be approximated by a weak plane wave (not just a plane-fronted wave). Then the radiation is produced in the scattering of this weak wave by the strong plane-fronted shock 2 at distances of $O(1)$ from black hole 2. Thus it is calculated by means of perturbation theory in a plane-fronted shock background. Expressions for this radiation have been found by the author, although they are not written down here. The radiation is to be found at all angles θ, and the news function in this region scales as $O(\gamma^{-2})$ with a time-scale of $O(1)$ when we fix $\theta(\neq 0$ or $\pi)$ and let $\gamma \to \infty$; it carries only a negligible fraction of the total energy emitted.

Third, another type of radiation will be generated at distances of $O(\gamma^{-1})$ away from black hole 2 when it passes through shock 1, and similarly at distances of $O(\gamma^{-1})$ away from black hole 1 when it passes through shock 2. The gravitational field at distances of $O(\gamma^{-1})$ away from black hole 2 (say) cannot be accurately represented by a plane-fronted wave; instead one must take account of the details of the Schwarzschild geometry on the length-scales of $O(M) = O(\gamma^{-1})$ comparable with the radius of the event horizon. The radiation generated in this region is most easily calculated by working in the rest frame of black hole 2. In this frame, the approaching shock 1 will again resemble a plane wave, and the Riemann curvature of shock 1 will be much less than the $O(\gamma^2)$ Riemann curvature of black hole 2 on these length-scales. Thus the radiation should be calculated from the scattering of weak plane gravitational waves in a Schwarzschild background; this is discussed, for example, by Futterman *et al.* (1987). There will not be a simple analytical form for the resulting scattered radiation, since the incoming plane waves must be Fourier-analysed and decomposed into angular harmonics before the scattering is found mode by mode. When we boost back into the centre-of-mass frame, we find that this radiation is beamed, with structure on angular scales of $O(\gamma^{-1})$, close to $\theta = \pi$. The news function scales as $O(1)$, and the time-scale for the radiation is $O(\gamma^{-2})$. Thus the radiation from this region attains a large power/solid angle for a very brief time. To an observer situated at a large radius, near $\theta = \pi$, this radiation will appear as some rapid variations superimposed on the much smoother news function generated in the far-field curved shock 2. In Fig. 5.12, this radiation is to be found in the regions denoted by III; it has not yet had time to move far from the black holes.

Fourth, there will be some radiation with longer wavelengths of $O(\gamma)$, comparable with the impact parameter $b\gamma$. Like the strong radiation in the far-field curved shocks, it will be generated by the collision of the incoming shocks at distances of $O(\gamma)$ from the black holes; but unlike that radiation, it travels out at all angles θ of $O(1)$. It can be found by working with the expansion

$$g_{ab}(t, x, y, z, \gamma) = \eta_{ab} + \gamma^{-2} f_{ab}^{(2)}(\gamma^{-1}t, \gamma^{-1}x, \gamma^{-1}y, \gamma^{-1}z) + \cdots \tag{5.7.4}$$

appropriate to the structure of the field in region IV of Fig. 5.12, at t, x, y, z values of $O(\gamma)$. The leading term $\gamma^{-2} f_{ab}^{(2)}$ is calculated from the scattering of two weak impulsive plane-fronted waves; such a scattering has been discussed by Curtis (1978a). The news function is of $O(\gamma^{-2})$ over the time-scale of $O(\gamma)$, characteristic of this radiation, and so the energy carried by such radiation is small compared to the energy in the beams.

We now return to the strongest part of the radiation, generated in the far-field curved shocks. Just as in the axisymmetric case, the detailed structure of the far-field curved shock 1 will be given by an asymptotic expansion

$$g_{ab}(t, x, y, z, \gamma) =$$
$$\eta_{ab} + \gamma^{-2} g_{ab}^{(2)}(\gamma^{-2}t, \gamma^{-1}x, \gamma^{-1}y, z - t - 8\mu \log \gamma)$$

$$+ \gamma^{-3} g_{ab}^{(3)} (\gamma^{-2} t, \gamma^{-1} x, \gamma^{-1} y, z - t - 8\mu \log \gamma) + \cdots. \quad (5.7.5)$$

If we also use the previous notation

$$U = \gamma^{-2} t, \qquad X = \gamma^{-1} x, \qquad Y = \gamma^{-1} y,$$
$$V = z - t - 8\mu \log \gamma, \tag{5.7.6}$$

then we again find that the 'transverse' components $g_{xx}^{(2)}$, $g_{xy}^{(2)}$, and $g_{yy}^{(2)}$, which determine the leading radiation field near the axis, satisfy flat-space wave equations

$$2g_{xx,UV}^{(2)} + g_{xx,XX}^{(2)} + g_{xx,YY}^{(2)} = 0,$$
$$2g_{xy,UV}^{(2)} + g_{xy,XX}^{(2)} + g_{xy,YY}^{(2)} = 0, \tag{5.7.7}$$
$$2g_{yy,UV}^{(2)} + g_{yy,XX}^{(2)} + g_{yy,YY}^{(2)} = 0.$$

The characteristic initial data at the 'null surface' $\{U = 0\}$ are found as in Section 5.6 by following shock 1 as it passes through shock 2. Thus suppose that the coordinates are arranged to make the incoming black hole 1 aproach with $x = -b\gamma$, $y = 0$ at early times, while black hole 2 initially has $x = 0$, $y = 0$. The only effect of shock 2 on the dominant structure of shock 1 is to displace the incoming data along the generators of the characteristic initial surface. This leads to

$$g_{xx}^{(2)}(U = 0, X, Y, V) =$$
$$[Y^2 - (X + b)^2] f\left([(X + b)^2 + Y^2]^{\frac{1}{2}}, V - 4\mu \log(X^2 + Y^2)\right),$$
$$g_{xy}^{(2)}(U = 0, X, Y, V) =$$
$$- 2(X + b)Y f\left([(X + b)^2 + Y^2]^{\frac{1}{2}}, V - 4\mu \log(X^2 + Y^2)\right), \tag{5.7.8}$$
$$g_{yy}^{(2)}(U = 0, X, Y, V) =$$
$$[(X + b)^2 - Y^2] f\left([(X + b)^2 + Y^2]^{\frac{1}{2}}, V - 4\mu \log(X^2 + Y^2)\right),$$

where f is the function defined in eqn (5.3.16). The scaling of the impact parameter as $b\gamma$ has been designed to make an appreciable alteration to the characteristic initial data of eqns (5.6.10) and (5.6.14) and hence to the structure of the radiation on angular scales of $O(\gamma^{-1})$ close to the axis.

The characteristic initial data (5.7.8) demonstrate a significant difference between the axisymmetric collision problem and the encounter with $b > 0$. Since the leading Riemann curvature at $O(\gamma^{-2})$ is linear in $\partial^2 g_{xx}^{(2)}/\partial V^2$, $\partial^2 g_{xy}^{(2)}/\partial V^2$, and $\partial^2 g_{yy}^{(2)}/\partial V^2$, and since

$$\partial^2 f(p, q)/\partial q^2 = -6\mu(p^2 + q^2)^{-\frac{5}{2}}, \tag{5.7.9}$$

we see that the leading curvature terms become singular at $X = -b$, $Y = 0$, $V = 8\mu \log b$ when $b > 0$, but that they have no singularities at finite X, Y, V on the initial surface when $b = 0$. On the other hand, the $O(\gamma^{-2})$ curvature terms die away like $(X^2 + Y^2 + V^2)^{-\frac{3}{2}}$ as we move off to infinity in any direction on the initial surface when $b > 0$, whereas the curvature grows indefinitely as we let $V \to -\infty$ and stay close to $X = 0$, $Y = 0$ in the axisymmetric problem. Thus the singularity appearing at $X = -b$, $Y = 0, V = 8\mu \log b$, where black hole 1 emerges through shock 2, has been pushed to $V = -\infty$ by the infinite logarithmic delay when $b = 0$. It was this behaviour of the initial data at large negative V which made the news function tend to a limiting shape as $\psi \to \infty$ in Section 5.6, showing that the radiation in the axisymmetric problem is not beamed. In the non-axisymmetric problem with $b > 0$, we shall find that the radiation is beamed, with the leading term in the news function tending to zero as $\psi \to \infty$, because of the rapid decrease of the initial curvature as $X^2 + Y^2 + V^2 \to \infty$.

In Section 5.6 there was no need to consider the motion of black hole 1 after it hits shock 2, because of the infinite delay, but here for $b > 0$ we must subtract out the singular part of the initial data representing the non-radiative far field of black hole 1, which continues to travel almost in the z direction behind shock 2. To find this part of the initial data, note that black hole 1 will move essentially along a geodesic of the gravitational field produced by black hole 2, using arguments as in Chapter 3 (D'Eath 1975a). Hence for our purposes we can regard black hole 1 as moving along a straight line before and after it passes through shock 2, with the deflection in the line found from the geodesic equation in an impulsive shock background. The incoming line is given by

$$X = -b, \qquad Y = 0, \qquad z = (1 - \frac{1}{2}\gamma^{-2})t, \tag{5.7.10}$$

and hence the outgoing line is

$$
\begin{aligned}
&X = -b + 8\mu t / b\gamma^2, \\
&Y = 0, \\
&z = (1 - \frac{1}{2}\gamma^{-2} - 32\mu^2 / b^2\gamma^2)t + 8\mu \log(b\gamma),
\end{aligned}
\tag{5.7.11}
$$

disregarding corrections of higher order in γ^{-1}. Thus black hole 1 has been deflected by an angle of approximately $8\mu \log(b\gamma)$ due to the attraction of the other black hole, and it remains in the region of the far-field curved shock thereafter, since X, Y, and $(z - t - 8\mu \log \gamma)$ remain of $O(1)$ when t is of $O(\gamma^2)$ in eqn (5.7.11). In (U, X, Y, V) coordinates the outgoing line is

$$
\begin{aligned}
&X = -b + 8\mu U / b, \\
&Y = 0, \\
&V = 8\mu \log b - (\frac{1}{2} + 32\mu^2 / b^2)U,
\end{aligned}
\tag{5.7.12}
$$

with $U \geq 0$.

By considering the incoming field (5.6.3) and (5.6.4) of black hole 1, we can find the solutions of the wave equation which represent a boosted Schwarzschild field with no radiation, where the black hole moves along the line $X = 0, Y = 0, V = -\frac{1}{2}U$. These are just

$$g_{xx}^{(2)}(U, X, Y, V) = -g_{yy}^{(2)}(U, X, Y, V)$$
$$= (Y^2 - X^2)f\left((X^2 + Y^2)^{\frac{1}{2}}, V + \frac{1}{2}U\right),\qquad (5.7.13)$$
$$g_{xy}^{(2)}(U, X, Y, V) = -2XYf\left((X^2 + Y^2)^{\frac{1}{2}}, V + \frac{1}{2}U\right).$$

By constructing a Poincaré transformation in the (U, X, Y, V) space which maps the line $X = 0, Y = 0, V = -\frac{1}{2}U$ into the line (5.7.12), we arrive at the solutions which represent the uniform motion of black hole 1 in $\{U > 0\}$:

$$g_{xx}^{(2)}(U, X, Y, V) = -g_{yy}^{(2)}(U, X, Y, V)$$
$$= [Y^2 - (X + b - 8\mu U/b)^2]f\left([(X + b - 8\mu U/b)^2 + Y]^{\frac{1}{2}},\right.$$
$$\left. V + \frac{1}{2}U - 32\mu^2 U b^{-2} + 8\mu X b^{-1} + 8\mu - 8\mu \log b\right),$$
$$g_{xy}^{(2)}(U, X, Y, V) = -2(X + b - 8\mu U/b)Y f\left([(X + b - 8\mu U/b)^2 + Y^2]^{\frac{1}{2}},\right.$$
$$\left. V + \frac{1}{2}U - 32\mu^2 U b^{-2} + 8\mu X b^{-1} + 8\mu - 8\mu \log b\right).$$
$$(5.7.14)$$

Since we only wish to calculate the radiation generated by the passage of shock 1 through shock 2, these solutions must be subtracted off from the complete $g_{xx}^{(2)}, g_{xy}^{(2)}$, and $g_{yy}^{(2)}$. This leaves us with the initial data

$$g_{xx}^{(2)}(U = 0, X, Y, V) = -g_{yy}^{(2)}(U = 0, X, Y, V)$$
$$= [Y^2 - (X + b)^2]\{f\left([(X + b)^2 + Y^2]^{\frac{1}{2}}, V - 4\mu \log(X^2 + Y^2)\right)$$
$$- f\left([(X + b)^2 + Y^2]^{\frac{1}{2}}, V + 8\mu X b^{-1} + 8\mu - 8\mu \log b\right)\},$$
$$g_{xy}^{(2)}(U = 0, X, Y, V) = -2(X + b)Y\{f\left([(X + b)^2 + Y^2]^{\frac{1}{2}}, V - 4\mu \log(X^2 + Y^2)\right)$$
$$- f\left([(X + b)^2 + Y^2]^{\frac{1}{2}}, V + 8\mu X b^{-1} + 8\mu \log b\right)\}$$
$$(5.7.15)$$

for the radiative part of the field. These initial data are somewhat better behaved than those in eqn (5.7.8), since the curvature again decreases like $(X^2 + Y^2 + V^2)^{-\frac{3}{2}}$ as $X^2 + Y^2 + V^2 \to \infty$ on the initial surface, but now the curvature only grows like $[(X + b)^2 + Y^2 + (V - 8\mu \log b)^2]^{-\frac{1}{2}}$ as the point $X = -b, Y = 0, V = 8\mu \log b$ is approached. Correspondingly, the first derivatives $\partial g_{xx}^{(2)}/\partial V, \partial g_{xy}^{(2)}/\partial V$ and $\partial g_{yy}^{(2)}/\partial V$ are now bounded on the surface $\{U = 0\}$.

The solutions of the flat-space wave equation with these characteristic initial data can be found as in Section 5.6 by using the representation (5.6.16); we are requiring, as before, that $g_{xx}^{(2)}, g_{xy}^{(2)}$ and $g_{yy}^{(2)}$ tend to zero rapidly when we let $V \to +\infty$ and move well ahead of the far-field curved shock 1. The dominant behaviour as $\gamma \to \infty$ of the complex news function $c_\tau(\tau, \theta = \psi\gamma^{-1}, \phi)$, characterizing the gravitational radiation near the axis, is again found by considering $g_{xx}^{(2)}, g_{xy}^{(2)}$ and $g_{yy}^{(2)}$ as $w \to \infty$ with

$$U = w, \qquad X = \psi w \cos \phi, \qquad Y = \psi w \sin \phi,$$
$$V = -\tau - \frac{1}{2}\psi^2 w. \tag{5.7.16}$$

Using the definition of the news function in (Sachs 1962), we find

$$c_\tau(\tau, \theta = \psi\gamma^{-1}, \phi) = \frac{1}{2}e^{-2i\phi} \lim_{w \to \infty} \left[w \left(\frac{\partial g_{xx}^{(2)}}{\partial \tau} + i\frac{\partial g_{xy}^{(2)}}{\partial \tau} \right) \right]$$
$$+ O(\gamma^{-1}) \tag{5.7.17}$$

as $\gamma \to \infty$ with τ, ψ, ϕ fixed. Hence $\partial g_{xx}^{(2)}/\partial \tau$ and $\partial g_{xy}^{(2)}/\partial \tau$ provide the leading term $Q_0(\tau, \psi, \phi)$ of an expansion of the form (5.7.2) describing the radiation near $\theta = 0$. For a non-axisymmetric problem, the news function then determines the rate of mass loss in gravitational waves at retarded time τ according to

$$\frac{d(\text{mass})}{d\tau} = \frac{-1}{4\pi} \int_0^\pi d\theta \int_0^{2\pi} d\phi \sin \theta \mid c_\tau(\tau, \theta, \phi) \mid^2 . \tag{5.7.18}$$

From eqn (5.7.15) and the representation (5.7.16), we obtain the asymptotic behaviour of $\partial g_{xx}^{(2)}/\partial \tau$ and $\partial g_{xy}^{(2)}/\partial \tau$ needed for the calculation of $Q_0(\tau, \psi, \phi)$:

$$\frac{\partial g_{xx}^{(2)}}{\partial \tau} \sim \frac{3\mu}{\pi w} \int_{-\infty}^\infty \int_{-\infty}^\infty dX\,dY[(X + b)^2 - Y^2]\times$$
$$\times \left(\{(X + b)^2 + Y^2 + [\tau + X\psi \cos \phi + Y\psi \sin \phi + 4\mu \log(X^2 + Y^2)]^2\}^{-\frac{5}{2}} \right.$$
$$- [(X + b)^2 + Y^2 + (\tau + X\psi \cos \phi + Y\psi \sin \phi - 8\mu b^{-1}X$$
$$\left. + 8\mu \log b - 8\mu)^2]^{-\frac{5}{2}} \right),$$

$$\tag{5.7.19a}$$

$$\frac{\partial g_{xy}^{(2)}}{\partial \tau} \sim \frac{6\mu}{\pi w} \int_{-\infty}^{\infty} \int_{-\infty}^{\infty} dX \, dY \, (X+b) Y \times$$

$$\times \left(\{(X+b)^2 + Y^2 + [\tau + X\psi \cos \phi + Y\psi \sin \phi + 4\mu \log(X^2 + Y^2)]^2\}^{-\frac{5}{2}} \right.$$

$$- [(X+b)^2 + Y^2 + (\tau + X\psi \cos \phi + Y\psi \sin \phi - 8\mu b^{-1} X$$

$$\left. + 8\mu \log b - 8\mu)^2]^{-\frac{5}{2}} \right)$$

$$(5.7.19b)$$

as $w \to \infty$.

It can be shown that the second term,

$$[(X+b)^2 + Y^2 + (\tau + X\psi \cos \phi + Y\psi \sin \phi - 8\mu b^{-1} X + 8\mu \log b - 8\mu)^2]^{-\frac{5}{2}},$$

in the heavy parentheses of eqns (5.7.19a) and (5.7.19b) does not actually contribute to the integrals, provided that $\tau \neq b\psi \cos \phi - 8\mu \log b$. This term has appeared because we subtracted out the initial data for eqn (5.7.14), representing the boosted Schwarzschild field of black hole 1 after the encounter. Its only role is to ensure that the integrals (5.7.19a) and (5.7.19b) are well behaved near $\tau = b\psi \cos \phi - 8\mu \log b$, the retarded time at which future null infinity is intersected by the future null cone of the point $(U = 0, X = -b, Y = 0, V = 8\mu \log b)$, where black hole 1 has just emerged through shock 2.

To obtain more detailed understanding of this radiation, one should compute the integrals (5.7.19a,b) numerically. So far this has not been carried out because of the amount of computation required. An accurate calculation of the news function for just one value of the parameter b/μ would involve computing two integrals for a range of values of the three quantities (τ, ψ, ϕ), rather than only the one integral depending on the two parameters (τ, ψ), computed in the case $b = 0$.

However, the integrals (5.7.19a,b) do permit us to see that the radiation is beamed when $b > 0$, so that the strong-field intensity, carrying power/solid angle comparable with c^5/G, only reaches out to θ values of $O(\gamma^{-1})$. For as $\psi \to \infty$, we find that the leading term $Q_0(\tau, \psi, \phi)$ decreases like ψ^{-2}. When ψ is large, the main contribution to the integrals comes from the region in which X and Y are of order ψ^{-1}. By transforming to integration variables $X\psi, Y\psi$ one obtains

$$Q_0(\tau\psi, \phi) = \frac{3\mu b^2 e^{-2i\phi}}{2\pi \psi^2} \int_{-\infty}^{\tau} d\tau' \int_0^{\infty} \rho \, d\rho \int_0^{2\pi} d\phi' \frac{\partial}{\partial \tau'} \times$$

$$\times [b^2 + (\tau' - 8\mu \log \psi + \rho \cos \phi' + 8\mu \log \rho)^2]^{-\frac{5}{2}}$$

$$+ o(\psi^{-2})$$

$$(5.7.20)$$

as $\psi \to \infty$, where the leading ψ^{-2} behaviour arises in $\partial g_{xx}^{(2)}/\partial \tau$, while $\partial g_{xy}^{(2)}/\partial \tau$ decreases as ψ^{-3} for large ψ.

Note also the complicated dependence of $Q_0(\tau, \psi, \phi)$ on the black-hole masses, through the dimensionless quantity μ/b. This differs from the quadratic dependence (news function $\propto M_1 M_2$) found by other authors for encounters so distant as to be tractable by 'fast-motion' perturbation theory based on Minkowski space (Peters 1970; Kovács and Thorne 1978). Once again, this difference emphasizes that the methods used here can deal with much closer high-speed encounters—generating much more radiation—than are amenable to other perturbation approaches. The validity of the 'fast-motion' methods is limited by their requirement that the perturbations of flat space be everywhere weak. For an encounter of equal masses M (say) at large γ, viewed in a centre-of-mass frame, the condition becomes $M\gamma^2 \ll$ impact parameter, found by considering the metric component g_{tt}. But the calculations of this section are still valid when the impact parameter is comparable to $M\gamma^2$.

Both types of method should give an accurate description of the beamed radiation when we take $b \gg 1$. Thus our results should match as $b \to \infty$ onto the $\gamma \to \infty$ results of the other methods. In this limit we can actually carry out the integrals involved in the leading behaviour of $Q_0(\tau, \psi, \phi)$. For large b, the main contribution to the integrals (5.7.19a,b) arises from X and Y values of order b. Then the 'delay' term $4\mu \log(X^2 + Y^2)$ becomes a small correction when compared with the terms $X\psi \cos \phi$ and $Y\psi \sin \phi$. By transforming to integration variables Xb^{-1}, Yb^{-1}, and expanding out the small 'delay' corrections in a Taylor series, we find the leading behaviour

$$
Q_0(\tau, \psi, \phi) = \frac{6\mu^2 e^{-2i\phi}}{\pi b^2} \int_{-\infty}^{\infty} \int_{-\infty}^{\infty} dX\,dY (X + iY + 1)^2 \times
$$
$$
\times \log(X^2 + Y^2) \frac{\partial}{\partial(\tau/b)}[(X + 1)^2 + Y^2 +
$$
$$
(\tau b^{-1} + X\psi \cos \phi + Y\psi \sin \phi)^2]^{-\frac{5}{2}} + o(b^{-2}) \quad (5.7.21)
$$

as $b \to \infty$. This already exhibits the quadratic factor (μ/b^2) which sets the order of magnitude of the news function, and also shows that the characteristic timescale of the radiation is provided by $b =$ impact parameter$/\gamma$ in this régime.

The integral in eqn (5.7.21) can either be evaluated directly, or computed with the help of the Green function techniques in weakly curved space–time, set up by Kovács and Thorne (1977). The result for the beamed part of the radiation is

$$
Q_0(\tau, \psi, \phi) =
$$
$$
8\mu^2 e^{-2i\phi}(\psi^2 + 1)^{-1}(\tau^2 + b^2)^{-3/2} \times
$$
$$
[4\psi^2\tau^2 + 4b\psi \cos \phi(1 - \psi^2)\tau + b^2(1 + 2\psi^2 - 4\psi^2 \cos^2 \phi + \psi^4)]^{-2} \times
$$
$$
\times \Big((\psi^2 + 1)\{8\psi^2(\psi^2 - 1)(2\cos^2 \phi - 1)\tau^5
$$
$$
+ 8b\psi \cos \phi[-\psi^4 + (4\cos^2 \phi - 2)\psi^2 - 1]\tau^4
$$
$$
+ 2b^2(\psi^2 - 1)[\psi^4 + (8\cos^2 \phi - 4)\psi^2 + 1]\tau^3
$$

$$+ 4b^3 \psi \cos \phi [-3\psi^4 + (12\cos^2 \phi - 4)\psi^2 - 3]\tau^2$$
$$+ b^4 (\psi^2 - 1)[3\psi^4 - (12\cos^2 \phi + 2)\psi^2 + 3]\tau$$
$$+ 2b^5 \psi \cos \phi [\psi^4 + (4\cos^2 \phi - 6)\psi^2 + 1]\}$$
$$\pm 2(\tau^2 + b^2)^{\frac{3}{2}} \{ 4\psi^2 (\psi^4 + 1)(1 - 2\cos^2 \phi)\tau^2$$
$$+ 4b\psi \cos \phi (\psi^2 - 1)[\psi^4 + 4(\sin^2 \phi)\psi^2 + 1]\tau$$
$$+ b^2 [-\psi^8 + (4\cos^2 \phi - 2)\psi^6 \tag{5.7.22}$$
$$+ (16\cos^2 \phi \sin^2 \phi - 2)\psi^4 + (4\cos^2 \phi - 2)\psi^2 - 1]\}$$
$$+ 4i(\psi^2 + 1)\psi \sin \phi (\tau^2 + b^2)\{4\psi \cos \phi (\psi^2 - 1)\tau^3$$
$$+ 2b[-\psi^4 + 4(\cos^2 \phi)\psi^2 - 1]\tau^2$$
$$+ b^3 [-\psi^4 + (4\cos^2 \phi - 2)\psi^2 - 1]\}$$
$$\pm 8i\psi \sin \phi (\tau^2 + b^2)^{\frac{1}{2}} \{ -2\psi \cos \phi (\psi^4 + 1)\tau^2$$
$$+ b(\psi^2 - 1)[\psi^4 + (2 - 4\cos^2 \phi)\psi^2 + 1]\tau$$
$$+ b^2 \psi \cos \phi [\psi^4 + (2 - 4\cos^2 \phi)\psi^2 + 1]\} \Big)$$
$$+ o(b^{-2})$$

as $b \to \infty$. The upper sign applies when $\tau < b\psi \cos \phi$, and the lower when $\tau > b\psi \cos \phi$. The discontinuous behaviour given by eqn (5.7.22) at $\tau = b\psi \cos \phi$ has appeared as a result of fixing attention on τ values comparable with b, before letting $b \to \infty$. For any particular large finite b, the news function will be continuous near $\tau = b\psi \cos \phi$, but will vary there over time-scales short compared to b, which cannot be seen in the expansion (5.7.22). These more rapid variations are unimportant by comparison with the behaviour in eqn (5.7.22), in the sense that they carry a negligible fraction of the total energy radiated, in the limit $b \to \infty$. The form (5.7.22), which can be derived either from the high-speed methods or from the Thorne–Kovács methods, shows how these methods work together in the appropriate region.

When the black-hole masses are unequal, these calculations of the radiation field near the axis must again be modified slightly; near the forward direction γ should be replaced by γ_1, and near the backward direction γ should be replaced by γ_2. Thus

$$Q_0(\tau, \psi, \phi) = \lim_{\gamma \to \infty} c_\tau(\tau, \theta = \psi/\gamma_1, \phi)$$

will have the same functional form if we scale the impact parameter $= b\gamma_1$, while working in a centre-of-mass frame.

If one is only interested in the dominant beamed part of the radiation generated by a high-speed encounter, there is no need to restrict attention to black holes. The expressions for the leading radiation field should still be accurate when the proper radii R_1, R_2 of the two bodies satisfy $R_1 \ll M_1(\gamma_1)^2, R_2 \ll M_2(\gamma_2)^2$, while the impact parameter is comparable to (or larger than) $M_1(\gamma_1)^2$ and $M_2(\gamma_2)^2$. This is simply because the detailed structure of such bodies is not apparent on the length-scales of $O(\gamma)$ in the x and y directions, and of $O(1)$ in the z direction, appropriate to the far-field curved shocks where the radiation is generated.

5.8 Conclusions

We have seen in this chapter how certain parts of any space–time, formed by the high-speed collision or encounter of two black holes with comparable masses, can be described using perturbation methods. When the impact parameter is of order $\mu\gamma$ or larger, so that we have a fairly distant encounter, essentially all of the space–time geometry can be analysed by perturbation theory, as in Section 5.7. But even when the impact parameter is comparable to the energy μ, so that there is a highly non-linear collision, useful information about the space–time can be extracted from a description of the shock regions which form the boundary of the strong-field interaction zone. In the special case of an axisymmetric collision with zero impact parameter, these shock regions have been analysed in Sections 5.4, 5.5, and 5.6.

Arguably, the most significant outcome of this treatment has been the calculation of the gravitational radiation emitted near the forward and backward directions, found by considering the evolution of the far-field shocks after they have distorted each other during their interaction. By this means we have been able to find beamed bremsstrahlung radiation in a régime not accurately accessible by other methods, where the impact parameter is of order $\mu\gamma$. Such an encounter is sufficiently close that it generates radiation of the typical strong-field power/solid angle c^5G^{-1}, lasting a time of $O(\mu)$, which emerges in a beam at angles of $O(\gamma^{-1})$ from the direction of motion. This radiation is described by eqns (5.7.17) and (5.7.19a,b). Thus, the present treatment differs from most other perturbation calculations of gravitational wave generation, which work with asymptotic expansions valid only in the limit that the radiation becomes weak.

Similar techniques have been applied to find the detailed structure of the radiation on angular scales of $O(\gamma^{-1})$ near the direction of motion, when the impact parameter is of order μ and the black holes undergo a close encounter or a collision. The leading term in the asymptotic expansion for such radiation is given by eqn (5.6.24), and this news function is depicted in Fig. 5.7. Again, the radiation has the intensity characteristic of fully non-linear gravitational interactions, but it does not appear to be beamed. Instead, the news function tends to a limiting shape at angles θ satisfying $\gamma^{-1} \ll \sin\theta \ll 1$; this should allow the radiation near the axis to merge into a much smoother radiation pattern spread all over the celestial sphere. In particular, the limiting news function of eqn (5.6.28) should be precisely the news function at $\theta = 0$ produced by a collision at the speed of light. Thus we have found part of a strong-field news function while using only perturbation methods. Clearly, it is only the very special nature of the collision problem at large γ which has made this possible. This has allowed us to calculate the dominant radiation near the axis by examining the collision of the Coulomb far fields of the two black holes. As suggested in Section 5.6, one could go on from here to build up more information about the angular structure of the full $\gamma = \infty$ news function, by solving inhomogeneous wave equations in the far-field curved shocks and finding higher-order terms in the expansion (5.6.25), before matching out to the expansion (5.6.29). At each higher order, the calculation would dig further and further into the non-linear details of the strong-field interaction region, allowing for more powers of r^{-1} in the incoming fields. If carried to completion, this process would enable us to calculate the true efficiency for the conversion of incoming energy into gravitational waves during an axisymmetric collision at the speed of light. At present, our only guide as to that true efficiency is provided by the limiting news function, which would give an efficiency of 25% if it were part of an isotropic radiation pattern.

The next term, $a_2(\tau')$ in the speed-of-light expansion (5.6.31) for the news function will be calculated by studying speed-of-light collisions, at the next order of perturbation theory, in Chapters 6, 7, and 8. This will give a better idea of the angular distribution of radiation and of the total energy emitted.

6

AXISYMMETRIC BLACK-HOLE COLLISIONS AT THE SPEED OF LIGHT: PERTURBATION TREATMENT OF GRAVITATIONAL RADIATION

6.1 Introduction

As remarked in Chapter 5, in the many years since general relativity was originally formulated by Einstein, no-one has found, owing to the complexity and nonlinearity of the field equations, any physically realistic analytic solution which does not possess a large number of simplifying symmetries. However, if one is interested in studying the generation of gravitational radiation by realistic physical sources, then one must of necessity consider isolated gravitating systems that are time-dependent and which can have no simplifying features apart from axisymmetry. General reviews of the subject of gravitational waves are given by Thorne (1977, 1987).

An exact treatment of such problems is, at present, quite out of the question, and one must therefore seek recourse to approximation procedures (Thorne 1987). There are two alternatives: the first is numerical simulation, whereby one replaces the space–time continuum by a discrete grid and the differential field equations by a finite difference scheme. One sets up appropriate boundary conditions on some initial surface, and then one 'constructs' the space–time to the future of this surface by evolving the initial data on a computer. This approach has been used to study, amongst other problems, the axisymmetric collision of two black holes starting from rest at a finite separation (Smarr et al. 1976; Smarr 1977b, 1979; Anninos et al. 1993, 1995a, b). All these papers are concerned with the evolution of time-symmetric initial data for a pair of equal-mass black holes. An analytic estimate of $E = 0.0025M$, corrected by certain factors, where M is the mass of one black hole, is obtained for the energy loss in gravitational waves for the case when the black holes start with large initial separation. This estimate is obtained by a certain extrapolation from the calculation of Davis et al. (1971), where a small black hole is dropped into a large black hole, exciting the normal modes (Detweiler 1979; Chandrasekhar 1983) of the large black hole. This estimate agrees well with the numerical results (Anninos et al. 1993, 1995a, b). But note that this estimate is not based on any systematic perturbation theory, since there is no obvious small parameter in the non-linear calculation.

A second analytic estimate is obtained in the limit that the two black holes are initially very close together. In that case, a common horizon surrounds both holes, and the geometry exterior to the horizon can be treated as a non-spherical perturbation of a single Schwarzschild black hole. Here one has a genuine perturbation problem, in which the small perturbation parameter measures the distance between the initial black holes. But the initial data is not realistic; nor is the radiation strong. Other papers concerning gravitational radiation generated by the motion of a small mass (which might be a small black hole) in the field of a large Schwarzschild or Kerr black hole are Ruffini (1973), Detweiler (1979), Kojima and Nakamura (1983), Nakamura and Haugan (1983), Oohara and Nakamura (1983), and Oohara (1984). Scattering by a black hole is described by Futterman et al. (1987).

The other method of treatment is perturbation theory, as just indicated. Here one assumes that the space–time metric differs only very slightly from some fixed background (which is taken to be one of the highly symmetric exact solutions mentioned above). The field equations for the metric perturbations are linear to lowest order, and often prove mathematically tractable, owing to the (relatively) simple nature of the background metric.

However, since the time-dependent perturbations must be small, the gravitational radiation produced is almost always correspondingly weak (in the sense that the energy carried off by the waves is only a small fraction of the total energy of the system). To deduce the behaviour of gravitating systems when the perturbations are not small, one is obliged to extrapolate from the weak-field limit, which can provide physical insight, but no strict quantitative results.

In fact, there is only one physical process in which perturbation methods have proved successful in describing truly strong-field gravitational radiation: namely, the high-speed collision of two black holes, as seen in Chapter 5. Based on the work of the present chapter, it will be seen in Chapters 7 and 8 that the angular distribution of gravitational radiation up to the term $a_2(\tau)$ of eqn (5.6.31) can be found analytically. The success of perturbation theory in these space–times is due to certain special features of their geometry, found in Chapter 5, which we now briefly recapitulate.

Owing to special-relativistic effects, the gravitational field of a black hole travelling close to the speed of light becomes concentrated in the vicinity of its trajectory, which lies close to a null plane in the surrounding nearly Minkowskian space–time. At precisely the speed of light, the black hole turns into a particular sort of impulsive gravitational plane-fronted wave (eqn (5.3.2) or (5.3.5)) (as the speed increases one must scale down the rest mass appropriately, in order that the energy be finite). The curvature is then zero except on the null plane of its trajectory, and there is a massless particle travelling along the axis of symmetry at the centre of this null plane.

An important property of this sort of gravitational shock wave is that geodesics crossing it are not only bent inwards, but also undergo an instantaneous translation along the null surface that describes the trajectory of the wave. The nature of this translation is such that geodesics crossing the shock close to the axis of symmetry are delayed relative to those which cross the shock far out from the axis. Hence, when two such waves pass through each other in a head-on collision, the far-field region of each wave (i.e. the region far from the axis of symmetry) is given a large head start over its near-field counterpart, in addition to being bent slightly inwards. Because of this, the self-interaction of the far field of each wave as it propagates out towards null infinity takes place without interference from the highly non-linear region near the axis of symmetry; and because gravity is weak in the far-field region, perturbation theory can be used to study this process. However, the radiation produced in the forward and backward null directions is not weak, for although the far fields contain only a fraction of the total energy, the solid angle into which they are focused is small, and so the energy flux per unit solid angle in these directions is not small (i.e. the news function (5.7.17), which characterizes the gravitational radiation, is of order unity in dimensionless units). Thus perturbation methods can successfully describe the generation of truly strong-field gravitational radiation in these space–times (D'Eath 1978; D'Eath and Payne 1992a-c).

The collision of two impulsive plane-fronted waves studied here is not to be confused with the problem of colliding *plane* waves, which yields a space–time containing two commuting Killing vector fields. Many colliding plane-wave space–times have been found analytically, and there is a substantial literature on the subject. See, for example, Griffiths (1991) and references therein.

The physics is practically identical in collisions at just less than the speed of light—the main difference being that everything is slightly smoothed out, since the incoming shock waves are no longer impulsive (see Chapter 5).

For each of a range of impact parameters in the speed-of-light collision, Penrose (1974) has found an apparent horizon (a compact two-dimensional spacelike surface such that the outgoing null geodesics have zero divergence ρ (Hawking and Ellis 1973)) on the union of the two null planes that describe the trajectories of the incoming waves. If one makes the cosmic censorship hypothesis (see, for example, Hawking and Ellis 1973), then the area of the initial apparent horizon can be used to put a lower bound on the areas of the event horizons of any black holes formed by the collision. In this way, Penrose has shown that the total rest mass of the black hole (or holes) formed by the axisymmetric collision must be more than $100/\sqrt{2}\%$ of the initial energy. Conversely, if too much energy is carried off by gravitational waves then the hypothesis must be wrong. Thus, the function of energy carried away in an axisymmetric collision must be less than 29% of the initial energy 2μ. These high-speed collisions thus provide an interesting test of cosmic censorship.

There are two different perturbation methods that one can use to treat these high-speed collisions. In the preceding Chapter 5, the collision was studied at large but finite γ, where γ is the Lorentz factor of the incoming holes. It was shown that the metric of a single high-speed hole—and hence also the pre-collision metric in the high-speed collision—can be expressed as a perturbation series in γ^{-1}. A method of matched asymptotic expansions could then be used to investigate the space–time geometry to the future of the collision. It is necessary to use a number of different asymptotic expansions to allow for the various length- and time-scales characteristic of the gravitational field in different parts of the space–time. One expects that expansions holding in adjacent regions will match smoothly on to each other; the regions to the past thereby providing boundary conditions for those neighbouring regions to the future.

Following this approach, one may calculate the radiation on angular scales of $O(\gamma^{-1})$ produced by the focusing of the far fields of the waves as they pass through each other during the collision. It was found in Chapter 5 (eqn (5.6.31)) that in this region the news function has an asymptotic expansion of the form

$$c_0(\bar{\tau}, \hat{\theta} = \gamma^{-1}\psi) \sim \sum_{n=0}^{\infty} \gamma^{-2n} Q_{2n}(\bar{\tau}, \psi), \qquad (6.1.1)$$

valid as $\gamma \to \infty$ with $\bar{\tau}, \psi$ fixed, where $\bar{\tau}$ is a suitable retarded time coordinate and $\hat{\theta}$ is the angle from the symmetry axis in the centre-of-mass frame. In Chapter 5 (eqn (5.6.27)), the leading term $Q_0(\hat{\tau}, \psi)$ was calculated; this does not vanish and is a regular function of $\hat{\tau}$. Since $Q_0(\hat{\tau}, \psi)$ is not damped by any power of γ^{-1}, the news function is $O(1)$, and so describes truly strong-field gravitational radiation (the square of the news function is $4\pi x$ power radiated/unit solid angle). $Q_0(\hat{\tau}, \psi)$ and its first angular derivative, $\partial Q_0(\hat{\tau}, \psi)/\partial \psi$, both vanish at $\psi = 0$, as they must if the space–time is to be regular (Bondi *et al.* 1962). What is most interesting is that as ψ tends to infinity $Q_0(\hat{\tau}, \psi)$ approaches a non-zero limiting form, which is such that 25% of the incident energy would be carried off by gravitational waves if the radiation were emitted isotropically with the limiting power/solid angle.

It was further shown in Chapter 5 that on angular scales of $O(1)$ the news function should have an asymptotic expansion of the form

$$c_0(\hat{\tau}, \hat{\theta}) \sim \sum_{n=0}^{\infty} \gamma^{-n} S_n(\hat{\tau}, \hat{\theta}), \qquad (6.1.2)$$

valid as $\gamma \to \infty$ with $\hat{\tau}, \hat{\theta}$ fixed. (The retarded time variables used in eqns (6.1.1) and (6.1.2) are not the same, owing to the varying time delays suffered by different parts of the shocks when they collide.) Here $S_0(\hat{\tau}, \hat{\theta})$ must be the news function for the collision at the speed of light ($\gamma = \infty$). If the two asymptotic expansions (6.1.1) and (6.1.2) both hold in the intermediate region in which $\gamma^{-1} \ll \hat{\theta} \ll 1$, then matching enables one to gain information about the angular dependence of $S_0(\hat{\tau}, \hat{\theta})$ near the axis $\hat{\theta} = 0$. Moreover, if $S_0(\hat{\tau}, \hat{\theta})$ is sufficiently regular then it will possess a convergent series of the form

$$S_0(\hat{\tau}, \hat{\theta}) = \sum_{n=0}^{\infty} a_{2n}(\hat{\tau}) \sin^{2n} \hat{\theta}, \qquad (6.1.3)$$

(eqn (5.8.31)), since it is symmetrical about $\hat{\theta} = \pi/2$ (in the centre-of-mass frame). Since $\hat{\theta} = \gamma^{-1}\psi$ in eqn (6.1.1), the $\hat{\theta}^{2n}$ part of eqn (6.1.3) will be found from the $(\gamma^{-1}\psi)^{2n} = \gamma^{-2n}\psi^{2n}$ part of eqn (6.1.1), and thus finding $Q_{2n}(\hat{\tau}, \psi)$ enables one to determine the coefficient $a_{2n}(\hat{\tau})$ of $\sin^{2n} \hat{\theta}$ in eqn (6.1.3). In this way $a_0(\hat{\tau})$ was found (given by eqn (5.6.28)), given by the limiting form of $Q_0(\hat{\theta}, \psi)$ as $\psi \to \infty$. If these matching ideas are correct, then perturbation methods can be used to determine the entire news function of the highly non-linear speed-of-light collision. But to calculate higher-order $Q_2n(\hat{\tau}, \psi)$ requires the solution of inhomogeneous flat-space wave equations with extremely complicated source terms, and it is not a technically feasible way of determining the non-isotropic part of $S_0(\hat{\tau}, \hat{\theta})$.

There is another way of calculating $S_0(\hat{\tau}, \hat{\theta})$ using perturbation methods, which deals with the collision at precisely the speed of light. The method was used by Curtis (1978a, b), following a suggestion by Penrose. Curtis examined the result of scattering a weak shock wave off a fully non-linear one, using twistor methods (Penrose and Rindler 1984, 1986). The perturbation parameter here is the ratio of the energies of the two waves. Curtis derived an expression for the radiation pattern at lowest order in perturbation theory, valid over the whole celestial 2-sphere except very near $\hat{\theta} = \pi$. However, he did not use this expression to derive quantitative answers. It was pointed out in D'Eath (1978) that results in perturbation theory concerning the radiation pattern in this weak shock–strong shock system translate, when one makes a Lorentz boost to a centre-of-mass frame, to a description of the gravitational radiation in the neighbourhood of $\hat{\theta} = 0$ in the fully non-linear space–time formed by the collision of two shocks with equal energy. One can then match the expressions one derives for the news function close to $\hat{\theta} = 0$ with eqn (6.1.3) in order to find the entire news function $S_0(\hat{\tau}.\hat{\theta})$, just as in the finite-$\gamma$ collision. The collision of plane (not just plane-fronted) or spherical shocks has been studied by Dray and 't Hooft (1985a, 1986). The gravitational shock wave (6.2.1) of a massless particle has also been studied by Dray and 't Hooft (1985b). Quantum scattering off such a shock wave was first studied by 't Hooft (1987); this leads to an estimate of quantum scattering cross-sections at Planckian energies and above—see Chapter 9.

In this chapter, we describe a calculation similar to that outlined above, based in part on the work of Chapman (1979). Starting with the speed-of-light collision of two shocks which each have energy μ, one can then make a large Lorentz boost away from the centre-of-mass frame, so that one shock becomes much more energetic than the other. The metric describing the scattering of the weak shock off the strong one possesses a perturbation expansion in powers of λ/ν, where λ and ν are the energies of the weak and strong shock respectively. The news function can be found to lowest order in λ/ν in the boosted frame, and then matched to obtain an expression for $a_0(\hat{\tau})$. Pleasingly, the resulting expression is identical to that derived in Chapter 5 (eqn (5.6.28)) for the isotropic part of the news function in the finite-γ collision on angular scales of $O(1)$. By solving the second-order field equations in the boosted frame, which take the form of inhomogeneous flat-space wave equations with complicated source terms, one can go on to derive an integral expression for the first non-isotropic term $a_2(\hat{\tau})$ in eqn (6.1.3). In Chapters 7 and 8, we will show how the computation of $a_2(\hat{\tau})$ can be simplified by reducing the perturbation field equations to equations in two independent variables, and will discuss the implications for the energy emitted in gravitational radiation and the nature of the radiative space–time.

One expects the apparent and event horizons in the finite-γ collision to be very similar to those in the speed-of-light encounter. A plausible scenario for each process is that at the collision there is a burst of radiation accompanied by the formation of a black hole, which settles down asymptotically to a Schwarzschild geometry. However, since $a_0(\hat{\tau})$ is non-zero, the speed-of-light news function does not vanish on the axis of symmetry $\hat{\theta} = 0, \pi$, which indicates that this space–time is not smooth on the axis at future null infinity (Bondi et al. 1967). The logarithmic singularity (Fig. 5.10) in the news function is another indication that, strictly speaking, this space–time is not asymptotically flat. Further, the speed-of-light space–time is certainly not asymptotically flat in the past, since the null shocks extend to infinity. These properties show that one should be careful about considering the speed-of-light collision as an isolated radiating system, and that it is better to think of it as the limit of the perfectly regular finite-γ collisions; and of the speed-of-light news function $S_0(\hat{\tau}, \hat{\theta})$ as describing the radiation in the finite-γ collisions on angular scales of $O(1)$. This will be our attitude here; that is, we are principally interested in the speed-of-light collision as a calculational tool which we use to find the higher-order moments in the off-axis news function in the finite-γ encounters. Whenever we loosely refer to future null infinity $\mathcal{I}^+$ for the speed-of-light collision, the argument can always be re-phrased in terms of limiting properties of the finite-γ space–times, which are expected to have a regular $\mathcal{I}^+$.

It has been conjectured (Smarr 1977a; D'Eath 1979b) that the radiation pattern in the high-speed collision is isotropic, apart from the detailed structure near the axis of symmetry. This conjecture was motivated by the zero-frequency limit (ZFL) calculation of Smarr, who found that the zero-frequency limit of the gravitational energy spectrum does have this angular distribution. If valid, it would mean that all the $a_{2n}(\hat{\tau})$ vanish for $n \geq 1$, and that the relative mass loss is 25%. It has been shown by Payne (1983b) that Smarr's ZFL calculation is in fact a linearized approximation valid only when the gravitational radiation is weak, so that it cannot be applied to the head-on collision of two black holes. Further discussion of these ideas is given in Section 6.6. As will be seen from the results presented in Chapter 8, $a_2(\hat{\tau})$ is certainly non-zero, and a complicated angular distribution is expected for the gravitational radiation.

In Section 6.2, we review the geometry of a single black hole moving at the speed of light—giving an impulsive plane-fronted wave. The axisymmetric collision of two such shock waves is then studied (Section 6.3). After the initial collision, the shocks lie on curved surfaces to the future of regions of Minkowski space–time. To the future of both curved shocks lies the curved interaction region of the space–time, which contains the gravitational radiation. As already described, in the approach used here a large Lorentz boost is applied such that one incoming wave has an energy ν much greater than the energy λ of the other wave. The scattering of the weak shock off the strong shock is regarded as a characteristic-initial value problem for the perturbed space–time, with characteristic initial data known just to the future of the strong shock, which in lowest approximation appears as a null hyperplane between two portions of Minkowski space–time. In Section 6.3, the characteristic initial data is described, and the region of validity of the perturbation theory is studied. As with many strong-field problems in general relativity, when subjected to perturbation methods, the perturbation series is expected to be non-uniform, corresponding to a singular perturbation problem.

In Section 6.4, the first-order field equations are studied and the first-order news function found. On boosting back to the centre-of-mass frame, one finds $a_0(\hat{\tau})$. The second-order field equations are solved in terms of integrals in Section 6.5, leading to an integral expression for the second-order news function and hence ultimately for $a_2(\hat{\tau})$. This expression is, however, intractable numerically in this particular form. Sections 6.3–6.5 draw heavily on the unpublished work of Chapman (1979). In Section 6.6, we derive a new mass-loss formula for the axisymmetric finite-γ collision, which shows that if the product of the collision is a single black hole, and the burst of radiation the form of which we are calculating is the only gravitational radiation present in the space–time, then the final mass of this hole is determined, with fractional error tending to zero as $\gamma \to \infty$, by only the first two coefficients, $a_0(\hat{\tau})$ and $a_2(\hat{\tau})$, in eqn (6.1.3). This provides further motivation for computing $a_2(\hat{\tau})$. Section 6.7 provides a summary of the chapter.

In Chapter 7, we then go on to show that in the speed-of-light collision, in the boosted frame of reference, the metric possesses a conformal symmetry at each order in perturbation theory. This, along with the obvious axisymmetry, enables us to reduce all the field equations from four dimensions to two. This reduction to two dimensions makes the numerical computation of the second-order metric practicable. We analyse the reduced field equations, show that they are hyperbolic partial differential equations, and find their Green functions. The numerical calculation of the second-order metric perturbation entails extensive computation, and is not entirely straightforward. We also show that there are $\log r/r$ terms present in the second-order metric. The existence of such terms in second-order perturbation theory in harmonic gauges is well known (Fock 1964). We eliminate then by an explicit transformation to a Bondi coordinate system (Bondi *et al.* 1962).

The numerical results are presented in Chapter 8. We show that our mass-loss formula makes the unphysical prediction that the final mass of the residual hole will be approximately twice the initial energy of the colliding waves, indicating either that the collision product is not a single black hole, or else that there is some other gravitational radiation present in the space–time. The latter possibility is more likely, and we indicate how this radiation might be generated. We also give an estimated upper bound on the final mass of the residual hole, using the conventional formula of Bondi *et al.* (1962).

6.2 The boosted metric and the axisymmetric collision

As seen in Section 5.3, at the speed of light with energy μ fixed, the Schwarzschild metric approaches a limit, which can be put in the form

$$ds^2 = du\, dv + dx^2 + dy^2 - 4\mu \log(x^2 + y^2)\delta(u)du^2, \qquad (6.2.1)$$

where $u = z + t$ and $v = z - t$. The axisymmetric space–time described by eqn (6.2.1) is flat, except on the null hypersurface $u = o$ where the curvature has a delta-function singularity. It can be shown that the energy–momentum tensor has the form $T^{ab} = \mu\delta(u)\delta(x)\delta(y)\ell^a\ell^b$, where the null vector ℓ^a is orthogonal to the hypersurface $u = 0$. Hence there is a massless point-like particle of energy μ travelling at the speed of light along the axis of symmetry in $u = 0$. The impulsive plane-fronted shock wave represented by the metric (6.2.1) is the gravitational field generated by the null particle. Owing to the infinite boost the black hole properties have been lost and there is no event horizon present.

Making the discontinuous coordinate transformation (eqn (5.3.3))

$$x = \hat{x} - 4\mu\hat{u}\theta(\hat{u})\frac{\hat{x}}{\hat{x}^2 + \hat{y}^2}\ ,$$

$$y = \hat{y} - 4\mu\hat{u}\theta(\hat{u})\frac{\hat{y}}{\hat{x}^2 + \hat{y}^2}\ ,$$

$$u = \hat{u}, \qquad (6.2.2)$$

$$v = \hat{v} + 4\mu\theta(\hat{u})\log(\hat{x}^2 + \hat{y}^2) - \frac{16\mu^2\hat{u}\theta(\hat{u})}{\hat{x}^2 + \hat{y}^2}\ ,$$

where $\theta(\hat{u})$ is the Heaviside step function, the metric (6.2.1) becomes

$$ds^2 = d\hat{u}\, d\hat{v} + [1 + 4\mu\hat{u}\theta(\hat{u})\hat{\rho}^{-2}]^2 d\hat{\rho}^2 + \hat{\rho}^2[1 - 4\mu\hat{u}\theta(\hat{u})\hat{\rho}^{-2}]^2 d\phi^2, (6.2.3)$$

where $\hat{\rho}^2 = \hat{x}^2 + \hat{y}^2$ and $\phi = \arctan(\hat{y}/\hat{x})$. The half-space $u > 0$ in eqn (6.2.1) has been mapped into the region $\hat{u} > 0, 4\mu \leq \hat{\rho}^2$ by (6.2.2). The 'boundary' $4\mu\hat{u} = \hat{\rho}^2$ is in fact the axis of symmetry $\rho = 0$. The metric (6.2.3) is continuous but there is the disadvantage that the metric form is no longer Minkowskian behind the shock (in $\hat{u} > 0$).

Apart from a discontinuity at $\hat{u} = 0$ the Christoffel symbols Γ^a_{bc} are well-behaved functions of $\hat{u}, \hat{v}, \hat{x}, \hat{y}$. From the geodesic equation

$$\frac{d^2 x^a}{ds^2} + \Gamma^a_{bc}\frac{dx^b}{ds}\frac{dx^c}{ds} = 0, \qquad (6.2.4)$$

we immediately see that in the hatted coordinate system a geodesic crossing the shock $\hat{u} = 0$ will be continuous and have a continuous tangent vector. Returning to the unhatted coordinates u, v, x, y and using eqn (6.2.2), we see that the value of v will change discontinuously by $8\mu \log \rho$ on crossing the shock, and that the geodesic will simultaneously be bent inwards.

The metric (6.2.1) transforms very simply under Lorentz boosts along the z-axis. Define u', v', x', y' by

$$x = x', \qquad y = y',$$

$$u = e^{\alpha}u', \qquad v = e^{-\alpha}v', \qquad (6.2.5)$$

where $e^\alpha = [(1 + \beta)/(1 - \beta)]^{1/2}$ (the primed system is therefore moving with speed β in the $+z$ direction with respect to the unprimed system). Written in terms of u', v', x', y', eqn(6.2.1) becomes

$$ds^2 = du'dv' + dx'^2 + dy'^2 - 4\mu e^\alpha \log(x'^2 + y'^2)\delta(u')du'^2. \tag{6.2.6}$$

Thus the effect of the Lorentz boost is simply to scale the energy by a factor e^α, which from the Doppler formula

$$E' = E\left[\frac{1 + \beta}{(1 - \beta^2)^{\frac{1}{2}}}\right] \tag{6.2.7}$$

is exactly how we expect the energy of a massless particle to transform. It is obvious that in the boosted frame the metric in its continuous form will be

$$ds^2 = d\hat{u}'d\hat{v}' + [1 + 4\mu e^\alpha \hat{u}'\theta(\hat{u}')\hat{\rho}'^{-2}]^2 d\hat{\rho}'^2$$
$$+ \hat{\rho}'^2[1 - 4\mu e^\alpha \hat{u}'\theta(\hat{u}')\hat{\rho}'^{-2}]^2 d\phi^2. \tag{6.2.8}$$

To obtain the metric (in its $C^{(0)}$ form) describing an identical wave travelling in the opposite direction, we merely replace $\hat{z}$ by $-\hat{z}$ in eqn (6.2.3); or equivalently $\hat{u}$ by $-\hat{v}$ and $\hat{v}$ by $-\hat{u}$. Thus the metric will be

$$ds^2 = d\hat{u}\,d\hat{v} + [1 - 4\mu\hat{v}\theta(-\hat{v})\hat{\rho}^{-2}]d\hat{\rho}^2$$
$$+ \hat{\rho}^2[1 + 4\mu\hat{v}\theta(-\hat{v})\hat{\rho}^{-2}]^2 d\phi^2. \tag{6.2.9}$$

We now consider the head-on collision of two such plane waves. Initially, we work in the centre-of-mass frame and denote the energy of each wave by μ. The region ahead of each shock is flat and so before the waves collide they propagate freely, each unaware of the other's presence. We shall choose the origin of coordinates so that the trajectories of the waves before the collision are given by $\hat{u} = 0$ (call this shock 1) and $-\hat{v} = 0$ (shock 2) respectively. In addition to the region ahead of the waves, there will be two further flat regions in the space–time; one behind shock 1 before shock 2 comes by, and its 'mirror image' behind shock 2 before shock 1 makes its presence felt. The metric in the union of these various regions will be simply the 'sum' of the individual metrics for each wave. In its C^0 form this is

$$ds^2 = d\hat{u}\,d\hat{v} + [1 + 4\mu\hat{u}\theta(\hat{u})\hat{\rho}^{-2}]^2 d\hat{\rho}^2$$
$$+ [-8\mu\hat{v}\theta(-\hat{v})\hat{\rho}^{-2} + 16\mu^2\hat{v}^2\theta(-\hat{v})\hat{\rho}^{-4}]d\hat{\rho}^2$$
$$+ \hat{\rho}^2[1 - 4\mu\hat{u}\theta(\hat{u})\hat{\rho}^{-2}]^2 d\phi^2$$
$$+ \hat{\rho}^2[8\mu\hat{v}\theta(-\hat{v})\hat{\rho}^{-2} + 16\mu^2\hat{v}^2\theta(-\hat{v})\hat{\rho}^{-4}]d\phi^2. \tag{6.2.10}$$

Before the collision the null generators of shock 2 (say) are the lines

$$\hat{u} = \Lambda, \qquad -\hat{v} = 0, \qquad \hat{x} = \xi, \qquad \hat{y} = \eta, \tag{6.2.11}$$

where the affine parameter Λ is negative. These null geodesics will intersect shock 1 at the spacelike collision surface $\hat{u} = -\hat{v} = 0$. Their continuation into $\hat{u} > 0$ will mark the future boundary of one of the flat space–time regions mentioned above; the metric having the form (6.2.10) to its past. Since geodesics are $C^{(1)}$ when viewed in hatted coordinates, the null generators of shock 2 will still have $d\hat{v}/d\Lambda = d\hat{x}/d\Lambda = d\hat{y}/d\Lambda = 0$ at $\hat{u} = 0^+$. Inspection of eqn (6.2.10) now shows that their continuation into is simply given by eqn (6.2.11) with $\Lambda > 0$. A similar result holds for the shock generators of shock 1. Thus the metric form (6.2.10) is valid in the regions denoted by I, II, and III in Fig. 5.1. Region IV to the future of the shocks will be curved, and the metric there can be found only by integrating Einstein's equations in some appropriate way.

The geometry of the shocks can be most easily understood if we view them from regions II and III when using flat coordinates. Therefore let us transform to coordinates defined by eqn (6.2.2). In this coordinate system the null generators of shock 2 are clearly parametrized by

$$u = \Lambda,$$
$$v = 4\mu\theta(\Lambda)\log(\xi^2 + \eta^2) - \frac{16\mu^2\Lambda\theta(\Lambda)}{\xi^2 + \eta^2},$$
$$x = \xi\left[1 - \frac{4\mu\Lambda\theta(\Lambda)}{\xi^2 + \eta^2}\right],$$
$$y = \eta\left[1 - \frac{4\mu\Lambda\theta(\Lambda)}{\xi^2 + \eta^2}\right]. \tag{6.2.12}$$

Thus viewed in these coordinates a geodesic generator of shock 1 crossing shock 2 suffers an instantaneous translation of magnitude $4\mu\log(\xi^2 + \eta^2)$ along the hypersurface $u = 0$ and is simultaneously bent inwards and backwards. Following this geodesic into $\Lambda > 0$ we see that after a finite affine distance ($\lambda = (\xi^2 + \eta^2)/4\mu$) it crosses the axis of symmetry $\rho = 0$. This means that shock 2 intersects itself at a caustic, given by

$$\rho = 0, \ v = 4\mu\log(4\mu u) - 4\mu. \tag{6.2.13}$$

The geometrical configuration is shown in Fig. 5.3. An analogous (inverted) picture can, of course, be drawn for shock 1. These results were first found by Penrose (1974). In the finite-γ collision the gross features of the geometry will be similar to those shown in Fig. 5.3; the main difference being that the shock and caustic structures will taper off at distances of $O(\gamma)$ from the axis of symmetry on the initial surface and so will not extend out to infinity.

In the finite-γ collision it was shown (Section 5.4) that the curvature within each shock is $O(\gamma)$ before it reaches the caustic. In the limit $\gamma \to \infty$ this $O(\gamma)$ structure becomes the delta-function profile of the impulsive speed-of-light shock. It was also shown in Section 5.5 that in the caustic region itself the curvature rises to $O(\gamma^{3/2})$. This means that in the speed-of-light collision there will be a curvature singularity at each caustic, this singularity being, in some sense, worse than just a delta function. This has in fact been shown by Corkill (1983), who by integrating Einstein's equations has, in addition to deriving the metric form (6.2.10), shown that the delta-function part of the Weyl tensor is singular at $4\mu\hat{u} = \hat{\rho}^2$ on $-\hat{v} = 0$ and at $-4\mu\hat{v} = \hat{\rho}^2$ on $\hat{u} = 0$ (the former being simply the caustic equation (6.2.13) written in terms of careted coordinates, and the latter its equivalent for shock 1). One might have expected a breakdown in predictability to the future of each singularity, the Cauchy horizon coinciding with the continuation of the shock generators beyond the caustic. However, at finite γ it was found in Section 5.5 that although the shock's profile is altered by its wavelike self-interaction in the caustic region, it continues to travel in a nearly null direction beyond the caustic into the analogue of the curved region IV. This indicates that the singularities in the speed-of-light collision should be quite harmless, the space–time having a natural extension through the supposed Cauchy horizons alluded to above. The fact that we can do perturbation theory exactly as if such a continuation does exist and obtain sensible results using it (results that agree exactly with the radiation calculation of Chapter 5 at finite γ) supports this view, and we shall assume it is the case.

In the finite-γ collision, it follows from Section 5.5 that the curvature of each shock has the form $\gamma^{3/2}/(1 + A\rho^2)^{3/2}$ just before the caustic, where $\gamma\rho$ measures the distance from the peak of the shock and A is some constant. As mentioned above, this shock profile is the finite-γ analogue of the delta function in the speed-of-light collision. Owing to the shock's self-interaction, its curvature profile just beyond the caustic is approximately $\gamma^{3/2}/q$ when q is large, where again γq measures the distance from the centre of the shock. This indicates that the impulsive structure of the shocks in the speed-of-light collision will be destroyed at the singularities, the curvature having a rather odd $1/s$ form to leading order near the continuation of the shocks beyond their respective caustics. That this $1/s$ form continues all the way out to $\mathcal{I}^+$ can be seen from Section 5.6, for $a_0(\hat{\tau})$ has a has a logarithmic singularity and so the curvature, which is given by the time-derivative of the news function, has a $1/(\hat{\tau} - \hat{\tau}_0)$ term at the retarded time where the null shock generator passes through the caustic on its way to infinity.

One consequence of the self-intersection of each shock is that any point on a null shock generator to the future of that shock's caustic will be timelike connected with points on the same shock to the past of the caustic (Penrose 1972) and will lie within the curved space–time region IV in Fig. 5.1. This means that the null surfaces on which the shocks lie cannot be used as characteristic initial surfaces (for solving Einstein's equations) beyond their caustics, since null data cannot be freely given there. This is consistent with eqn (6.2.10), which provides initial data only for $4\mu\hat{u} \leq \hat{\rho}^2$ on $-\hat{v} = 0$ and for $-4\mu\hat{v} \leq \hat{\rho}^2$ on $\hat{u} = 0$ (i.e. up to the caustics, but not beyond). This difficulty with caustics is a common feature of the characteristic initial value problem in general relativity (for a general discussion and possible resolution, see Friedrich and Stewart 1983), but we shall see that it does not affect our perturbation problem.

One should note that the portion of each null hypersurface on which data can be given (and which is given by eqn (6.2.10)) does not intersect future null infinity. This is easily seen for shock 2 (say), since eqn (6.2.13) shows that the caustic does not reach $\mathcal{J}^+(-v \to \infty$ as $u \to -\infty$ on the caustic), and therefore neither does the pre-collision shock (see Fig. 5.3). This would appear to create problems if one were to try to construct the whole space–time numerically by evolving the initial data off the null shock surfaces (see Stewart and Friedrich (1982) and Corkill and Stewart (1983) for a general discussion of the numerical construction of space–times from characteristic initial surfaces), since one cannot reach $\mathcal{I}^+$ by making a series of finite jumps off these surfaces. But perhaps one could discover all the essential features of the space–time without going too far out from the centre.

Penrose (1974) has found an apparent horizon (Hawking and Ellis 1973) (i.e. a compact two-dimensional spacelike surface, such that the expansion of the outgoing null generators is zero) on the union of the two null planes that describe the trajectories of the incoming shock waves in the speed-of-light collision. The apparent horizon is formed by the union of two flat discs the common boundary of which is a circle $\rho = 4\mu$ in the collision surface $\hat{u} = -\hat{v} = 0$, having area $32\pi\mu^2$. If cosmic censorship holds there will be an event horizon outside this apparent horizon (Hawking and Ellis 1973), and its area cannot decrease, so that if the space–time eventually settles down to the Schwarzschild geometry, as seems likely; the area of the final black hole must be greater than $32\pi\mu^2$; or in other words it must have a mass greater than $(1/\sqrt{2}) \cdot 2\mu$. (A figure $(1/2) \cdot 2\mu$ was wrongly quoted in D'Eath (1978).) Thus, the energy emitted in gravitational waves must be less than 29% of the incoming energy. We expect a horizon of a very similar nature to be formed in the finite-γ collision, and the lower bound on the mass of the final hole there should be the same to within a relative error tending to zero as $\gamma \to \infty$. It is interesting that the lower bound which cosmic censorship places on the ratio (mass of final black hole)/(initial energy) should be the same for the ultra-relativistic encounter and the collision of two black holes starting from rest at infinity (Hawking 1971), corresponding to an upper bound of 29% efficiency. However, the upper bound on radiation is much better filled in the high-speed than in the low-speed case.

The null particles lie within the apparent horizon and are therefore trapped, and presumably will eventually run into an unpleasant spacelike singularity, which should be hidden from infinity by the event horizon, assuming cosmic censorship holds. The inner portions of each shock (certainly for $\rho \leq 4\mu$) should fold up into the same singularity. It is likely that this singularity will be formed at the point at which the particles collide ($\hat{u} = -\hat{v} = \hat{\rho} = 0$). This would certainly be the most satisfactory outcome, since it would remove any ambiguities concerning the particle–particle interaction. It is reasonable as the limit of the finite-γ collision, in which the small, fast-moving, black holes should stop each other when they collide.

In order to calculate the form of the gravitational radiation in the speed-of-light space–time we make a large Lorentz boost and observe the collision in a frame of reference moving with velocity β (where $(1-\beta) \ll 1$) in the $+z$ direction with respect to the centre-of-mass frame. From the discussion leading up to eqn

(6.2.8), it is clear that in the boosted frame the pre-collision metric takes the form

$$
\begin{aligned}
ds^2 = {} & d\hat{u}' d\hat{v}' + [1 + 4\nu\hat{u}'\theta(\hat{u}')\hat{\rho}'^{-2}]^2 d\hat{\rho}'^2 \\
& + [-8\lambda\hat{v}'\theta(-\hat{v}')\hat{\rho}'^{-2} + 16\lambda^2\hat{v}'^2\theta(-\hat{v}')\hat{\rho}'^{-4}]d\hat{\rho}'^2 \\
& + \hat{\rho}'^2[1 - 4\nu\hat{u}'\theta(\hat{u}')\hat{\rho}'^{-2}]^2 d\phi^2 \\
& + \hat{\rho}'^2[8\lambda\hat{v}'\theta(-\hat{v}')\hat{\rho}'^{-2} + 16\lambda^2\hat{v}'^2\theta(-\hat{v}')\hat{\rho}'^{-4}]d\phi^2, \quad (6.2.14)
\end{aligned}
$$

where $\nu = \mu e^\alpha$, $\lambda = \mu e^{-\alpha}$, and $e^\alpha = [(1 + \beta)/(1 - \beta)]^{1/2}$. The collision appears to be between a weak shock wave of energy λ and a strong shock of energy ν, where $\lambda/\nu \ll 1$.

We now consider the evolution of the weak shock in the region behind the strong shock. For convenience we drop the prime on coordinates in the boosted frame, since we will be working exclusively in this frame of reference. The boundary data on $\hat{u} = 0$ for the characteristic initial value problem the solution of which describes the propagation of the weak shock in $\hat{u} > 0$ is

$$
g_{ab} = \eta_{ab} + \lambda\hat{h}^{(1)}_{ab} + \lambda^2\hat{h}^{(2)}_{ab}, \quad (6.2.15)
$$

where the only non-zero components of $\hat{h}^{(1)}_{ab}$ and $\hat{h}^{(2)}_{ab}$ are

$$
\begin{aligned}
\hat{h}^{(1)}_{\hat{\rho}\hat{\rho}} &= -\hat{\rho}^{-2}\hat{h}^{(1)}_{\phi\phi} = -8\lambda\hat{v}\theta(-\hat{v})\hat{\rho}^{-2}, \\
\hat{h}^{(2)}_{\hat{\rho}\hat{\rho}} &= \hat{\rho}^{-2}\hat{h}^{(2)}_{\phi\phi} = 16\lambda^2\hat{v}^2\theta(-\hat{v})\hat{\rho}^{-4}.
\end{aligned} \quad (6.2.16)
$$

On the initial surface $\hat{u} = 0$, in the region where $\hat{\nu}$ and $\hat{\rho}$ are of $O(\nu)$ (call this region R_ν), the contribution to eqn (6.2.15) from the weak shock appears as a small perturbation of $O(\lambda/\nu)$ to the 'background' metric of the strong shock. Therefore, in that region of space–time to the future of the initial surface which can be influenced only by R_ν, the metric should possess a perturbation expansion in powers of λ/ν. We shall demonstrate this explicitly below.

The geometry is easier to visualize if we transform to a coordinate system (u, v, x, y) in which the background metric of the strong shock in $\hat{u} > 0$ is manifestly Minkowskian. An appropriate coordinate transformation is given by eqn (6.2.2) with μ replaced by ν:

$$
\begin{aligned}
x &= \hat{x}[1 - 4\nu\hat{u}\theta(\hat{u})\hat{\rho}^{-2}], \\
y &= \hat{y}(1 - 4\nu\hat{u}\theta(\hat{u})\hat{\rho}^{-2}), \\
u &= \hat{u}, \\
v &= \hat{v} + 8\nu\theta(\hat{u})\log\hat{\rho} - 16\nu^2\hat{u}\theta(\hat{u})\hat{\rho}^{-2}.
\end{aligned} \quad (6.2.17)
$$

The metric transforms as

$$
g_{ab} = \left[\frac{\partial\hat{x}^c}{\partial x^a}\right]\left[\frac{\partial\hat{x}^d}{\partial x^b}\right]\hat{g}_{cd}. \quad (6.2.18)
$$

On $u = 0^+$ one easily finds

$$\frac{\partial \hat{x}^a}{\partial x^b} = \begin{pmatrix} 1 & 0 & 0 & 0 \\ -16\nu^2\rho^{-2} & 1 & -8\nu x\rho^{-2} & -8\nu y\rho^{-2} \\ 4\nu x\rho^{-2} & 0 & 1 & 0 \\ 4\nu y\rho^{-2} & 0 & 0 & 1 \end{pmatrix}, \tag{6.2.19}$$

where $\hat{x}^a = (\hat{u}, \hat{v}, \hat{x}, \hat{y})$ and $(x^b = u, v, x, y)$. If we rescale the coordinates by a factor ν (to exhibit more clearly the perturbative behaviour mentioned above), letting

$$X^a_{\text{old}} = \nu X^a_{\text{new}} \tag{6.2.20}$$

and redefine u and v through $u = z + t/\sqrt{2}$ and $v = z - t/\sqrt{2}$, then the form of the metric in the unhatted coordinate system on $u = 0^+$ is

$$g_{ab} = \nu^2 \left[\eta_{ab} + \left(\frac{\lambda}{\nu}\right) h^{(1)}_{ab} + \left(\frac{\lambda}{\nu}\right)^2 h^{(2)}_{ab} \right]. \tag{6.2.21}$$

Here

$$\begin{aligned} h(1)_{uu} &= A, & h^{(1)}_{vv} &= 0, & h^{(1)}_{xx} &= (y^2 - x^2)\rho^{-2}E, \\ h^{(1)}_{yy} &= (x^2 - y^2)\rho^{-2}E, & h^{(1)}_{uv} &= 0, & h^{(1)}_{ux} &= 0, \\ h^{(1)}_{xy} &= -2xy\rho^{-2}E, & h^{(1)}_{ux} &= x\rho^{-1}B, & & \\ h^{(1)}_{uy} &= 0, & h^{(1)}_{uy} &= y\rho^{-1}B, & & \end{aligned} \tag{6.2.22}$$

where

$$\begin{aligned} A &= 32\rho^{-4}f(8\log(\nu\rho) - \sqrt{2}v), \\ B &= 4\sqrt{2}\rho^{-3}f(8\log(\nu\rho) - \sqrt{2}v), \\ E &= -\rho^{-2}f(8\log(\nu\rho) - \sqrt{2}v), \end{aligned} \tag{6.2.23}$$

and $f(x) = 8x\theta(x)$. Also

$$\begin{aligned} h^{(2)}_{uu} &= H^{(2)}, & h^{(2)}_{vv} &= 0, & h^{(2)}_{xx} &= D^{(2)}, & h^{(2)}_{yy} &= D^{(2)}, \\ h^{(2)}_{uv} &= 0, & h^{(2)}_{ux} &= 0, & h^{(2)}_{xy} &= 0, & & \\ h^{(2)}_{ux} &= x\rho^{-1}I^{(2)}, & h^{(2)}_{vy} &= 0, & h^{(2)}_{uy} &= y\rho^{-1}I^{(2)}, & & \end{aligned} \tag{6.2.24}$$

where

$$\begin{aligned} H^{(2)} &= 32\rho^{-6}g(8\log(\nu\rho) - \sqrt{2}v), \\ I^{(2)} &= 4\sqrt{2}\rho^{-5}g(8\log(\nu\rho) - \sqrt{2}v), \\ D^{(2)} &= \rho^{-4}g(8\log(\nu\rho) - \sqrt{2}v), \end{aligned} \tag{6.2.25}$$

and $g(x) = 16x^2\theta(x)$. The flat background metric η_{ab} is given by

$$ds^2 \eta_{ab}dx^a dx^b = 2du\, dv + dx^2 + dy^2. \tag{6.2.26}$$

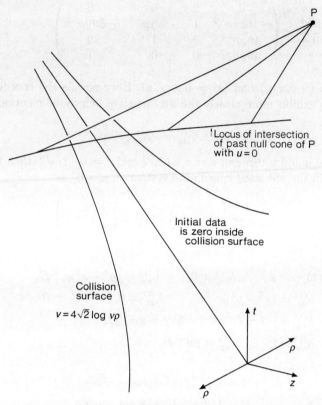

FIG. 6.1. This shows the locus of intersection of the past null cone of a point P with the initial surface $u = 0$, on which the incoming shock 1 lies. The initial data for the perturbation problem are zero inside the collision surface $v = 4\sqrt{2}\log(\nu\rho)$.

To the future of $u = 0$ the metric will possess the perturbation expansion

$$g_{ab} \sim \nu^2 \left[\eta_{ab} + \sum_{i=1}^{\infty} \left(\frac{\lambda}{\nu} \right)^i h_{ab}^{(i)} \right]. \tag{6.2.27}$$

In principle we can find $h_{ab}^{(1)}$, $h_{ab}^{(2)}$, ... by successively solving the linearized field equations at first, second, ... order in λ/ν.

We now comment on the region of validity of this expansion. In Fig. 6.1 we show the locus of intersection of the past null cone of a point $P \equiv (u, v, \rho, \phi)$ with the initial surface $(0, v', \rho', \phi')$. It is a paraboloid, given by

$$v' = v + \frac{(\rho')^2 - 2\rho\rho'\cos(\phi - \phi') + \rho^2}{2u}. \tag{6.2.28}$$

We shall see in the next section that, as might be expected, the gravitational radiation in the space–time is concentrated in the region surrounding the continuation of the weak shock generators beyond the caustic. From eqn (6.2.12), it is clear that these generators are parameterized by

$$u = \sqrt{2}\Lambda,$$

$$v = 4\sqrt{2}\log(\nu\sqrt{\xi^2 + \eta^2}) - \frac{16\sqrt{2}}{\xi^2 + \eta^2}\Lambda,$$

$$x = \xi\left[1 - \frac{8\Lambda}{\xi^2 + \eta^2}\right],$$

$$y = \eta\left[1 - \frac{8\Lambda}{\xi^2 + \eta^2}\right].$$

(6.2.29)

Therefore let u, v, and p be given by

$$u = \sqrt{2}\Lambda,$$

$$v = 4\sqrt{2}\log(\nu\rho_0) - \frac{16\sqrt{2}}{\rho_0^2}\Lambda - \frac{K}{\sqrt{2}},$$

$$\rho = \left[\frac{8\Lambda}{\rho_0^2} - 1\right]\rho_0,$$

(6.2.30)

where future null infinity is reached by letting $\Lambda \to \infty$. (As K increases the paraboloid sweeps up the initial surface in the $-v$ direction.) Then

$$\tan\theta \equiv \frac{\rho}{z} = \frac{2(4/\rho_0)}{1 - (4/\rho_0)^2} + O(\Lambda^{-1}),$$

(6.2.31)

and so $\tan\theta/2 = 4\rho_o + O(\Lambda^{-1})$, from the standard trigonometric formula. For the point (6.2.30), eqn (6.2.28) reduces to

$$v' = 4\sqrt{2}\left[\log(\nu\rho_0) - 1 - \frac{\rho'}{\rho_0}\cos(\phi - \phi')\right] - \frac{K}{\sqrt{2}}.$$

(6.2.32)

Using eqns (6.2.22)–(6.2.25), we find that, at this locus of intersection,

$$h_{ab}^{(1)} \propto S\theta(S)(\rho')^{-m}, \qquad m = 2, 3, \text{ or } 4,$$

$$h_{ab}^{(2)} \propto S^2\theta(S)(\rho')^{-n}, \qquad n = 4, 5, \text{ or } 6,$$

(6.2.33)

where S is defined by

$$S = 8\log\left[\frac{\rho'}{\rho_0}\right] + 8 + 8\left[\frac{\rho'}{\rho_0}\right]\cos(\phi - \phi') + K.$$

(6.2.34)

Because of the form of the initial data (6.2.22)–(6.2.25), the metric perturbations become small at early retarded times or, equivalently, as $K \to -\infty$. For simplicity then, let us consider the worst case in which K is large and positive. Differentiating eqn (6.2.33), we find that $|h_{ab}^{(1)}|$ and $|h_{ab}^{(2)}|$ take their respective maximum values when

$$
\begin{aligned}
\frac{8}{\rho'} + \frac{8}{\rho_0}\cos(\phi - \phi') - \frac{mS}{\rho'} &= 0, \\
\frac{16}{\rho'} + \frac{16}{\rho_0}\cos(\phi - \phi') - \frac{nS}{\rho'} &= 0.
\end{aligned}
\tag{6.2.35}
$$

At these points

$$
h_{ab}^{(1)} \propto \left[\tan\frac{\theta}{2}e^{K/8}\right]^m, \qquad h_{ab}^{(2)} \propto \left[\tan\frac{\theta}{2}e^{K/8}\right]^n.
\tag{6.2.36}
$$

Roughly speaking, the metric perturbations decrease as we move away from the initial surface along the past null cone of P. We see that the perturbation series (6.2.27) will converge everywhere in this null cone only if $\lambda/\nu[\tan(\theta/2)\exp(K/8)]^4$ is small. (If desired, this property can be checked more carefully using the integral representation of the metric perturbations, as employed in Sections 6.3 and 6.4.) But since λ/ν can be made arbitrarily small, our perturbation theory is valid in the region of interest surrounding $K = 0$.

Note that these rough estimates indicate the possibility that the metric perturbations may grow exponentially with time at late retarded times. Such non-uniform behaviour is typical of singular perturbation theory, encountered in a variety of strong-field problems in general relativity, as seen, for example, in Chapters 3 and 4. In fact it seems that $h_{ab}^{(1)}$ and $h_{ab}^{(2)}$ are well-behaved at late times, at least as far as is borne out by the behaviour of the corresponding parts $a_0(\hat{\tau})$ and $a_2(\hat{\tau})$ of the news function, computed in Sections 6.3 and 6.4, and in the following chapters 7 and 8. However, it is very likely that the higher-order perturbations $h_{ab}^{(3)}, \ldots$ do grow exponentially at late times, for reasons to be discussed further in Chapter 8.

It is easy to show that the caustic in the initial surface $u = 0$ is at

$$
8\log(\nu\rho) - \sqrt{2}v = \frac{\rho^2}{4(\lambda/\nu)}.
\tag{6.2.37}
$$

This lies well beyond the region in which perturbation theory is valid, and so can be ignored in our present calculation.

6.3   The first-order calculation

The field equations for $h_{ab}^{(1)}$ in $u \geq 0$ are

$$
-\bar{h}^{(1)}{}_{ab,c}{}^{c} - \eta_{ab}\bar{h}^{(1)}{}_{cd,}{}^{cd} + \bar{h}^{(1)}{}_{ac,}{}^{c}{}_{b} + \bar{h}^{(1)}{}_{bc,}{}^{c}{}_{a} = 0,
\tag{6.3.1}
$$

where $\bar{h}_{ab}^{(1)} = h_{ab}^{(1)} - \frac{1}{2}\eta_{ab}h^{(1)}{}_{c}{}^{c}$, $\partial/\partial x^c$ is denoted by $_{,c}$, and indices are raised and lowered using η_{ab}. We solve eqn (6.3.1) in the usual way, by making a gauge transformation $x^a = x^{Na} + (\lambda/\nu)\xi^a$, such that the new first-order perturbation $h_{ab}^{N(1)} = h_{ab}^{(1)} + 2\xi_{(a,b)}$ satisfies the de Donder gauge condition

$$\bar{h}^{N(1)}{}_{ab,}{}^b = 0 \qquad (6.3.2)$$

in $u \geq 0$. Then eqn (6.3.1) reduces to

$$\Box \bar{h}_{ab}^{N(1)} \equiv \left[2\frac{\partial^2}{\partial u \, \partial v} + \frac{\partial^2}{\partial x^2} + \frac{\partial^2}{\partial y^2} \right] \bar{h}_{ab}^{N(1)} = 0. \qquad (6.3.3)$$

We now prove that for eqn (6.3.2) to hold, it is sufficient that $h_{ab}^{N(1)}$ satisfy

$$\bar{h}^{N(1)}{}_{ab,}{}^b{}_v \mid_{u=0} = 0. \qquad (6.3.4)$$

The general solution to the wave equation $\Box F = 0$ in $u \geq 0$, with F given on $u = 0$ (subject to the restriction $F \to 0$ sufficiently rapidly as $v \to \infty$), is (Penrose 1980; Penrose and Rindler 1984)

$$F(u,v,x,y) = \frac{-1}{2\pi u} \int_0^\infty \int_0^{2\pi} \rho' d\rho' d\phi' \frac{\partial}{\partial v'} F(0,v',x',y'), \qquad (6.3.5)$$

where $x = \rho \cos\phi$, $y = \rho \sin\phi$, $x' = \rho' \cos\phi'$, $y' = \rho' \sin\phi'$, and v' is determined as a function of x' and y' through

$$v' = v + [\rho^2 - 2\rho\rho' \cos(\phi - \phi') + \rho'^2]/2u. \qquad (6.3.6)$$

The functional form of eqn (6.3.6) is such that $(0,v',x',y')$ lies in the past null cone of (u,v,x,y). It can be seen from eqn (6.3.5) that if $\partial F/\partial v \mid_{u=0} = 0$, then $F = 0$ everywhere in $u \geq 0$. Now if $\Box \bar{h}_{ab}^{N(1)} = 0$, then also $\Box \bar{h}^{N(1)}{}_{ab,}{}^b = 0$. Hence, $\bar{h}^{N(1)}{}_{ab,}{}^b{}_v \mid_{u=0} = 0$ will ensure that eqn (6.3.2) holds (our initial data (6.2.22) and (6.2.23) do go to zero sufficiently rapidly as $v \to \infty$). Thus, any solution to the combined system $\Box \bar{h}_{ab}^{N(1)} = 0, \bar{h}^{N(1)}{}_{ab,}{}^b{}_v \mid_{u=0} = 0$ will also be a solution to the field equations (6.3.1).

On rewriting eqn (6.3.4) in terms of ξ_a and $h_{ab}^{(1)}$, we find that ξ_a must satisfy

$$-\Box \xi_{[a,v]} - \frac{1}{2}\eta_{av}\Box \xi_{c,}{}^c = \bar{h}_{ax,xv}^{(1)} + \bar{h}_{ay,yv}^{(1)} + \bar{h}_{au,vv}^{(1)}$$

$$- \frac{1}{2}(\bar{h}_{av,xx}^{(1)} + \bar{h}_{av,yy}^{(1)}), \qquad (6.3.7)$$

where the $\bar{h}_{av,uv}^{(1)}$ term has been eliminated using eqn (6.3.3). For $a = v$ both sides of eqn (6.3.7) vanish identically; the right-hand side by virtue of eqn (6.3.7) with $a = b = v$. Substituting the initial data (6.2.22) and (6.2.23), we find that the other three equations (6.3.7) for ξ_a on $u = 0$ are

$$\Box \xi_{[x,v]} = 0, \qquad \Box \xi_{[y,v]} = 0,$$
$$-\frac{1}{2}\Box(\xi_{x,x} + \xi_{y,y} + 2\xi_{u,v}) = 128\rho^{-4}\theta(8\log(\nu\rho) - \sqrt{2}v). \qquad (6.3.8)$$

We look for a solution which has a power series expansion near $u = 0$:

$$\xi_a(u,v,x,y) = \xi_a^{(0)}(v,x,y) + u\xi_a^{(1)}(v,x,y) + \cdots. \qquad (6.3.9)$$

One possibility is

$$\xi_a^{(0)} = 0, \qquad \xi_v^{(1)} = \xi_x^{(1)} = \xi_y^{(1)} = 0,$$
$$\xi_u^{(1)} = -16\rho^{-4}(8\log(\nu\rho) - \sqrt{2}v)^2\theta(8\log(\nu\rho) - \sqrt{2}v). \qquad (6.3.10)$$

The gauge transformation (6.3.10) retains the form (6.2.22) for the metric coefficients on $u = 0$, but (6.2.23) becomes

$$A = 32\rho^{-4}f(8\log(\nu\rho) - \sqrt{2}v)$$
$$- 4\rho^{-4}(8\log(\nu\rho) - \sqrt{2}v)f(8\log(\nu\rho) - \sqrt{2}v),$$
$$B = 4\sqrt{2}\rho^{-3}f(8\log(\nu\rho) - \sqrt{2}v), \qquad (6.3.11)$$
$$E = -\rho^{-2}f(8\log(\nu\rho) - \sqrt{2}v).$$

It can be verified that eqn (6.3.4) is satisfied in this gauge.

The metric perturbation $h_{ab}^{N(1)}$ is traceless on the initial surface:

$$h^{N(1)} \mid_{u=0} \equiv \eta^{ab}h_{ab}^{N(1)} \mid_{u=0} = 0, \qquad (6.3.12)$$

and it is therefore traceless everywhere in $u \geq 0$ by virtue of eqn (6.3.3). Thus eqn (6.3.3) reduces to

$$\Box h_{ab}^{N(1)} = 0. \qquad (6.3.13)$$

It is clear that (6.3.13) preserves the metric form (6.2.22) in $u > 0$.

Following Section 5.6, and Bondi et al. (1962), the gravitational radiation in an axisymmetric, reflection-symmetric space–time is described by a single real function $c_0(\tau, \theta)$ of retarded time and polar angle, known as the news function. The news function is an invariant quantity, but for convenience we use the definition

$$c_0(\tau,\theta) = -\frac{1}{2}\lim_{r\to\infty}\left[r^{-1}(\sin\theta)^{-2}\frac{\partial g_{\phi\phi}}{\partial\tau}\right]. \qquad (6.3.14)$$

(Strictly speaking, eqn (6.3.14) is valid only when the metric is written in Bondi coordinates in the radiation zone. However, it gives the correct answer here, even for our Minkowskian coordinates. We defer an explicit verification of this technical point to the following Chapter 7.) Here (τ, r, θ, ϕ) are defined in terms of (u, v, x, y) through

$$r^2 = \nu^2(x^2 + y^2 + z^2),$$
$$\tau = \nu t - r, \quad (6.3.15)$$
$$\theta = \arctan\left[\frac{\rho}{z}\right], \qquad \phi = \arctan\left[\frac{y}{x}\right].$$

Using eqn (6.2.22), we find that the first-order news function in the boosted frame is

$$c_0^{(1)}(\tau, \theta) = -\frac{1}{2}\left[\frac{\lambda}{\nu}\right] \lim_{r \to \infty} \left[r\frac{\partial E}{\partial \tau}\right]. \quad (6.3.16)$$

Since $\Box h_{xx}^{(1)} = \Box h_{yy}^{(1)} = 0$, the metric function $E(u, v, \rho)$ satisfies

$$\Box(e^{2i\phi} E) = 0,$$

where

$$e^{2i\phi} E \mid_{u=0} = 8\rho^{-2}e^{2i\phi}[8\log(\nu\rho) - \sqrt{2}\nu]\theta(8\log(\nu\rho) - \sqrt{2}\nu). \quad (6.3.17)$$

Using the integral representation (6.3.5), we find

$$e^{2i\phi} E = \frac{-4\sqrt{2}}{\pi u} \int_0^\infty \int_0^{2\pi} \frac{d\rho'}{\rho'}d\phi' e^{2i\phi'} \times$$
$$\times \theta\left[8\log(\nu\rho') - \sqrt{2}\left(v + \frac{\rho^2 - 2\rho\rho'\cos(\phi - \phi') + \rho'^2}{2u}\right)\right]. \quad (6.3.18)$$

Eliminating u, v, ρ using eqn (6.3.15), we find

$$E(\tau, r, \theta) = \frac{-8\nu}{\pi[\tau + 2r\cos^2(\theta/2)]} \int_0^\infty \int_0^{2\pi} \frac{d\rho'}{\rho'}d\omega \cos(2\omega) \times$$
$$\times \theta\left[8\log(\nu\rho') + \frac{\tau}{\nu}\sec^2(\theta/2) + 2\tan(\theta/2)\rho'\cos\omega\right]$$
$$+ O(1/r). \quad (6.3.19)$$

Defining $s = 2\rho'\tan(\theta/2)$ and letting $r \to \infty$, the first-order news function is found to be

$$c_0^{(1)}(\tau, \theta) = \left[\frac{\lambda}{\nu}\right]\frac{2}{\pi}\sec^4(\theta/2) \int_0^\infty \int_0^{2\pi} \frac{ds}{s}d\omega \cos(2\omega) \times$$
$$\times \delta\left[8\log s + s\cos\omega + \frac{\tau}{\nu}\sec^2(\theta/2) - 8\log\left(\frac{2\tan(\theta/2)}{\nu}\right)\right] \quad (6.3.20)$$

Performing the angular integration, eqn (6.3.20) reduces to

$$c_0^{(1)}(\tau, \theta) = \left[\frac{\lambda}{\nu}\right]\sec^4(\theta/2)H_0(T), \quad (6.3.21)$$

where $T = (\tau/\nu)\sec^2(\theta/2) - 8\log[2\tan(\theta/2)/\nu]$ and

$$H_0(T) = \frac{4}{\pi} \int_D \frac{ds}{s^2} \left[2\left(\frac{T + 8\log s}{s}\right)^2 - 1 \right] \times$$

$$\times \left[1 - \left(\frac{T + 8\log s}{s}\right)^2 \right]^{-\frac{1}{2}} \tag{6.3.22}$$

Here D is the domain in which

$$-s \le (8\log s + T) \le s. \tag{6.3.23}$$

The reason for the appearance of the $8\log[2\tan(\theta/2)/\nu]$ term in the expression for T can be traced to the logarithmic delay across the strong shock. In fact, if we look at the null geodesic generators of the weak shock in $u \ge 0$, as given by eqn (6.2.29), and let the affine parameter $\Lambda \to \infty$, then we find that at their intersection with $\mathcal{I}^+, T = 8 - 8\log 8$. The news function has a logarithmic singularity at this value of T, and is significantly non-zero only in the surrounding region, dying away asymptotically on either side of the weak shock.

We now transform back to the centre-of-mass frame to see what eqn (6.3.21) tells us about the $\sin^2\hat{\theta}$ series (6.1.3) for the news function. Let $\hat{\tau}, \hat{r}, \hat{\theta}$ denote the centre-of-mass coordinates. Then (Bondi *et al.* 1962)

$$\tau = \hat{\tau}/K, \qquad r = K\hat{r},$$
$$\tan(\theta/2) = e^a \tan(\hat{\theta}/2), \tag{6.3.24}$$

while the news function transforms as

$$\hat{c}_0 = c_0/K^2. \tag{6.3.25}$$

Here $K(\hat{\theta}) = \cosh\alpha - \sinh\alpha\cos\hat{\theta}$. Using eqns (6.3.24) and (6.3.25), we find that (6.3.21) transforms to

$$\hat{c}_0(\hat{\tau}, \hat{\theta}) = \frac{2}{1 + \cos\hat{\theta}} H_0(T), \tag{6.3.26}$$

where now

$$T = [2\hat{\tau}/\mu(1 + \cos\hat{\theta})] - 8\log\{2[\tan(\hat{\theta}/2)]/\mu\}.$$

By making a supertranslation,

$$\hat{\tau} = \hat{\tau}' + 4\log[2\tan(\hat{\theta}/2)/\mu]/\mu(1 + \cos\hat{\theta}),$$

which leaves the news function invariant (Bondi *et al.* 1962)—that is, by adding a function of angles only to the retarded time coordinate—we may eliminate the logarithmic term from T. In terms of $\hat{\tau}'$, one has $T = 2\hat{\tau}'/\mu(1 + \cos\hat{\theta})$, but for convenience we shall omit the prime on $\hat{\tau}'$ and merely write $T = 2\hat{\tau}/\mu(1+\cos\hat{\theta})$ in what follows.

The expression (6.3.21) for the first-order news function in the boosted frame is valid only for values of θ not too close to π (i.e. for $(\pi - \theta) \sim 1$; but not, for example, for $(\pi - \theta) = O(\lambda/\nu)$). It is easy to see this from the parametric representation (6.2.29) for the weak shock generators; for if $(\pi - \theta) \ll 1$ out near $\mathcal{I}^+$, then $(\xi^2 + \eta^2) \ll 1$ on $u = 0$, and the initial data (6.2.22) will then not be a small perturbation. This implies that, in the centre-of-mass frame, eqn (6.3.26) is valid only in the neighbourhood of $\hat{\theta} = 0$ (and by symmetry near $\hat{\theta} = \pi$). Thus, the right-hand side of eqn (6.3.26) should really be written as

$$H_0\left[\frac{\hat{\tau}}{\mu}\right] + \sin^2\hat{\theta}\left[\frac{1}{4}H_0\left(\frac{\hat{\tau}}{\mu}\right) + \frac{1}{4}\left(\frac{\hat{\tau}}{\mu}\right)H_0'\left(\frac{\hat{\tau}}{\mu}\right)\right] + \cdots. \tag{6.3.27}$$

Hence all that eqn (6.3.26) tells us is that the isotropic term $a_0(\hat{\tau})$ in eqn (6.1.3) is $H_0(\hat{\tau}/\mu)$. We cannot say at this stage what the $\sin^2\hat{\theta}, \sin^4\hat{\theta}, \ldots$ terms are, since there will be contributions to these from the second, third,... order news functions in the boosted frame.

The expression $H_0(\hat{\tau}/\mu)$ for $a_0(\hat{\tau})$ agrees with that derived previously in eqn (5.6.28) as the isotropic part of the news function in the finite-γ collisions on angular scales of order 1. This is pleasing, for it indicates that the matching ideas of Chapter 5 outlined in Section 6.1 are working. The form of $a_0(\hat{\tau})$ is shown in Fig. 5.10. The singularity is logarithmic. The magnitude of $a_0(\hat{\tau})$ is such that if the radiation were isotropic (that is, if $a_{2n}(\hat{\tau}) = 0$, $\forall n \geq 1$) then the total energy carried off by gravitational waves would be 25% of the initial energy 2μ.

6.4   The second-order calculation

We now show that, by finding the news function to second order in λ/ν in the boosted frame, we may determine the coefficient $a_2(\hat{\tau})$ of $\sin^2\hat{\theta}$ in eqn (6.1.3), and thereby get some idea of the angular dependence of the total news function.

Near the axis, the series expansion for the news function in the centre-of-mass frame is

$$\hat{c}_0(\hat{\tau}, \hat{\theta}) = a_0\left[\frac{\hat{\tau}}{\mu}\right] + a_2\left[\frac{\hat{\tau}}{\mu}\right]\sin^2\hat{\theta} + O(\sin^4\hat{\theta}), \tag{6.4.1}$$

in which we have so far found the first term $a_0(\hat{\tau}/\mu) = H_0(\hat{\tau}/\mu)$. (For convenience we write $\hat{\tau}/\mu$ instead of $\hat{\tau}$; and in fact $\hat{\tau}$ and μ will appear in this combination at every order. That is to say, in this section we make the replacement $a_{2n}(\hat{\tau}) \rightarrow a_{2n}(\hat{\tau}/\mu)$.) In the boosted frame, eqn (6.4.1) becomes

$$c_0(\tau, \theta) = K^{-2}\left[a_0\left(\frac{\tau}{\mu K}\right) + \frac{\sin^2\theta}{K^2}a_2\left(\frac{\tau}{\mu K}\right)\right.$$
$$\left. + O\left(\frac{\sin^4\theta}{K^4}\right)\right], \tag{6.4.2}$$

where

$$K(\theta) = \cosh\alpha + \sinh\alpha\cos\theta$$
$$= e^{\alpha}\cos^2(\theta/2)[1 + e^{-2\alpha}\tan^2(\theta/2)].\tag{6.4.3}$$

Combinging eqns (6.4.3) and (6.4.2), we find

$$c_0(\tau,\theta) = e^{-2\alpha}\sec^4(\theta/2)H_0[(\tau/\nu)\sec^2(\theta/2)]$$
$$+ e^{-4\alpha}\{4\tan^2(\theta/2)\sec^4(\theta/2)a_2[(\tau/\nu)\sec^2(\theta/2)]$$
$$- 2\tan^2(\theta/2)\sec^4(\theta/2)H_0[(\tau/\nu)\sec^2(\theta/2)]$$
$$- (\tau/\nu)\sec^6(\theta/2)\tan^2(\theta/2)H_0'[(\tau/\nu)\sec^2(\theta/2)]\} + \cdots.\tag{6.4.4}$$

Thus the $e^{-4\alpha}$ term depends solely on $H_0(\hat{\tau}/\mu)$ and $a_2(\hat{\tau}/\mu)$. Conversely, by finding the news function to second order in the boosted frame we may determine the first two coefficients in eqn (6.4.1).

To find the field equations satisfied by $h_{ab}^{(2)}$, write eqn (6.2.27) as

$$g_{ab} \sim \nu^2(\eta_{ab} + e^{-2\alpha}h_{ab}^{(1)} + e^{-4\alpha}h_{ab}^{(2)} + \cdots).\tag{6.4.5}$$

Then

$$g^{ab} \sim \frac{1}{\nu^2}[\eta^{ab} - e^{-2\alpha}h^{(1)ab}$$
$$- e^{-4\alpha}(h^{(2)ab} - h^{(1)ad}h^{(1)}{}_d{}^b) + \cdots]\tag{6.4.6}$$

(where indices on the right-hand side are raised and lowered with η_{ab}). Using the properties $h^{(1)}{}_c{}^c = h^{(1)}{}_{ab,}{}^b = 0$, it is straightforward to show that the $e^{-4\alpha}$ term in the Ricci tensor is

$$R_{ab}^{(2)} = \frac{1}{2}(h^{(2)c}{}_{a,bc} + {}^{(2)c}{}_{b,ac} - h^{(2)}{}_{ab,}{}^c{}_c - h^{(2)}{}_c{}^c{}_{,ab})$$
$$- \frac{1}{2}h^{(1)cd}(h_{ca,bd}^{(1)} + h_{cb,ad}^{(1)} - h_{ab,cd}^{(1)} - h_{cd,ab}^{(1)})$$
$$+ \frac{1}{4}h^{(1)cd}{}_{,a}h_{cd,b}^{(1)} - \frac{1}{2}h^{(1)c}{}_{a,d}(h^{(1)d}{}_{b,c} - h^{(1)}{}_{cb,}{}^d).\tag{6.4.7}$$

Rearranging the terms slightly, we may write the second-order field equation $R_{ab}^{(2)} = 0$ as

$$\Box h_{ab}^{(2)} - \bar{h}^{(2)}{}_{ac,}{}^c{}_b - \bar{h}^{(2)}{}_{bc,}{}^c{}_a = T_{ab}(h_{cd}^{(1)}),\tag{6.4.8}$$

where $\bar{h}_{ab}^{(2)} = h_{ab}^{(2)} - \frac{1}{2}\eta_{ab}h^{(2)}{}_c{}^c$ and T_{ab} is equal to twice the sum of all the $h_{ab}^{(1)}$ terms on the right-hand side of eqn (6.4.7). Equations (6.4.8) are similar in form to those of linearized theory with a source term, the first-order metric perturbations $h_{ab}^{(1)}$ giving rise to an effective energy–momentum tensor T_{ab}. It may be easily verified that T_{ab} satisfies the conservation equation

$$\bar{T}_{ab,}{}^{b} = 0. \tag{6.4.9}$$

As in the first-order case, we look for a gauge transformation $x^a = x^{Na} + e^{-4\alpha}\delta^a$ such that $h_{ab}^{N(2)} = h_{ab}^{(2)} + 2\delta_{(a,b)}$ satisfies the de Donder condition

$$\bar{h}^{N(2)}{}_{ab,}{}^{b} = 0 \qquad \text{in } u \geq 0.$$

As before, a sufficient condition for this to hold is $\bar{h}^{N(2)}{}_{ab,}{}^{b}{}_{v} \mid_{u=0} = 0$. The argument is identical to that used in Section 6.3, since by eqn (6.4.9), we again have $\Box \bar{h}^{N(2)}{}_{ab,}{}^{b} = 0$.

Before proceeding, we must first find the change in the second-order initial data induced by the first-order gauge transformation with parameter ξ^a. This takes the form

$$2\xi^{d}{}_{,(a}h_{b)d}^{(1)} + 2\xi_{d,(a}\xi^{d}{}_{,b)} + \xi_{d,a}\xi^{d}{}_{,b}. \tag{6.4.10}$$

However, from eqn (6.4.10) we see that the only non-zero $\xi^c{}_{,d}$ on $u = 0$ is

$$\xi^{v}{}_{,u} = -16\rho^{-4}[8\log(\nu\rho) - \sqrt{2}v]^2\theta(8\log(\nu\rho) - \sqrt{2}v). \tag{6.4.11}$$

Thus, although the first-order gauge transformation will change $h_{ab}^{(2)}$ on $u = 0$, it will not affect the radiative components of the field, $h_{xx}^{(2)}, h_{yy}^{(2)}$, and $h_{xy}^{(2)}$.

The equations that δ^a must satisfy on $u = 0$ are identical to those in the first-order case, except for an extra term in T_{ab}:

$$-\Box\delta_{[a,v]} - \frac{1}{2}\eta_{av}\Box\delta_{c,}{}^{c} = \bar{h}_{ax,xv}^{(2)} + \bar{h}_{ay,yv}^{(2)} + \bar{h}_{au,vv}^{(2)}$$
$$- \frac{1}{2}(\bar{h}_{av,xx}^{(2)} + \bar{h}_{av,yy}^{(2)}) + \frac{1}{2}\bar{T}_{av}. \tag{6.4.12}$$

As before, we look for a solution

$$\delta_a = uf_a(\rho, v) + O(u^2). \tag{6.4.13}$$

When $a = v$, both sides of eqn (6.4.12) vanish identically. We therefore choose $f_v = 0$. For $a = x$, and $a = y$, eqn (6.4.12) has the form

$$\frac{\partial^2}{\partial v^2}f_x(\rho, v) = g_x(\rho, v),$$
$$\frac{\partial^2}{\partial v^2}f_y(\rho, v) = g_y(\rho, v), \tag{6.4.14}$$

which can be integrated directly to find f_x and f_y. Using these functions, the remaining equation can be written $\partial^2 f_u(\rho, v)/\partial v^2 = g_u(\rho, v)$, determining f_u. Thus, a solution of the form (6.4.13) does exist. However, there is no need to calculate it explicitly, since it will not alter the initial data for the radiative components of the field. In this gauge, the field equations for $h_{ab}^{(2)}$ reduce to

$$\Box h_{ab}^{(2)} = T_{ab}. \tag{6.4.15}$$

The most general form that $h_{ab}^{(2)}$ can have in $u \geq 0$ is

$$\begin{aligned}
h_{tt}^{(2)} &= A^{(2)}, & h_{tx}^{(2)} &= \rho^{-1}xB^{(2)}, \\
h_{ty}^{(2)} &= \rho^{-1}yB^{(2)}, & h_{tz}^{(2)} &= C^{(2)}, \\
h_{zz}^{(2)} &= G^{(2)}, & h_{zx}^{(2)} &= \rho^{-1}xF^{(2)}, \\
h_{zy}^{(2)} &= \rho^{-1}yF^{(2)}, & h_{xx}^{(2)} &= D^{(2)} + (y^2 - x^2)\rho^{-2}E^{(2)}, \\
h_{xy}^{(2)} &= -2xy\rho^{-2}E^{(2)}, & h_{yy}^{(2)} &= D^{(2)} + (x^2 - y^2)\rho^{-2}E^{(2)}.
\end{aligned} \tag{6.4.16}$$

The initial data for the radiative part of the field are given in eqns (6.2.24) and (6.2.25):

$$\begin{aligned}
D^{(2)}\,|_{u=0} &= \rho^{-4}g[8\log(\nu\rho) - \sqrt{2}v], \\
E^{(2)}\,|_{u=0} &= 0.
\end{aligned} \tag{6.4.17}$$

From the definition (6.3.14) of the news function we find

$$c_0^{(2)}(\tau, \theta) = -\frac{1}{2}\left[\frac{\lambda}{\nu}\right]^2 \lim_{r \to \infty}\left[r\frac{\partial}{\partial\tau}(D^{(2)} + E^{(2)})\right]. \tag{6.4.18}$$

(We shall see in Chapter 7 that this is not quite correct. All the information about the second-order news function *is* contained in $D^{(2)} + E^{(2)}$. In addition, there are some spurious gauge terms which must be eliminated by transforming to Bondi coordinates. This will be carried out in Chapter 7.) A straightforward, though tedious, analysis of eqn (6.4.15) shows that $D^{(2)}$ satisfies

$$\begin{aligned}
\Box D^{(2)} =&\, \rho^{-1}BE_{,v} + BE_{,v\rho} - \frac{1}{2}(B_{,v})^2 + 2E_{,u}E_{,v} \\
&+ 4\rho^{-2}E^2 + 2\rho^{-1}EE_{,\rho} + (E_{,\rho})^2 + E_{,v}E_{,\rho},
\end{aligned} \tag{6.4.19}$$

and $E^{(2)}$ satisfies

$$\begin{aligned}
\Box(e^{2i\phi}E^{(2)}) =&\, e^{2i\phi}[AE_{,vv} + \rho^{-1}BE_{,v} + BE_{,v\rho} - E_{,v}B_{,\rho} \\
&+ \frac{1}{2}(B_{,v})^2 - EE_{,\rho\rho} - \rho^{-1}EE_{,\rho} \\
&+ 4\rho^{-2}E^2].
\end{aligned} \tag{6.4.20}$$

These equations are each of the form

$$\Box F = H, \qquad F(0, v, x, y) \text{ known.} \tag{6.4.21}$$

The solution to eqn (6.4.21) at P with coordinates (u, v, x, y) may be expressed as the sum of two integrals (Penrose 1980; Penrose and Rindler 1984): (1) a surface term given by eqn (6.3.5) which arises from integrating over the two-surface, where the past null cone of P intersects $u = 0$; and (2) a volume term, which comes from integrating down the past null cone of P to the initial surface $u = 0$. It is given by

$$F_{\text{vol}}(t, \mathbf{r}) = \frac{1}{4\pi} \int_{u' \geq 0} \frac{H(t - |\mathbf{r} - \mathbf{r}'|, \mathbf{r}')}{|\mathbf{r} - \mathbf{r}'|} d^3 r'. \tag{6.4.22}$$

The contribution to $\partial D^{(2)}/\partial \tau$ in the far field from the surface integral is easily found to be

$$\frac{\partial D^{(2)}_{\text{surf}}}{\partial \tau} = \frac{16 \sec^2(\theta/2) \tan^2(\theta/2)}{\pi r} \int_0^\infty \int_0^{2\pi} \frac{ds}{s^3} d\omega$$

$$\times \theta \left[8 \log s + s \cos \omega + \frac{\tau}{\nu} \sec^2(\theta/2) - 8 \log \left(\frac{2 \tan(\theta/2)}{\nu} \right) \right]. \tag{6.4.23}$$

The volume terms, however, are by no means easy to compute. If the source terms in eqns (6.4.19) and (6.4.20) are inserted into eqn (6.4.22), then we see that to find the functional forms of $D^{(2)}$ and $E^{(2)}$ would require the evaluation of triple integrals, the integrands of which are themselves products of single integrals. To do this analytically is clearly out of the question, since the single integrals are themselves not analytically tractable. On the other hand, any numerical computation would suffer from two difficulties. First, one would have to identify the regions of integration which contribute most to the total integral. Second, and perhaps more importantly, the accurate numerical computation of what are essentially four-dimensional integrals with infinite ranges of integration, would require an exorbitant amount of computer time: indeed, it is quite impractical. However, we shall see in Chapter 7 that the perturbative field equations can be reduced to equations in only two independent variables rather than three. This allows an accurate numerical computation of the second-order news function, or equivalently of $a_2(\hat{\tau})$. The results are presented in Chapter 8.

6.5 A new mass-loss formula for the axisymmetric collision

To provide further motivation for the computation of the function $a_2(\hat{\tau})$ in eqn (6.1.3), in this section we shall show that if the product of the high-speed black-hole collision at finite γ is a single black hole at rest, plus outgoing gravitational radiation the form of which close to the axis of symmetry is fully described by eqn (6.1.1), then the final mass of this residual black hole is determined by $Q_0(\bar{\tau}, \psi)$ and $Q_2(\bar{\tau}, \psi)$, up to small corrections of $O(\gamma^{-2})$. Further, if eqns (6.1.1) and (6.1.2) match smoothly, then an alternative, more useful formula relating the final mass to the first two coefficients $a_0(\hat{\tau})$ and $a_2(\hat{\tau})$ in eqn (6.1.3) can be obtained. It is the latter that we derive first, and the arguments used are similar to those employed in Payne (1983b), for studying Smarr's zero-frequency limit (Smarr 1977a). Related work is described by Hobill (1984).

Since this system is axisymmetric and reflection symmetric, we may use the results of Bondi et al. (1962), concerning the form of the geometry near $\mathcal{I}^+$. In particular, we shall make use of the supplementary condition of equation (34) in Bondi et al. (1962):

$$\frac{\partial M}{\partial \hat{\tau}} = -\left[\frac{\partial c}{\partial \hat{\tau}}\right]^2 + \frac{1}{2}\frac{\partial}{\partial \hat{\tau}}\left[\frac{\partial^2 c}{\partial \hat{\theta}^2} + 3\cot\hat{\theta}\frac{\partial c}{\partial \hat{\theta}} - 2c\right]. \qquad (6.5.1)$$

Here $M(\hat{\tau}, \hat{\theta})$ is the mass aspect of the system, and $\partial c(\hat{\tau}, \hat{\theta})/\partial \hat{\tau}$ is the news function. In the Bondi metric, M appears in

$$g_{\hat{\tau}\hat{\tau}} = 1 - \frac{2M(\hat{\tau}, \hat{\theta})}{r} + \cdots, \qquad (6.5.2)$$

while

$$g_{\hat{\theta}\hat{\theta}} = r^2\left[1 + \frac{2c}{r} + \cdots\right],$$
$$g_{\hat{\phi}\hat{\phi}} = r^2\sin^2\hat{\theta}\left[1 - \frac{2c}{r} + \cdots\right]. \qquad (6.5.3)$$

M is thus a generalized mass suitable for non-static systems. The mass aspect of a particle of rest mass m moving with speed v (and Lorentz factor $\gamma = (1 - v^2)^{-\frac{1}{2}} = \cosh\lambda$) in the $\hat{\theta} = \pi$ direction is (Bondi et al. (1962), Appendix 3)

$$M(\hat{\theta}) = \frac{m}{(\cosh\lambda + \cos\hat{\theta}\sinh\lambda)^3}. \qquad (6.5.4)$$

Reexpressing this in terms of v, we find

$$M = \frac{m(1 - v^2)^{3/2}}{(1 + v\cos\hat{\theta})^3}. \qquad (6.5.5)$$

Let the rest mass of each body now be m. Before the collision one particle moves with speed v in the $\hat{\theta} = 0$ direction, while the other moves with the same speed in the $\hat{\theta} = \pi$ direction. The respective mass aspects of these two particles are thus

$$M_1 = \frac{m(1 - v^2)^{3/2}}{(1 - v\cos\hat{\theta})^3}, \qquad M_2 = \frac{m(1 - v^2)^{3/2}}{(1 + v\cos\hat{\theta})^3}. \qquad (6.5.6)$$

In the distant past the total mass aspect of the system is simply the linear superposition of M_1 and M_2:

$$M(-\infty, \hat{\theta}) = M_1(\hat{\theta}) + M_2(\hat{\theta}). \qquad (6.5.7)$$

If the final product of the collision is a single black hole, then

$$M(\hat{\tau} = \infty, \hat{\theta}) = m_{\text{final}}. \qquad (6.5.8)$$

Integrating eqn (6.5.1) over $\hat{\tau}$ we find that

$$m_{\text{final}} = M(-\infty, \hat{\theta}) - \int_{-\infty}^{\infty} (c_0)^2 \, d\hat{\tau}$$
$$+ \frac{1}{2} \left[\frac{\partial^2}{\partial \hat{\theta}^2} + 3 \cot \hat{\theta} \frac{\partial}{\partial \hat{\theta}} - 2 \right] (c \mid_{-\infty}^{\infty}), \qquad (6.5.9)$$

under the assumption (6.5.8). We shall examine eqn (6.5.9) in the region in which (6.1.1) is presumed to match with (6.1.2); that is, in which $\gamma^{-1} \ll \hat{\theta} \ll 1$.

From eqns (6.5.4) and (6.5.7) we have

$$M(-\infty, \hat{\theta}) = \mu \gamma^{-4} \{[1 + \cos \hat{\theta}(1 - \gamma^{-2})^{\frac{1}{2}}]^{-3}$$
$$+ [1 - \cos \hat{\theta}(1 - \gamma^{-2})^{\frac{1}{2}}]^{-3}\}, \qquad (6.5.10)$$

with $\mu = m\gamma$. One can easily show from the above that $M(-\infty, \hat{\theta}) = O(\gamma^{-1})$ when $\hat{\theta}$ is $O(\gamma^{-1/2})$.

The news function is given by eqn (6.1.1), which, reformulated, says that

$$c_0(\bar{\tau}, \psi = \gamma \hat{\theta}) - \sum_{n=0}^{m} \gamma^{-2n} Q_{2n}(\bar{\tau}, \psi) = O[\gamma^{-(2m+2)}], \qquad (6.5.11)$$

$\forall m$ as $\gamma \to \infty$ with $\bar{\tau}, \psi$ fixed. It can be seen from eqn (6.5.9) that $\int_{-\infty}^{\infty} c_0(\bar{\tau}, \psi) d\bar{\tau}$ and its first two angular derivatives must exist; otherwise, $M(\infty, \hat{\theta})$ will not be well defined. In order to make use of eqn (6.5.9), we must assume that eqn (6.1.1) satisfies a kind of uniformity condition, by which we mean that in addition to eqn (6.5.11), it will be assumed that the news function satisfies (for each fixed ψ)

$$\left| \gamma^{2m+2} \left[c_0(\bar{\tau}, \psi) - \sum_{n=0}^{m} \gamma^{-2n} Q_{2n}(\bar{\tau}, \psi) \right] \right| \leq f_m(\bar{\tau}, \psi), \qquad (6.5.12)$$

$\forall m, \forall \gamma > \Gamma$ where Γ is some constant, for some functions $f_m(\bar{\tau}, \psi)$, where every $\int_{-\infty}^{\infty} f_m(\bar{\tau}, \psi) d\bar{\tau}$ exists. If this is the case, then eqn (6.5.11) may be integrated to yield

$$\int_{-\infty}^{\infty} c_0(\bar{\tau}, \psi) d\bar{\tau} - \sum_{n=0}^{m} \gamma^{-2n} \int_{-\infty}^{\infty} Q_{2n}(\bar{\tau}, \psi) d\bar{\tau} = O[\gamma^{-(2m+2)}], \qquad (6.5.13)$$

$\forall m$ as $\gamma \to \infty$, with ψ fixed. Clearly, $\int_{-\infty}^{\infty} Q_{2n}(\bar{\tau}, \psi) d\bar{\tau}$ has to exist if eqn (6.5.12) is to hold, so that each $Q_{2n}(\bar{\tau}, \psi)$ must $\to 0$ as $\bar{\tau} \to \infty$ (presumably according to some inverse power law, if $Q_0(\bar{\tau}, \psi)$ is any guide). But, more importantly, in order that the news function satisfy eqn (6.5.12), it is necessary that the burst of radiation described by eqn (6.5.11), and its continuation to large angular scales, be the only gravitational radiation in the space–time, the system approaching isotropy asymptotically after it comes by. Another way of looking at this is to say that, if eqn (6.5.12) is to be valid, the news function must have the form (6.1.1), even in the limit $\bar{\tau} \to \infty$ (with γ large but fixed) when the perturbation theory breaks down near the initial surface. It is plausible only if the $Q_{2n}(\bar{\tau}, \psi)$ do all fall off according to various inverse power laws as $\bar{\tau} \to \infty$.

We also assume that (6.1.1) matches smoothly to (6.1.2), so that (6.1.1) is still a good asymptotic expansion in the intermediate region in which $1 \ll \psi \ll \gamma$. The analogue of eqn (6.5.13) when ψ is $O(\gamma^{1/2})$ will then be

$$\int_{-\infty}^{\infty} c_0(\bar{\tau}\psi)d\bar{\tau} - \sum_{n=0}^{m} \gamma^{-2n} \int_{-\infty}^{\infty} Q_{2n}(\bar{\tau}, \psi)d\bar{\tau} = O(\gamma^{-(m+1)}), \quad (6.5.14)$$

$\forall m$ as $\gamma \to \infty$ with $\gamma^{-\frac{1}{2}}\psi$ fixed. The reason why $\gamma^{-(m+1)}$ appears on the right-hand side instead of $\gamma^{-(2m+2)}$ is that the $Q_{2n}(\bar{\tau}, \psi)$ must grow as ψ^{2n} when $\psi \to \infty$, in order to match to the $\sin^{2n}\hat{\theta}$ terms in eqn (6.1.3), and so $Q_{2n}(\bar{\tau}, \psi)$ will be $O(\gamma^n)$ when ψ is $O(\gamma^{\frac{1}{2}})$.

Indeed, if matching works, $Q_{2n}(\bar{\tau}, \psi)$ will have the form given by

$$\gamma^{-2n} Q_{2n}(\bar{\tau} = \tau + 8\mu \log(\psi/\mu), \psi)$$
$$\sim \sin^{2n}\hat{\theta}\left[a_{2n}(\tau) + \frac{fn(\tau)}{\psi} + \frac{fn(\tau)}{\psi^2} + \cdots\right] \quad (6.5.15)$$

as $\gamma \to \infty$ with $\tau, \gamma^{1/2}\psi$ fixed (the origin of the $8\mu \log(\psi/\mu)$ term can be traced to the logarithmic delay across the shocks of Chapter 5, and

$$\gamma^{-2n} \int_{-\infty}^{\infty} Q_{2n}(\bar{\tau}, \psi)d\bar{\tau}$$
$$\sim \sin^{2n}\hat{\theta}\left[\int_{-\infty}^{\infty} a_{2n}(\tau)d\tau + \frac{\lambda_{n1}}{\psi} + \frac{\lambda_{n2}}{\psi^2} + \cdots\right] \quad (6.5.16)$$

as $\gamma \to \infty$ with $\gamma^{-1/2}\psi$ fixed. This is consistent with the calculation of D'Eath (1979b), who showed that

$$\int_{-\infty}^{\infty} Q_0(\bar{\tau}, \psi)d\bar{\tau} = [\psi^2/(1+\psi^2)] \int_{-\infty}^{\infty} a_0(\tau)d\tau.$$

It is clear from the form of eqn (6.5.16) that only

$$\int_{-\infty}^{\infty} Q_0(\bar{\tau}, \psi)d\bar{\tau} = \left[\int_{-\infty}^{\infty} a_0(\tau)d\tau\right]\left[1 - \frac{1}{\psi^2} + \cdots\right] \quad (6.5.17)$$

and

$$\gamma^{-2} \int_{-\infty}^{\infty} Q_2(\bar{\tau}, \psi)d\bar{\tau} \sim \sin^2\hat{\theta}\left[\int_{-\infty}^{\infty} a_2(\tau)d\tau + \frac{\lambda_{21}}{\psi} + \cdots\right] \quad (6.5.18)$$

can make an order 1 contribution to the last term in eqn (6.5.9) as $\gamma \to \infty$ with $\gamma^{-1/2}\psi$ fixed. Differentiating these expressions with respect to $\hat{\theta}$ we find

$$\frac{1}{2}\left[\frac{\partial^2}{\partial\hat{\theta}^2} + 3\cot\hat{\theta}\frac{\partial}{\partial\hat{\theta}} - 2\right](c\mid^{\infty}_{-\infty})$$

$$= \int_{-\infty}^{\infty}[4a_2(\tau) - a_0(\tau)]d\tau + O(\gamma^{-1/2}) \qquad (6.5.19)$$

as $\gamma \to \infty$ with $\gamma^{-1/2}$ constant. (Note that it is fortunate, or perhaps significant, that λ_{01} vanishes, since it would otherwise give a γ^2/ψ^3 contribution to eqn (6.5.19).) Clearly,

$$\int_{-\infty}^{\infty}(c_0)^2 d\hat{\tau} = \int_{-\infty}^{\infty}a_0(\tau)^2 d\tau + O(\gamma^{-1/2})$$

in the same limit, so collecting the various results we find that

$$m_{\text{final}} = -\int_{-\infty}^{\infty}[a_0(\tau)]^2 d\tau$$

$$+ \int_{-\infty}^{\infty}[4a_2(\tau) - a_0(\tau)]d\tau + O(\gamma^{-1/2}) \qquad (6.5.20)$$

as $\gamma \to \infty$ with $\gamma^{1/2}\psi$ fixed, proving our assertion at the beginning of this section, subject to the uniformity assumption (6.5.12).

One derives the equation relating m_{final} to $Q_0(\bar{\tau}, \psi)$ and $Q_2(\bar{\tau}, \psi)$ by examining eqn (6.5.9) in the limit $\psi \to 0$. It is

$$m_{\text{final}} = \mu(8\gamma^2 - 2)$$

$$+ \frac{1}{2}\lim_{\psi\to 0}\left[\gamma^2\frac{\partial^2}{\partial\psi^2} + \frac{\gamma}{\psi}\frac{\partial}{\partial\psi}\right]\left[\int_{-\infty}^{\infty}[Q_0(\bar{\tau}, \psi) + \gamma^{-2}Q_2(\bar{\tau}, \psi)]\bar{\tau}\right]$$

$$+ O(\gamma^{-2}), \qquad (6.5.21)$$

subject to the uniformity assumption (6.5.12). As noted previously, it is rather hard to calculate $Q_2(\bar{\tau}, \psi)$, which is why we shall use eqn (6.5.20), rather than eqn (6.5.21), to calculate m_{final}.

Clearly, a formula similar to eqn (6.5.21) could be derived for any process in which the initial momenta are known and the final product is a single body at rest.

The first term in eqn (6.5.20) has been calculated numerically (D'Eath 1978; Chapman 1979), and found to be $\mu/2$ to three significant figures (here 2μ is the initial energy). The term $\int_{-\infty}^{\infty}a_0(\tau)d\tau$ can be calculated exactly. From eqns (6.3.22) and (6.3.27) we have

$$a_0(\tau) = \frac{4}{\pi}\int_0^{\infty}\int_0^{\pi}\frac{ds}{s}d\omega\cos(2\omega)\delta\left[8\log s + s\cos\omega + \frac{\tau}{\mu}\right]. \qquad (6.5.22)$$

Hence

$$\int_{-\infty}^{\infty} a_0(\tau)d\tau = \frac{4\mu}{\pi} \int_0^{\infty} \int_0^{\pi} \frac{ds}{s} d\omega \cos(2\omega)[\theta(8\log s + s \cos\omega + x)]\mid_{x=-\infty}^{x=\infty}$$

$$= \frac{4\mu}{\pi} \left[\int_{|\cos\phi_0|\leq 1} \frac{ds}{s} \sin\phi_0 \cos\phi_0 \right] \Bigg|_{x=-\infty}^{x=\infty}, \qquad (6.5.23)$$

where

$$\cos\phi_0 = - \left[\frac{8\log s + x}{s} \right]. \qquad (6.5.24)$$

When $x \ll -1$, the domain in which $|\cos\phi_0| \leq 1$ is connected, bounded below by $S_L \simeq -x$, and unbounded above. Hence

$$\lim_{x \to -\infty} \int_{|\cos\phi_0|\leq 1} \frac{ds}{s} \sin\phi_0 \cos\phi_0$$

$$= \lim_{x \to -\infty} \int_{-x}^{\infty} \frac{ds}{s} \left[1 - \left(\frac{x}{s}\right)^2 \right]^{1/2} \left[\frac{-x}{s} \right]$$

$$= \int_1^{\infty} \frac{d\omega}{w^3} \sqrt{\omega^2 - 1} = \frac{\pi}{4}. \qquad (6.5.25)$$

When $x \gg 1$, the range of integration has two disconnected regions. One is bounded below by $S_L \simeq x$ and unbounded above, and clearly contributes $-\pi/4$ to the total integral. The other is approximately $[e^{-x/8}(1 - \frac{1}{8}e^{-x/8}), e^{-x/8}(1 + \frac{1}{8}e^{-x/8})]$ and so does not contribute to eqn (6.5.22). Therefore

$$\left(\int_{|\cos\phi_0|\leq 1} \frac{ds}{s} \sin\phi_0 \cos\phi_0 \right) \Bigg|_{x=-\infty}^{x=\infty} = \frac{-\pi}{2} \qquad (6.5.26)$$

and $\int_{-\infty}^{\infty} a_0(\tau)d\tau = -2\mu$. Substituting this into eqn (6.5.20), we find

$$m_{\text{final}} = \frac{3\mu}{2} + 4 \int_{-\infty}^{\infty} a_2(\tau)d\tau + o(1), \qquad (6.5.27)$$

where $o(1)$ denotes a term tending to zero as $\gamma \to \infty$, subject to the uniformity condition (6.5.12).

6.6 Summary

We have studied in this chapter the axisymmetric collision of two black holes at the speed of light, with a view to understanding the more physically realistic collision of two black holes at large but finite γ. Following earlier work of Curtis (1978a) and Chapman (1979), the curved region IV of the space–time depicted in Fig. 5.1, resulting from this collision of two impulsive plane-fronted waves, has been treated by means of perturbation theory. A large Lorentz boost applied to the incoming states (each with energy μ) yields two null particles with energies λ, ν, where $\lambda \ll \nu$. The metric of the curved region IV can be found as a perturbation of flat space–time, in powers of the small parameter λ/ν, by solving a sequence of characteristic initial-value problems with initial data given just to the future of the strong shock with energy ν. The perturbation theory is expected to be singular: it should give a good description of the parts of the space–time near the forward and backward directions in which the incoming shocks have been delayed and deflected by small angles during the interaction at large distances from the symmetry axis. But it will give a less and less accurate description of the geometry as one examines regions further into the centre of the space–time, where formation of a single final Schwarzschild black hole should take place, with associated further emission of gravitational radiation.

In Section 6.3 the metric was calculated to first order in λ/ν. On boosting back to the centre-of-mass frame this yielded the contribution $a_0(\hat{\tau})$ (shown in Fig. 5.10) to the series conjectured for the news function in eqn (6.1.3). This agrees with the form found in Section 5.6 for the gravitational radiation at angles $\hat{\theta}$ fairly close to the axis, obeying $\gamma^{-1} \ll \hat{\theta} \ll 1$, in the finite-$\gamma$ collision. The form of $a_0(\hat{\tau})$ is such that 25% of the initial energy 2μ would be emitted in gravitational waves, if the radiation were isotropic. The calculation was continued to second order in Section 6.4, leading to an integral expression for the next coefficient $a_2(\hat{\tau})$ in the angular expansion (6.1.3) of the news function near the axis. Further motivation for the computation of $a_2(\hat{\tau})$ was provided in Section 6.5, which showed that if *all* the gravitational radiation near the axis in the finite-γ space–times is accurately described (in a certain precise sense) by eqn (6.1.1), and if eqns (6.1.1) and (6.1.2) match smoothly at angles obeying $\gamma^{-1} \ll \hat{\theta} \ll 1$, then the mass of the (assumed) final static black hole can be found from a knowledge only of $a_0(\hat{\tau})$ and $a_2(\hat{\tau})$, up to corrections which tend to zero as $\gamma \to \infty$. Since Penrose (1974) has found an apparent horizon for the speed-of-light collision on the union of the two incoming shocks, the collision space–time is thus providing an interesting test of the assumption of cosmic censorship, which gives a lower bound of $\sqrt{2}\mu$ for the final mass.

In the following Chapter 7, we show how the perturbative field equations can be reduced to equations in only two independent variables, because of a conformal symmetry at each order of perturbation theory. This yields an alternative integral expression for $a_2(\hat{\tau})$ which allows us to calculate this quantity numerically. The results are presented in Chapter 8, where it is found that the mass-loss formula of Section 6.5 makes the unphysical prediction that the mass of the assumed final static black hole is approximately twice the initial energy of the colliding waves. The most likely explanation for this apparently surprising result is that there is some other gravitational radiation present in the space–time. A 'second burst' of radiation produced deep inside the space–time will be delayed by an amount proportional to $\log \hat{\theta}$ relative to the 'first burst' described by eqn (6.1.3), which is in part produced at very large radii. As discussed further in Chapter 8, this will have the consequence that the expansion (6.1.3) is not valid uniformly with respect to retarded time, so that the assumptions made in Section 6.5 fail to hold. Nevertheless, knowledge of $a_2(\hat{\tau})$ together with $a_0(\hat{\tau})$ does give some further information about the angular distribution of radiation, and allows a rough estimate of the emitted energy, following the conventional formula of Bondi *et al.* (1962).

7

AXISYMMETRIC BLACK-HOLE COLLISIONS AT THE SPEED OF LIGHT: REDUCTION TO TWO INDEPENDENT VARIABLES AND CALCULATION OF THE SECOND-ORDER NEWS FUNCTION

7.1 Introduction

Here further progress will be described concerning the gravitational radiation emitted in the axisymmetric collision of two black holes at the speed of light. The preceding chapter was concerned with describing the problem and with setting up an analytical treatment using a perturbation approach. In this chapter we describe analytical simplifications which make feasible the numerical calculation of the second-order news function, which gives partial information about the angular distribution of gravitational radiation. Results and conclusions concerning the radiation emitted and consequent mass loss are presented in Chapter 8.

In Chapter 6, the axisymmetric collision of two black holes travelling at the speed of light, each described in the centre-of-mass frame before the collision by an impulsive plane-fronted shock wave with energy μ, was investigated in a new frame to which a large Lorentz boost had been applied. There the energy $\nu = \mu e^{\alpha}$ of the incoming shock 1, which initially lies on the hyperplane $z + t = 0$ between two portions of Minkowski space, obeys $\nu \gg \lambda$, where $\lambda = \mu e^{-\alpha}$ is the energy of the incoming shock 2, which initially lies on the hypersurface $z - t = 0$. In the boosted frame, to the future of the strong shock 1, the metric possesses the perturbation expansion (eqn (6.2.27))

$$g_{ab} \sim \nu^2 \left[\eta_{ab} + \sum_{i=1}^{\infty} \left(\frac{\lambda}{\nu} \right)^i h_{ab}^{(i)} \right] \tag{7.1.1}$$

with respect to suitable coordinates, where η_{ab} is the Minkowski metric. The problem of solving the Einstein field equations becomes a (singular) perturbation problem of finding $h^{(1)}{}_{ab}$, $h^{(2)}{}_{ab}$,... by successively solving the linearized field equations at first, second,... order in λ/ν, given characteristic initial data on the surface $u = 0$ just to the future of the strong shock 1.

On boosting back to the centre-of-mass frame, one finds that the perturbation series (7.1.1) gives an accurate description of the space–time geometry in the region in which gravitational radiation propagates at small angles away from the forward symmetry axis $\hat{\theta} = 0$. By reflection symmetry, an analogous series also gives a good description near the backward axis $\hat{\theta} = \pi$. The news function (Bondi *et al.* 1962) (eqn (6.3.14)) $c_0 \left(\hat{\tau}, \hat{\theta} \right)$, which describes the gravitational radiation arriving at future null infinity $\mathcal{I}^+$ in the centre-of-mass frame, is expected to have the convergent series expansion (6.1.3)

$$c_0 \left(\hat{\tau}, \hat{\theta} \right) = \sum_{n=0}^{\infty} a_{2n} \left(\hat{\tau}/\mu \right) \sin^{2n} \hat{\theta} \,, \qquad (7.1.2)$$

where $\hat{\tau}$ is a suitable retarded time coordinate. (The series of eqn (6.1.3) has been modified here by making the replacement $a_{2n} \left(\hat{\tau} \right) \rightarrow a_{2n} \left(\hat{\tau}/\mu \right)$, since $\hat{\tau}$ will always appear as an argument in the dimensionless combination $\left(\hat{\tau}/\mu \right)$ (see Section 6.4).) The first-order perturbation calculation of $h_{ab}^{(1)}$ in Section 6.3, on boosting back to the centre-of-mass frame, yielded $a_0 \left(\hat{\tau}/\mu \right)$, in agreement with the expression found in eqn (5.6.28) by studying the collision of two black holes at large but finite incoming Lorentz factor γ. This is such that, if the radiation were isotropic (i.e. if $a_{2n} \left(\hat{\tau}/\mu \right)$ were zero for $n \geq 1$), 25% of the initial energy 2μ would be emitted in gravitational waves. The second-order calculation of $h_{ab}^{(2)}$ in Section 6.4, on boosting back to the centre-of-mass frame, gave an integral expression for the next coefficient $a_2 \left(\hat{\tau}/\mu \right)$ which unfortunately was so complicated as to be intractable numerically. In the present chapter we shall show how the calculation of $a_2 \left(\hat{\tau}/\mu \right)$ can be simplified analytically so as to enable us to compute this function numerically. As was shown in Section 6.5, if all the gravitational radiation in the space–time is (in a certain precise sense) accurately described by eqn (7.1.2), then the mass of the (assumed) final static Schwarzschild black hole remaining after the collision can be determined from knowledge only of $a_0 \left(\hat{\tau}/\mu \right)$ and $a_2 \left(\hat{\tau}/\mu \right)$. Further, Penrose (1974) has found an apparent horizon on the union of the two null planes on which the incoming shocks lie; if the Cosmic Censorship Hypothesis (Hawking and Ellis 1973) is correct, this gives a lower bound of $\sqrt{2}\mu$ for the energy of the final black hole (or holes). This also gives an upper bound of (29%) for the efficiency of gravitational wave generation, slightly greater than the above crude estimate of 25% based on the knowledge of $a_0(\hat{\tau}/\mu)$ only. The computation of $a_2 \left(\hat{\tau}/\mu \right)$ in addition is thus linked to an interesting test of cosmic censorship, via eqn (6.5.27).

In Section 7.2, we begin the process of finding a simpler form for $a_2 \left(\hat{\tau}/\mu \right)$ by noting that, because of a conformal symmetry at each order in perturbation theory, the field equations obeyed by the metric perturbations $h_{ab}^{(1)}$, $h_{ab}^{(2)}$, ... in eqn (7.1.1) may all be reduced to equations in only two independent variables. The resulting reduced differential equations are studied in Section 7.3; the equations are shown to be hyperbolic, and their characteristics are found. The retarded Green function for the reduced differential operator is found in Section 7.4 by reduction from the retarded flat-space Green function in four dimensions. This allows the transverse components of the second-order metric perturbation $h_{ab}^{(2)}$ (from which $a_2 \left(\hat{\tau}/\mu \right)$ can found) to be expressed in two-dimensional form (Section 7.5). The resulting integral expressions are considerably simpler than those found from a four-dimensional approach in Section 6.4, thus making feasible the numerical computation of a_2 $(\hat{\tau}/\mu)$, of which the results will be presented in Chapter 8.

In order to extract a_2 $(\hat{\tau}/\mu)$ from the metric perturbations, one has to deal with certain terms introduced in the metric as a result of the choice of the harmonic gauge, employed in the calculation of $h_{ab}^{(1)}$ and $h_{ab}^{(2)}$. As is well known (Fock 1964), this gauge leads to the appearance of $(\log r)/r$ terms in the metric tensor at second and higher orders in perturbation theory, where r is a radial coordinate. In Section 7.4, we calculate the $(\log r)/r$ term in the transverse part of $h_{ab}^{(2)}$, and show how to eliminate this term by finding an explicit coordinate transformation to a Bondi coordinate system (Bondi *et al.* 1962) at first order in perturbation theory. In Section 7.7, we show that, while the construction of this coordinate transformation can be carried on to second order, knowledge of the full second-order gauge transformation is not needed in order to calculate the second-order news function, which describes the gravitational radiation at this order. Section 7.8 discusses the ambiguity in the second-order news function caused by the freedom to make supertranslations (Bondi *et al.* 1962); use of this freedom is in fact essential in order to put the news function in a form which is square-integrable at each order in perturbation theory. (The complete news function must be square-integrable, in order that the mass loss in gravitational waves be finite.) Some comments are included in Section 7.9.

7.2 Reduction to two dimensions

We shall now show that the (four-dimensional) field equations satisfied by the metric perturbations $h_{ab}^{(1)}, h_{ab}^{(2)}, \ldots$ in eqn (7.1.1) may all be reduced to two-dimensional form.

Consider the C^0 form of the infinitely boosted black-hole metric (eqn (6.2.3)):

$$ds^2 = d\hat{u}\, d\hat{v} + \left[1 + 4\mu\hat{u}\theta\,(\hat{u})\,\hat{\rho}^{-2}\right]^2 d\hat{\rho}^2$$
$$+ \hat{\rho}^2 \left[1 - 4\mu\hat{u}\theta\,(\hat{u})\,\hat{\rho}^{-2}\right]^2 d\phi^2, \tag{7.2.1}$$

where $\theta\,(\hat{u})$ is the Heaviside step function. On using the discontinuous coordinate transformation

$$x = \hat{x} - 4\mu\hat{u}\theta\,(\hat{u})\,\hat{x}\hat{\rho}^{-2},$$
$$y = \hat{y} - 4\mu\hat{u}\theta\,(\hat{u})\,\hat{y}\hat{\rho}^{-2},$$
$$u = \hat{u},$$
$$v = \hat{v} + 8\mu\theta\,(\hat{u})\,\log\hat{\rho} - 16\mu^2\hat{u}\theta\,(\hat{u})\,\hat{\rho}^{-2}, \tag{7.2.2}$$

where $x = \rho\cos\phi$, $y = \rho\sin\phi$, $\hat{x} = \hat{\rho}\cos\phi$, $\hat{y} = \hat{\rho}\sin\phi$, this may be put in the form (eqn (6.2.1))

$$ds^2 = du\, dv + dx^2 + dy^2 - 4\mu\log\left(x^2 + y^2\right)\delta(u)du^2, \tag{7.2.3}$$

describing an impulsive plane-fronted wave between two portions of Minkowski space–time.

Let L denote the Lorentz transformation

$$(\hat{u}, \hat{v}, \hat{\rho}, \phi) \rightarrow (\hat{u}', \hat{v}', \hat{\rho}', \phi') = (e^{-\beta}\hat{u}, e^{\beta}\hat{v}, \hat{\rho}, \phi) \qquad (7.2.4)$$

(using eqn (7.2.2) it can be shown that L *is* a Lorentz transformation, even though the hatted coordinates are not Minkowskian in $\hat{u} > 0$), and let C denote the conformal transformation

$$(\hat{u}, \hat{v}, \hat{\rho}, \phi) \overset{C}{\rightarrow} (\hat{u}', \hat{v}', \hat{\rho}', \phi') = (e^{-\beta}\hat{u}, e^{-\beta}\hat{v}, e^{-\beta}\hat{\rho}, \phi). \qquad (7.2.5)$$

Then, under CL,

$$(\hat{u}, \hat{v}, \hat{\rho}, \phi) \overset{CL}{\rightarrow} (\hat{u}', \hat{v}', \hat{\rho}', \phi') = (e^{-2\beta}\hat{u}, \hat{v}, e^{-\beta}\hat{\rho}, \phi) \qquad (7.2.6)$$

and

$$d\hat{u}\, d\hat{v} + [1 + 4\mu\hat{v}\theta\hat{u})\hat{\rho}^{-2}]^2 d\hat{\rho}^2 + \hat{\rho}^2[1 - 4\mu\hat{u}\theta(\hat{u})\hat{\rho}^{-2}]d\phi^2$$
$$\overset{CL}{\rightarrow} e^{2\beta}\{d\hat{u}'d\hat{v}' + [1 + 4\mu\hat{u}'\theta(\hat{u}')\hat{\rho}'^{-2}]^2 d\hat{\rho}'^2 + \hat{\rho}'^2[1 - 4\mu\hat{u}'\theta(\hat{u}')\hat{\rho}'^{-2}]^2 d\phi^2\}. \qquad (7.2.7)$$

Thus the transformation CL is a conformal symmetry of (2.1). (This is easy to understand physically: the Lorentz transformation L increases the apparent energy of the wave from μ to μe^{β}; but this energy provides the only length-scale present in the metric. If, therefore, using C, we scale down all lengths by a factor e^{β}, then the apparent energy of the wave is reduced to μ again.) For a wave travelling in the opposite direction, the effect of CL is

$$d\hat{u}\, d\hat{v} + [1 - 4\mu\hat{v}\theta(-\hat{v})\hat{\rho}^{-2}]^2 d\hat{\rho}^2 + \hat{\rho}^2[1 + 4\mu\hat{v}\theta(-\hat{v})\hat{\rho}^{-2}]^2 d\phi^2$$
$$\overset{CL}{\rightarrow} e^{2\beta}\{d\hat{u}'d\hat{v}' + [1 - 4\mu e^{-2\beta}\hat{v}'\theta(-\hat{v}')\hat{\rho}'^{-2}]^2$$
$$d\hat{\rho}'^2 + \hat{\rho}'^2[1 + 4\mu e^{-2\beta}\hat{v}'\theta(-\hat{v}')\hat{\rho}'^{-2}]^2 d\phi^2\}. \qquad (7.2.8)$$

Now consider the axisymmetric collision of two such waves, viewed in the 'boosted frame' in which the waves have energies $\nu = \mu e^{\alpha}$ and $\lambda = \mu e^{-\alpha}$, and in which the pre-collision metric is given by eqn (6.2.14)

$$ds^2 = d\hat{u}\, d\hat{v} + \left[1 + 4v\hat{u}\theta(\hat{u})\hat{\rho}^{-2}\right]^2 d\hat{\rho}^2$$
$$+ \left[-8\lambda\hat{v}\theta(-\hat{v})\hat{\rho}^{-2} + 16\lambda^2\hat{v}^2\theta(-\hat{v})\hat{\rho}^{-4}\right]d\hat{\rho}^2$$
$$+ \hat{\rho}^2\left[1 - 4u\hat{v}\theta(\hat{u})\hat{\rho}^{-2}\right]^2 d\phi^2$$
$$+ \hat{\rho}^2\left[8\lambda\hat{v}\theta(-\hat{v})\hat{\rho}^{-2} + 16\lambda^2\hat{v}^2\theta(-\hat{v})\hat{\rho}^{-4}\right]d\phi^2. \qquad (7.2.9)$$

Let us denote by $g_{ab}(\nu, \lambda, \hat{X})$ this explicit form (7.2.9) of the pre-collision metric in the boosted frame ($\hat{X} \equiv (\hat{u}, \hat{v}, \hat{\rho}, \phi)$). From eqns (7.2.7) and (7.2.8), we see that $g_{ab}(\nu, \lambda, \hat{X})$ (and hence also the initial data on $\hat{u} = 0^+$) transforms as

$$g_{ab}(\nu, \lambda, \hat{X}) \overset{CL}{\rightarrow} e^{2\beta}g_{ab}(\nu, \lambda e^{-2\beta}, \hat{X}') \qquad (7.2.10)$$

under CL. The map CL has a natural continuation into the region in $\hat{u} > 0$ where the weak shock appears a small perturbation to the flat background of the strong shock: namely eqn (7.2.6) with the coordinates being those of the strong shock background. In the unhatted coordinate system, which is related to the hatted system through eqn (7.2.2) with μ replaced by ν, the metric possesses the perturbation expansion (7.1.1):

$$g_{ab}(X) \sim \nu^2 \left[\eta_{ab} + \sum_{i=1}^{\infty} \left(\frac{\lambda}{\nu} \right)^i h_{ab}^{(i)}(X) \right]. \tag{7.2.11}$$

(The coordinates have been rendered dimensionless, using $X \to X/\nu$ [as in eqn (6.2.20)], to obtain eqn (7.2.11).) Since the metric to the future of the strong shock is determined solely by the initial data on $\hat{u} = 0^+$, and since this initial data transforms as eqn (7.2.10), the effect of CL on eqn (7.2.11) is

$$\nu^2 \left[\eta_{ab} + \sum_{i=1}^{\infty} \left(\frac{\lambda}{\nu}^i \right) h_{ab}^{(i)}(X) \right]$$

$$\overset{CL}{\Rightarrow} e^{2\beta} \nu^2 \left[\eta_{ab} + \sum_{i=1}^{\infty} \left(\frac{\lambda e^{-2\beta}}{\nu} \right)^i h_{ab}^{(i)}(X') \right] \tag{7.2.12}$$

(where the explicit forms of the $h_{ab}^{(i)}$ are identical in the two expansions). Hence the transformation CL does not effect the intrinsic nature of the perturbation problem: it merely alters the value of the perturbation parameter. In other words, the space–time possesses a conformal symmetry at each order in perturbation theory:

$$h_{ab}^{(i)}(X) \overset{CL}{\Rightarrow} e^{2(i-1)\beta} h_{ab}^{(i)}(X') \tag{7.2.13}$$

(where, of course, $X \overset{CL}{\Rightarrow} X'$).

We can use eqn (7.2.13) to determine something of the behaviour of the $h_{ab}^{(i)}(X)$. From eqn (7.2.6) we deduce

$$g_{\hat{u}'\hat{u}'} = e^{4\beta} g_{\hat{u}\hat{u}}, \qquad g_{\hat{u}'\hat{\rho}'} = e^{3\beta} g_{\hat{u}\hat{\rho}},$$

$$g_{\underset{\hat{u}'\hat{v}'}{\tilde{\rho}'\tilde{\rho}'}} = e^{2\beta} g_{\underset{\hat{u}\hat{v}}{\hat{\rho}\hat{\rho}}}, \tag{7.2.14}$$

$$g_{\hat{v}'\tilde{\rho}'} = e^{\beta} g_{\hat{v}\hat{\rho}}, \qquad g_{\underset{\phi'\phi'}{\hat{v}'\hat{v}'}} = g_{\underset{\phi\phi}{\hat{v}\hat{v}}}.$$

Using the coordinate transformation of eqn (7.2.2) with μ replaced by ν, we can show that identical relationships hold between the unhatted coordinate systems:

$$g_{u'u'} = e^{4\beta} g_{uu}, \qquad g_{u'\rho'} = e^{3\beta} g_{u\rho},$$

$$g_{\underset{u'v'}{\rho'\rho'}} = e^{2\beta} g_{\underset{uv}{\rho\rho}}, \tag{7.2.15}$$

$$g_{\hat{v}'\tilde{\rho}'} = e^{\beta} g_{v\rho}, \qquad g_{\underset{\phi'\phi'}{v'v'}} = g_{\underset{\phi\phi}{vv}}.$$

Combining eqn (7.2.15) with eqn (7.2.13), we deduce the following:

$$h_{uu}^{(i)}(X') = e^{2(1+i)\beta} h_{uu}^{(i)}(X),$$
$$h_{u\rho}^{(i)}(X') = e^{(2i+1)\beta} h_{u\rho}^{(i)}(X),$$
$$h_{\substack{\rho\rho \\ vv}}^{(i)}(X') = e^{2\beta i} h_{\substack{\rho\rho \\ vv}}^{(i)}(X), \qquad (7.2.16)$$
$$h_{v\rho}^{(i)}(X') = e^{(2i-1)\beta} h_{v\rho}^{(i)}(X),$$
$$h_{\substack{vv \\ \phi\phi}}^{(i)}(X') = e^{(2i-2)\beta} h_{\substack{vv \\ \phi\phi}}^{(i)}(X).$$

From eqn (7.2.6) we note that the values of $\hat{v}$, $\hat{u}\hat{\rho}^{-2}$ and ϕ are left unchanged by the map CL. Using eqn (7.2.2) with μ replaced by ν we can show that the corresponding combinations of unhatted coordinates that are left invariant by CL are

$$r \equiv 8\log(\nu\rho) - \sqrt{2}v, \qquad q \equiv u\rho^{-2}, \qquad \phi \qquad (7.2.17)$$

(where we have removed a factor of ν from the coordinates, as in eqn (6.2.20), and redefined u and v by $u = \frac{1}{\sqrt{2}}(z+t)$ and $v = \frac{1}{\sqrt{2}}(z-t)$). The lines on which q, r and ϕ are constant may be interpreted geometrically as the orbits of the conformal symmetry CL.

If we express each $h_{ab}^{(i)}$ as $h_{ab}^{(i)}(q,r,\rho)$ (ϕ is ignorable) then the only coordinate that changes in value when CL is applied is ρ $\left(\rho \overset{CL}{\to} \rho' = e^{-\beta}\rho\right)$. Used in conjunction with eqn (7.2.16) this tells us that

$$h_{uu}^{(i)} = fn(q,r)\rho^{-(2i+2)},$$
$$h_{u\rho}^{(i)} = fn(q,r)\rho^{-(2i+1)},$$
$$h_{\substack{\rho\rho \\ vv}}^{(i)} = fn(q,r)\rho^{-2i}, \qquad (7.2.18)$$
$$h_{v\rho}^{(i)} = fn(q,r)\rho^{-(2i-1)},$$
$$h_{\substack{vv \\ \phi\phi}}^{(i)} = fn(q,r)\rho^{-(2i-2)}.$$

Thus each metric perturbation has a very simple dependence on ρ.

In an appropriate gauge the field equations for the $h_{ab}^{(i)}$ are all of the form $\Box h_{ab}^{(i)} = S_{ab}^{(i)}$, where $S_{ab}^{(i)}$ is a function of $h_{ab}^{(i-1)}, \ldots, h_{ab}^{(1)}$ and their derivatives ($S_{ab}^{(1)} = 0$). Since each $h_{ab}^{(i)}$ is of the form $fn(q,r)\rho^{-k}$, its corresponding $S_{ab}^{(i)}$ must be of the form $fn(q,r)\rho^{-(k+2)}$. This indicates that we can eliminate ρ from the field equations by separation of variables, thereby reducing them to two-dimensional differential equations.

Let us now perform the reduction to two dimensions explicitly, starting with the first-order perturbations $h_{ab}^{(1)}$. Consider the flat-space wave equation

$$\Box\psi \equiv 2\frac{\partial^2\psi}{\partial u\partial v} + \frac{1}{\rho}\frac{\partial}{\partial\rho}\left[\rho\frac{2\psi}{\partial\rho}\right] + \frac{1}{\rho^2}\frac{\partial^2\psi}{\partial\phi^2} = 0, \qquad (7.2.19)$$

where the boundary condition is

$$\psi\mid_{u=0} = e^{im\phi}\rho^{-n}f[8\log(v\rho) - \sqrt{2}v],$$
$$f(x) = 0, \qquad \forall x < 0 \tag{7.2.20}$$

(here m and n are integers and apart from the above restriction $f(x)$ is arbitrary). The field equations for $h_{ab}^{(1)}$ are special cases of the general system (7.2.19), (7.2.20). We know from our preceding arguments that ψ must be of the form $e^{im\phi}\rho^{-n}\chi(q,r)$ in $u \geq 0$ (where q and r are defined in eqn (7.2.17)). From eqn (7.2.17) we find

$$\left[\frac{\partial}{\partial u}\right]_{v,\rho,\phi} = \frac{1}{\rho^2}\left[\frac{\partial}{\partial q}\right]_{r,\rho,\phi},$$

$$\left[\frac{\partial}{\partial v}\right]_{u,\rho,\phi} = -\sqrt{2}\left[\frac{\partial}{\partial r}\right]_{q,\rho,\phi}, \tag{7.2.21}$$

$$\left[\frac{\partial}{\partial\rho}\right]_{u,v,\phi} = \left[\frac{\partial}{\partial\rho}\right]_{q,r,\phi} - \frac{2q}{\rho}\left[\frac{\partial}{\partial q}\right]_{r,\rho,\phi} - \frac{8}{\rho}\left[\frac{\partial}{\partial r}\right]_{q,\rho,\phi},$$

and therefore

$$2\frac{\partial^2}{\partial u\partial v} + \frac{1}{\rho}\frac{\partial}{\partial\rho}\left[\rho\frac{\partial}{\partial\rho}\right] + \frac{1}{\rho^2}\frac{\partial^2}{\partial\phi^2}$$
$$= \frac{1}{\rho^2}\left[-2\sqrt{2}\frac{\partial^2}{\partial q\partial r} + \left(\rho\frac{\partial}{\partial\rho} - 2q\frac{\partial}{\partial q} + 8\frac{\partial}{\partial r}\right) \times\right.$$
$$\left. \times \left(\rho\frac{\partial}{\partial\rho} - 2q\frac{\partial}{\partial q} + 8\frac{\partial}{\partial r}\right) + \frac{\partial^2}{\partial\phi^2}\right]. \tag{7.2.22}$$

Thus χ is the solution to

$$\mathcal{L}_{m,n}\chi$$
$$\equiv \left[-2\sqrt{2}\frac{\partial^2}{\partial q\partial r} + \left(-n - 2q\frac{\partial}{\partial q} + 8\frac{\partial}{\partial r}\right)\left(-n - 2q\frac{\partial}{\partial q} + 8\frac{\partial}{\partial r}\right) - m^2\right]\chi$$
$$= 0. \tag{7.2.23}$$

where the boundary condition is $\chi|_{q=0} = f(r)$.

Of course, for the homogeneous wave equation (7.2.19), where the solution has the simple integral form given in eqn (6.3.5)(Penrose 1980; Penrose and Rindler 1984) there is really no point in eliminating ρ and ϕ from the differential equation. However, consider the field equation for any one of the higher-order metric coefficients (i.e. $h_{ab}^{(i)}, i \geq 2$). It is an inhomogeneous flat-space wave equation

$$\Box\psi = S, \tag{7.2.24}$$

in which the source term is $S = e^{im\phi}\rho^{-(n+2)}H(q,r)$. (The boundary condition may be taken to be $\psi|_{u=0} = 0$, since any contribution to the solution from non-zero boundary conditions can be evaluated separately using eqn (6.3.5).) In contrast to the homogeneous case, the benefits to be gained by reducing eqn (7.2.24) to

$$\mathcal{L}_{m,n}X = H, \tag{7.2.25}$$

(where, of course, $\psi = e^{im\phi}\rho^{-n}\chi$) are not insignificant. First, the geometrical configuration of the problem is now much easier to visualize. Previously, to calculate the solution at some space–time point P we would have had to integrate the source term S (suitably weighted) over the past null cone of P. Now we need simply integrate the product of H and the Green function for the differential operator $\mathcal{L}_{m,n}$ over some two-dimensional region in the (q,r) plane. This makes it much easier to estimate the various contributions to the solution from different parts of the integration region. Second, although we must now find and calculate the Green function for $\mathcal{L}_{m,n}$, it turns out that there is a considerable computational gain which makes the numerical calculation of the solution practicable, whereas before it would have required a prohibitive amount of computer time.

7.3   The reduced differential equation

We shall now demonstrate that the differential operator $\mathcal{L}_{m,n}$ is hyperbolic and find its characteristics. Define new coordinates

$$\xi = \xi(q,r), \qquad \eta = \eta(q,r). \tag{7.3.1}$$

Now,

$$\mathcal{L}_{m,n} = -(2\sqrt{2} + 32q)\frac{\partial^2}{\partial q \partial r} + 4q^2\frac{\partial^2}{\partial q^2} + 64\frac{\partial^2}{\partial r^2} + \cdots, \tag{7.3.2}$$

where the terms omitted are first and zeroth order in $\partial/\partial q$ and $\partial/\partial r$. We wish to choose ξ and η so that $\mathcal{L}_{m,n}$ is transformed to normal hyperbolic form (Mackie 1965), in which

$$\mathcal{L}_{m,n} = f(\xi,\eta)\frac{\partial^2}{\partial\xi\partial\eta} + \cdots, \tag{7.3.3}$$

where the terms omitted are now of first and zeroth order in $\partial/\partial\xi$ and $\partial/\partial\eta$. Expressing $\mathcal{L}_{m,n}$ in terms of $\partial/\partial\xi$ and $\partial/\partial\eta$ we find that

$$\mathcal{L}_{m,n} = \left[-(2\sqrt{2} + 32q) \left(\frac{\partial \xi}{\partial q} \right) \left(\frac{\partial \xi}{\partial r} \right) + 4q^2 \left(\frac{\partial \xi}{\partial q} \right)^2 + 64 \left(\frac{\partial \xi}{\partial r} \right)^2 \right] \frac{\partial^2}{\partial \xi^2}$$

$$+ \left[-(2\sqrt{2} + 32q) \left(\frac{\partial \eta}{\partial q} \right) \left(\frac{\partial \eta}{\partial r} \right) + 4q^2 \left(\frac{\partial \eta}{\partial q} \right)^2 + 64 \left(\frac{\partial \eta}{\partial r} \right)^2 \right] \frac{\partial^2}{\partial \eta^2}$$

$$+ \left\{ -(2\sqrt{2} + 32q) \left[\left(\frac{\partial \xi}{\partial q} \right) \left(\frac{\partial \eta}{\partial r} \right) + \left(\frac{\partial \xi}{\partial r} \right) \left(\frac{\partial \eta}{\partial q} \right) \right] \right.$$

$$\left. + 8q^2 \left(\frac{\partial \xi}{\partial q} \right) \left(\frac{\partial \eta}{\partial q} \right) + 128 \left(\frac{\partial \xi}{\partial r} \right) \left(\frac{\partial \eta}{\partial r} \right) \right\} \frac{\partial^2}{\partial \eta \partial \xi} + \cdots. \tag{7.3.4}$$

In order that eqn (7.3.3) be satisfied, we must have

$$- (2\sqrt{2} + 32q) \left(\frac{\partial \eta}{\partial q} \right) \left(\frac{\partial \eta}{\partial r} \right) + 4q^2 \left(\frac{\partial \eta}{\partial q} \right)^2 + 64 \left(\frac{\partial \eta}{\partial r} \right)^2 = 0,$$

$$- (2\sqrt{2} + 32q) \left(\frac{\partial \xi}{\partial q} \right) \left(\frac{\partial \xi}{\partial r} \right) + 4q^2 \left(\frac{\partial \xi}{\partial q} \right)^2 + 64 \left(\frac{\partial \xi}{\partial r} \right)^2 = 0. \tag{7.3.5}$$

In other words, $(\partial \xi / \partial q) / (\partial \xi / \partial r)$ and $(\partial \eta / \partial q) / (\partial \eta / \partial r)$ must be the two real roots of the quadratic equation

$$4q^2 x^2 - (2\sqrt{2} + 32q)x + 64 = 0. \tag{7.3.6}$$

The discriminant of this quadratic is positive, so $\mathcal{L}_{m,n}$ *is* hyperbolic, and its characteristic coordinates ξ and η satisfy

$$\left(\frac{\partial \xi}{\partial q} \right) = \left(\frac{1 + 8\sqrt{2}q + \sqrt{(1 + 16\sqrt{2}q)}}{2\sqrt{2}q^2} \right) \left(\frac{\partial \xi}{\partial r} \right) \tag{7.3.7}$$

and

$$\left(\frac{\partial \eta}{\partial q} \right) = \left(\frac{1 + 8\sqrt{2}q - \sqrt{(1 + 16\sqrt{2}q)}}{2\sqrt{2}q^2} \right) \left(\frac{\partial \eta}{\partial r} \right), \tag{7.3.8}$$

where we have arbitrarily assigned the plus sign to ξ and the minus sign to η. For ease of calculation we now choose

$$\partial \xi / \partial r = 1, \qquad \partial \eta / \partial r = 1. \tag{7.3.9}$$

Solving eqns (7.3.7) and (7.3,8), subject to eqn (7.3.9), we find

$$\xi = r + 8 \log \left(\frac{\sqrt{(1 + 16\sqrt{2}q)} - 1}{2} \right) - \frac{8}{[\sqrt{(1 + 16\sqrt{2}q)} - 1]} - 4 \tag{7.3.10}$$

and

$$\eta = r + 8\log\left(\frac{\sqrt{(1 + 16\sqrt{2}q)} + 1}{2}\right) + \frac{8}{[\sqrt{(1 + 16\sqrt{2}q)} + 1]} - 4, \qquad (7.3.11)$$

where the constants of integration have been chosen for future convenience. The characteristics of $\mathcal{L}_{m,n}$ are the two families of lines

$$\xi = \text{const.}, \qquad \eta = \text{const.} \qquad (7.3.12)$$

They play the usual role of limiting the speed of propagation of information, so that a point A cannot be influenced by another point B, if B lies outside the region bounded by the two past characteristics through A.

The explicit forms (7.3.10) and (7.3.11) for ξ and η have the following simple geometrical interpretations. In our dimensionless coordinates (u, v, ρ, ϕ), the null geodesic generators of the weak shock have the parametric representation (6.2.29). That is,

$$u = \sqrt{2}\Lambda,$$

$$v = 4\sqrt{2}\log(\nu\rho_0) - \frac{16\sqrt{2}\Lambda}{(\rho_0)^2},$$

$$x = x_0\left(1 - \frac{8\Lambda}{(\rho_0)^2}\right), \qquad (7.3.13)$$

$$y = y_0\left(1 - \frac{8\Lambda}{(\rho_0)^2}\right),$$

where x and y are the usual Cartesian coordinates, so that $\rho = \left(x^2 + y^2\right)^{\frac{1}{2}}$ and $\Lambda \geq 0$ is an affine parameter along each of the null geodesics. Let us now find the locus of intersection of these null geodesics with a surface S of constant ρ and ϕ, on which

$$\rho = k, \qquad \phi = \phi_0. \qquad (7.3.14)$$

Consider a geodesic which comes through the collision surface at $\rho = \rho_0$, $\phi = \pi + \phi_0$. This geodesic will pass through the caustic at $\rho = 0$ before hitting S (ϕ jumps from $\pi + \phi_0$ at $\rho = 0^-$ to ϕ_0 at $\rho = 0^+$). Hence at its intersection with S

$$\rho_0\left(\frac{8\Lambda}{(\rho_0)^2} - 1\right) = k, \qquad \phi = \phi_0. \qquad (7.3.15)$$

Solving for Λ we find

$$\Lambda = \left(1 + \frac{k}{\rho_0}\right)\frac{(\rho_0)^2}{8}. \qquad (7.3.16)$$

Substituting this into eqn (7.3.13), we find that at the point of intersection

$$\rho = k, \qquad \phi = \phi_0,$$

$$u = \frac{(\rho_0)^2}{4\sqrt{2}} \left(1 + \frac{k}{\rho_0}\right),$$

(7.3.17)

$$v = 4\sqrt{2}\log(\nu\rho_0) - 2\sqrt{2}\left(1 + \frac{k}{\rho_0}\right).$$

Expressing eqn (7.3.17) in terms of r and q one finds

$$q = \frac{1}{4\sqrt{2}}\left(\frac{k}{\rho_0} + 1\right)\left(\frac{\rho_0}{k}\right)^2,$$

(7.3.18)

$$r = 8\log\left(\frac{k}{\rho_0}\right) + 4\left(\frac{k}{\rho_0} + 1\right).$$

Now by eliminating ρ_0/k from eqn (7.3.18) it is easy to show that

$$r + 8\log\left(\frac{\sqrt{(1 + 16\sqrt{2}q)} - 1}{2}\right) - \frac{8}{[\sqrt{(1 + 16\sqrt{2}q)} - 1]} - 4 = 0 \qquad (7.3.19)$$

at the geodesic's intersection with S (here $q > 0$).

Now consider a geodesic generator which originates at $\rho = \rho_0 \geq k$, $\phi = \phi_0$. A geodesic of this type will hit S *before* passing through the caustic. By following a similar argument to that of eqns (7.3.15)–(7.3.19) it is easy to show that at the point of intersection

$$r + 8\log\left(\frac{\sqrt{(1 + 16\sqrt{2}q)} + 1}{2}\right) + \frac{8}{[\sqrt{(1 + 16\sqrt{2}q)} + 1]} - 4 = 0, \qquad (7.3.20)$$

where again $q > 0$.

Thus in the near-field region, before it passes through the caustic, the weak shock intersects S at the line $\eta = 0$ (see Fig. 7.1). This line also marks the boundary of the region between the weak shock and the collision surface, underneath the caustic, in which space–time is flat and all the metric perturbations are zero.

The line $\xi = 0$, though, marks the intersection of the weak shock with the surface S after it has passed through the caustic and is propagating out towards null infinity (again see Fig. 7.1). However, we saw in Sections 6.2 and 6.3 that where the null geodesic generators of the weak shock intersect $\mathcal{I}^+$ the first-order news function has a logarithmic singularity, and that the news function is only significantly non-zero in the region immediately surrounding the weak shock. In our two-dimensional (q, r) plane this region is that in which $|\xi|$ is small and $r \to \infty$, $q \to 0$ (see Fig. 7.2). Thus out near null infinity the pulse of gravitational radiation is centred around $\xi = 0$, and dies away asymptotically as we go far away on either side of this line. ξ is therefore a measure of retarded time.

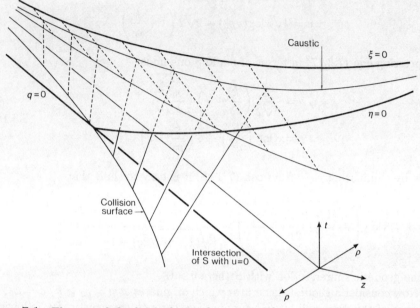

FIG. 7.1. The curved shock 2 is depicted, as viewed from the Minkowskian region III to its past, which lies above the incoming plane shock 1 $(u = 0)$. The heavy black lines all lie in the surface $S(\rho = k, \phi = \phi_0)$. The lines $\xi = 0$ and $\eta = 0$ mark the intersection of the null geodesic generators of shock 2 with S. The generators are drawn bold on the near side of S, and dashed on the far side.

The other characteristics (7.3.12) are obtained from $\xi = 0$ and $\eta = 0$ simply by translating these lines parallel to the q-axis. The line $q = 0$ is itself a characteristic, and so the boundary condition $\chi|_{q=0} = f(r)$ *is* sufficient to determine the solution uniquely in $q > 0$.

7.4 The Green function for the reduced equation

Let us now find the Green function for the differential operator $\mathcal{L}_{m,n}$. It is defined by Mackie (1965)

$$\mathcal{L}_{m,n} G_{m,n}(q, r; q_0, r_0) = \delta(q - q_0)\delta(r - r_0), \tag{7.4.1}$$

where $\mathcal{L}_{m,n}$ acts on the (q, r) part of $G_{m,n}$. Expressed in terms of the Green function $G_{m,n}$, the explicit solution to eqn (7.2.25) at a point (q, r) is

$$\mathcal{X}(q, r) = \int \int G_{m,n}(q, r; q_0, r_0) H(q_0, r_0) dq_0 dr_0, \tag{7.4.2}$$

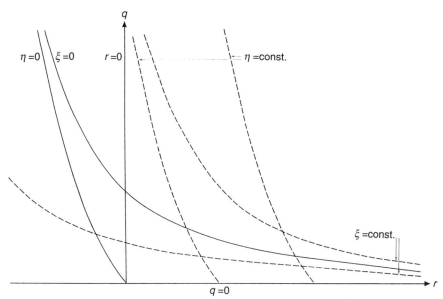

FIG. 7.2. Characteristics ξ = const. and η = const. are shown in the q–r plane. The lines $\xi = 0$ and $\eta = 0$ represent the curved surface of the weak shock 2. The caustic region has been mapped to their 'intersection' at infinite q. In $\eta < 0$, which corresponds to the flat space–time III underneath the curved shock, all the metric perturbations vanish. Non-zero initial data are set on the characteristic surface $q = 0$ for $r > 0$. The gravitational radiation is found from the metric perturbations in the region immediately surrounding $\xi = 0$, in the limit $r \to \infty$.

subject to suitable boundary conditions.

To solve eqn (7.4.1) we use a method of reduction. Because $\mathcal{L}_{m,n}$ is derived from the flat space d'Alembertian $\Box$ in the manner described in Section 7.2, it is clear that $G_{m,n}$ will satisfy

$$\Box[e^{im\phi}\rho^{-n}G_{m,n}(q,r;q_0,r_0)]$$
$$= e^{im\phi}\rho^{-(n+2)}\delta(q-q_0)\delta(r-r_0). \tag{7.4.3}$$

We can solve eqn (7.4.3) using the flat-space Green function $G(t,\mathbf{x};t_0,\mathbf{x_0})$, the explicit form of which is

$$G(t,\mathbf{x};t_0,\mathbf{x_0}) = \frac{-1}{4\pi}\frac{\delta(t-t_0-|\mathbf{x}-\mathbf{x_0}|)}{|\mathbf{x}-\mathbf{x_0}|}, \tag{7.4.4}$$

where t and $\mathbf{x}$ have their usual meanings as time and position vector, respectively. (In choosing the retarded flat-space Green function, we ensure that $G_{m,n}$ will be the retarded Green function of $\mathcal{L}_{m,n}$.) Thus

$$
e^{im\phi}\rho^{-n}G_{m,n}(q,r;q_0,r_0)
$$

$$
= \frac{-1}{4\pi} \iint\limits_{u_1>0} \int_0^\infty \int_0^{2\pi} \frac{\delta[t - t_1\sqrt{(z-z_1)^2 + \rho^2 - 2\rho\rho_1\cos(\phi-\phi_1) + \rho_1^2}]}{\sqrt{(z-z_1)^2 + \rho^2 - 2\rho\rho_1\cos(\phi-\phi_1) + \rho_1^2}}
$$

$$
\times e^{im\phi_1}\rho_1^{-(n+2)}\delta(q_1-q_0)\delta(r_1-r_0)\rho_1\,dt_1\,dz_1\,d\rho_1\,d\phi_1
$$

$$
= \frac{-e^{im\phi}}{4\pi} \iint\limits_{u_1>0} \int_0^\infty \int_0^{2\pi} \frac{\delta[t - t_1 - \sqrt{(z-z_1)^2 + \rho^2 - 2\rho\rho_1\cos\omega + \rho_1^2}]}{\sqrt{(z-z_1)^2 + \rho^2 - 2\rho\rho_1\cos\omega + \rho_1^2}}
$$

$$
\times \cos(m,\omega)\rho_1^{-(n+1)}\delta(q_1-q_0)\delta(r_1-r_0)\,dt_1\,dz_1\,d\rho_1\,d\omega. \tag{7.4.5}
$$

There is no $\sin(m\omega)$ term present in the second multiple integral of eqn (7.4.5) because such a term clearly integrates to zero.

If $f(\omega)$ is a function with n simple zeros $\omega_1,\ldots,\omega_n$, then

$$
\int_{-\infty}^\infty \delta[f(\omega)]g(\omega)\,d\omega = \sum_{i=1}^n g(\omega_i)/|f'(\omega_i)|. \tag{7.4.6}
$$

In eqn (7.4.5) let us first integrate over ω. This involves evaluating an integral of the form (7.4.6). For the particular case (7.4.5),

$$
f(\omega) = t - t_1 - \sqrt{(z-z_1)^2 + \rho^2 - 2\rho\rho_1\cos\omega + \rho_1^2} \tag{7.4.7}
$$

and

$$
g(\omega) = \frac{\cos(m\omega)}{\sqrt{(z-z_1)^2 + \rho^2 - 2\rho\rho_1\cos\omega + \rho_1^2}}. \tag{7.4.8}
$$

As a convenient shorthand, and for reasons which will become apparent, define $\cos\Omega_1$ by

$$
\cos\Omega_1 = \frac{(z-z_1)^2 + \rho^2 + \rho_1^2 - (t-t_1)^2}{2\rho\rho_1}. \tag{7.4.9}
$$

If the space–time coordinates (t,z,ρ) and (t_1,z_1,ρ_1) in eqn (7.4.9) are such that $|\cos\Omega_1| > 1$, then $f(\omega)$ will have no zero. On the other hand, if $|\cos\Omega_1| \le 1$ then $f(\omega)$ will have zeros at

$$
\omega = \pm\arccos(\cos\Omega_1). \tag{7.4.10}
$$

We deduce that

$$
e^{im\phi}\rho^{-n}G_{m,n}(q,r;q_0,r_0) = \frac{-e^{im\phi}}{2\pi} \iint\limits_{u_1>0} \int_0^\infty \frac{\cos(m\Omega_1)}{\rho\rho_1\sin\Omega_1} \times
$$

$$
\times \theta(1 - |\cos\Omega_1|)\rho_1^{-(n+1)}\delta(q_1-q_0)\delta(r_1-r_0)\,dt_1\,dz_1\,d\rho_1, \tag{7.4.11}
$$

where $\cos(m\Omega_1)$ and $\sin\Omega_1$ are related to $\cos\Omega_1$ through the standard trigonometric formulae and $\theta(x)$ is the Heaviside function, defined by

$$\theta(x) = \begin{pmatrix} 1, & x \geq 0, \\ 0, & x < 0. \end{pmatrix} \tag{7.4.12}$$

Now re-expressing $\cos\Omega_1$, first in terms of (u, v, ρ) and then (q, r, ρ), we find that

$$\cos\Omega_1 = \frac{2(u - u_1)(v - v_1) + \rho^2 + \rho_1^2}{2\rho\rho_1}$$

$$= \frac{1}{\sqrt{2}}\left(\frac{\rho}{\rho_1}\right)\left[q - q_1\left(\frac{\rho_1}{\rho}\right)^2\right]\left[8\log\left(\frac{\rho}{\rho_1}\right) - (r - r_1)\right]$$

$$+ \frac{1}{2}\left(\frac{\rho}{\rho_1} + \frac{\rho_1}{\rho}\right). \tag{7.4.13}$$

Also

$$dt_1\,dz_1 = \frac{\rho_1^2}{\sqrt{2}}dq_1\,dr_1. \tag{7.4.14}$$

Integrating out the two remaining delta functions in eqn (7.4.11), we find

$$e^{im\phi}\rho^{-n}G_{m,n}(q, r; q_0, r_0)$$

$$= \frac{-e^{im\phi}}{2\sqrt{2\pi}\rho}\int_0^\infty \frac{\cos(m\Omega_0)}{\sin\Omega_0}\theta(1 - |\cos\Omega_0|)\rho_1^{-n}d\rho_1, \tag{7.4.15}$$

where $\cos\Omega_0$ is defined as in eqn (7.4.9), except that (t_1, z_1, ρ) is replaced by (t_0, z_0, ρ_0). Making the substitution $\rho_1 = y\rho$ in eqn (7.4.15) we find that

$$G_{m,n}(q, r; q_0, r_0) = \frac{-1}{2\sqrt{2\pi}}\int_0^\infty \frac{\cos(m\epsilon)}{\sin\epsilon}\theta(1 - |\cos\epsilon|)\frac{dy}{y^n}, \tag{7.4.16}$$

where now

$$\cos\epsilon = \frac{1}{2y}[1 + y^2 - \sqrt{2}(q - q_0 y^2)(8\log y + r - r_0)]. \tag{7.4.17}$$

One can show that $G_{m,n}$ does vanish outside the region bounded by the past-directed characteristics through (q, r), as it should.

7.5 The second-order transverse metric functions in two-dimensional form

Let

$$e = \rho^2 E, \quad b = \rho^3 B, \quad a = \rho^4 A, \tag{7.5.1}$$

where E, B and A are the first-order metric functions defined by

$$h_{uu}^{(1)} = A, \qquad h_{vv}^{(1)} = 0 \,,$$
$$h_{xx}^{(1)} = -h_{yy}^{(1)} = \left(y^2 - x^2\right)\rho^{-2}E \,,$$
$$h_{uv}^{(1)} = 0 \,, \qquad h_{vx}^{(1)} = 0 \,, \qquad (7.5.2)$$
$$h_{xy}^{(1)} = -2xy\rho^{-2}E \,, \qquad h_{ux}^{(1)} = x\rho^{-1}B \,,$$
$$h_{vy}^{(1)} = 0, h_{uy}^{(1)} = y\rho^{-1}B \,.$$

When the first-order field equations are solved in harmonic (de Donder) gauge, subject to the characteristic initial data (6.2.23) on $u = 0^+$, one finds from eqn (6.3.18) and its obvious analogues for B and A that e, b and a are functions only of (q, r), given by

$$e = \frac{-4\sqrt{2}}{\pi q} \int_0^\infty \int_0^{2\pi} \cos(2\phi')\theta(\mathrm{Arg}) \frac{dy}{y} d\phi' \,,$$

$$b = \frac{32}{\pi q} \int_0^\infty \int_0^{2\pi} \cos\phi'\,\theta(\mathrm{Arg}) \frac{dy}{y^2} d\phi' \,,$$

$$a = \frac{128\sqrt{2}}{\pi q} \int_0^\infty \int_0^{2\pi} \theta(\mathrm{Arg}) \frac{dy}{y^3} d\phi' \,, \qquad (7.5.3)$$

$$- \frac{32}{\pi q^2} \int_0^\infty \int_0^{2\pi} \mathrm{Arg}\,\theta(\mathrm{Arg}) \frac{dy}{y^3} d\phi' \,,$$

where $\theta(x)$ is Heaviside's function and

$$\mathrm{Arg} = \sqrt{2}(8\log y + r) - (1 + y^2) + 2y\cos\phi' \,. \qquad (7.5.4)$$

Also define $d^{(2)}$ and $e^{(2)}$ by

$$d^{(2)} = \rho^4 D^{(2)}, \qquad e^{(2)} = \rho^4 E^{(2)} \,, \qquad (7.5.5)$$

where $D^{(2)}$ and $E^{(2)}$ are the second-order transverse metric functions introduced in eqn (6.4.16):

$$h_{xx}^{(2)} = D^{(2)} + \left(y^2 - x^2\right)\rho^{-2}E^{(2)} \,,$$
$$h_{yy}^{(2)} = D^{(2)} + \left(x^2 - y^2\right)\rho^{-2}E^{(2)} \,,$$
$$h_{xy}^{(2)} = -2xy\rho^{-2}E^{(2)} \,. \qquad (7.5.6)$$

Here $d^{(2)}$ and $e^{(2)}$ are functions of q and r only. We recall that the second-order news function in the boosted frame is defined in terms of $D^{(2)}$ and $E^{(2)}$ through

$$c_0^{(2)} = -\frac{1}{2}\left(\frac{\lambda}{\nu}\right)^2 \lim_{|\mathbf{r}|\to\infty} \left[|\mathbf{r}| \frac{\partial}{\partial\tau}\left(D^{(2)} + E^{(2)}\right)\right] \,, \qquad (7.5.7)$$

once the spurious gauge terms contributing to this equation are eliminated by transforming to Bondi coordinates (Sections 7.6 and 7.7). In harmonic gauge, the second-order metric functions $D^{(2)}$ and $E^{(2)}$ obey the inhomogeneous flat-space wave equations (6.4.19) and (6.4.20). On reduction, these imply that

$$
\begin{aligned}
\mathcal{L}_{0,4}d^{(2)} =&\, 4\sqrt{2}be_{,r} + 2\sqrt{2}qbe_{,qr} - 8\sqrt{2}qbe_{,rr} - (b_{,r})^2 \\
&+ 2\sqrt{2}qb_{,q}e_{,r} - 8\sqrt{2}e_{,r}b_{,r} - 2\sqrt{2}e_{,q}e_{,r} \\
&+ 4e^2 + 4qee_{,q} - 16ee_{,r} - 32qe_{,q}e_{,r} + 4q^2(e_{,q})^2 + 64(e_{,r})^2 \equiv S(q,r)
\end{aligned}
$$
$$(7.5.8)$$

and

$$
\begin{aligned}
\mathcal{L}_{2,4}e^{(2)} =&\, 2ae_{,rr} - 2\sqrt{2}be_{,r} + 2\sqrt{2}qbe_{,qr} - 8\sqrt{2}be_{,rr} \\
&- 2\sqrt{2}qe_{,r}b_{,q} + 8\sqrt{2}e_{,r}b_{,r} + (b_{,r})^2 \\
&- 12qee_{,q} + 32ee_{,r} - 4q^2ee_{,qq} + 32qee_{,rq} - 64ee_{,rr} \equiv T(q,r).
\end{aligned}
$$
$$(7.5.9)$$

From eqn (6.4.17) we find that the boundary conditions on $d^{(2)}$ and $e^{(2)}$ are

$$
d^{(2)}\,|_{q=0} = 16r^2\theta(r), \qquad e^{(2)}\,|_{q=0} = 0. \tag{7.5.10}
$$

It is not difficult to show that the contribution to $d^{(2)}$ from this surface term is

$$
d^{(2)}_{\text{surf}} = \frac{16}{\pi q^2} \int_0^\infty \int_0^{2\pi} \text{Arg}\,\theta(\text{Arg})\frac{dy}{y^3}d\phi', \tag{7.5.11}
$$

where Arg is as in eqn (7.5.4).

The Green function for $d^{(2)}$ is, from eqn (7.4.16),

$$
G_I(q,r;q_0,r_0) = \frac{-1}{2\sqrt{2\pi}} \int_0^\infty \theta(1 - |\cos\epsilon|)\frac{dy}{y^4\sin\epsilon} \tag{7.5.12}
$$

and that for $e^{(2)}$ is

$$
G_{II}(q,r;q_e,r_0) = \frac{-1}{2\sqrt{2\pi}} \int_0^\infty \theta(1 - |\cos\epsilon|)\frac{\cos(2\epsilon)dy}{y^4\sin\epsilon}, \tag{7.5.13}
$$

where, from eqn (7.4.17),

$$
\cos\epsilon = \frac{1}{2y}[1 + y^2 - \sqrt{2}(q - q_0y^2)(8\log y + r - r_0)]. \tag{7.5.14}
$$

We saw in Section 7.3 that to calculate the news function we must take the field point (q,r) out to the region in which r is very large and $|\xi|$ is $O(1)$ (and so q is small). From the definition (7.3.10) for ξ we find

$$
\xi = r + 8\log(4\sqrt{2}q) - \frac{1}{\sqrt{2q}} - 8 + O(q). \tag{7.5.15}
$$

Now let $x = y/4\sqrt{2}q$. Then when x, q_0 and r_0 are $O(1)$ and r is very large,

$$\cos\epsilon = \frac{1}{8\sqrt{2}qx}\left[1 + 32q^2x^2 - \sqrt{2}(q - 32x^4q^2q_0)\times\right.$$

$$\left.\times\left(8\log x + \frac{1}{\sqrt{2}q} + 8 + \xi - r_0 + O(q)\right)\right]$$

$$= \frac{16\sqrt{2}q_0x^2 - (8\log x + \xi - r_0 + 8)}{8x} + O(q), \qquad (7.5.16)$$

and thus

$$G_I = \frac{-1}{2\sqrt{2}\pi}\frac{1}{(4\sqrt{2}q)^3}\int_0^\infty \theta(1 - |\cos\epsilon|)\frac{dx}{x^4\sin\epsilon}$$

$$+ O(q^{-2}), \qquad (7.5.17)$$

and

$$G_{II} = \frac{-1}{2\sqrt{2}\pi}\frac{1}{(4\sqrt{2}q)^3}\int_0^\infty \theta(1 - |\cos\epsilon|)\frac{\cos(2\epsilon)dx}{x^4\sin\epsilon}$$

$$+ O(q^{-2}), \qquad (7.5.18)$$

where $\cos\epsilon$ is equal to the first term on the right-hand side of eqn (7.5.16). We can in fact ignore the $O(q^{-2})$ terms in eqns (7.5.17) and (7.5.18), since they do not contribute to the news function.

In Section 6.5, we derived a mass-loss formula (eqn (6.5.27)), which showed that if the gravitational radiation obeyed certain uniformity conditions, then the mass m_{final} of the final black hole (assumed to be a static Schwarzschild geometry) produced by the speed-of-light collision must be $m_{\text{final}} = \frac{3}{2}\mu + 4\int_{-\infty}^\infty a_2(\tau/\mu)d\tau$. Since it is the time-integral of the news function which is required in this formula, and not the news function itself, the quantity that we shall compute directly (as described further in Chapter 8) will be the combination $d^{(2)} + e^{(2)}$ of metric components, and not its time derivative. When we require the news function in Chapter 8, we shall differentiate numerically.

The surface term (7.5.11) contributes

$$\frac{\sqrt{2}}{\pi q^3}\int_0^\infty\int_0^{2\pi} (8\log x + \xi + 8 + 8x\,\cos\phi')\times$$

$$\times\,\theta(8\log x + \xi + 8 + 8x\,\cos\phi')\frac{dx\,d\phi'}{x^3} \qquad (7.5.19)$$

(plus irrelevant higher-order terms) to $d^{(2)} + e^{(2)}$ when $r \to \infty$ with ξ constant. There is, of course, also the source term

$$\iint_{\xi_0 < \xi, \eta_0 < \eta, q_0 > 0} [G_I(q, r; q_0, r_0)S(q_0, r_0)$$

$$+ G_{II}(q, r; q_0, r_0)T(q_0, r_0)]dq_0dr_0 \qquad (7.5.20)$$

to be added to this, where the source functions S and T are defined in eqns (7.5.8) and (7.5.9).

7.6 Eliminating logarithmic terms from the second-order transverse metric coefficients

It has been known for a long time (see Fock 1964) that harmonic gauges are complicated by the appearance of $(\log|\mathbf{r}|)/|\mathbf{r}|$ terms in the metric tensor at second and higher orders in perturbation theory ($|\mathbf{r}|$ is the radial coordinate). Initially it was not clear what, if any, physical significance these terms had, nor if gravitational radiation theory could be properly defined in such coordinate systems. (Naive calculations, using

$$\frac{dE}{d\Omega dt} = \frac{r_{ret}^2}{32\pi} \langle h_{jk}^{TT} h_{jk}^{TT} \rangle_{av}$$

(see Misner et al. 1973), of the power radiated per unit solid angle predict an infinite quantity of gravitational radiation.) However, it has been shown by Isaacson and Winicour (1968) in the axisymmetric case, and by Madore (1970a, b) for the general case, that these $(\log|\mathbf{r}|)/|\mathbf{r}|$ terms are coordinate artefacts which can be eliminated by transforming to a Bondi gauge, so that the news function is still well-defined. In this section we calculate the $(\log|\mathbf{r}|)/|\mathbf{r}|$ term in the transverse part of the second-order metric perturbation $h_{ab}^{(2)}$, and then show how to eliminate it by finding an explicit coordinate transformation to a Bondi gauge. The $(\log|\mathbf{r}|)/|\mathbf{r}|$ terms in the metric are produced, in the source integral (7.5.20), by the region of q_0–r_0 space corresponding to a source point near future null infinity $\mathcal{I}^+$, where (in particular) $q_0 \ll 1$. It is thus necessary to estimate the magnitudes of the Green functions G_I, G_{II} and of the source functions $S(q_0, r_0)$, $T(q_0, r_0)$ when (q_0, r_0) lies in this region.

From eqns (7.5.17) and (7.5.18):

$$G_{[I,II]}(q, r; q_0, r_0) = \frac{-1}{2\sqrt{2\pi}} \cdot \frac{1}{(4\sqrt{2}q)^3} \times$$
$$\times \int_0^\infty \theta(1 - |\cos \epsilon|) \left\{ \begin{pmatrix} 1 \\ \cos 2\epsilon \end{pmatrix} \right\} \frac{dx}{x^4 \sin \epsilon} \qquad (7.6.1)$$

plus corrections negligible in the radiation zone, where

$$\cos \epsilon = \frac{16\sqrt{2}q_0 x^2 - (8 \log x + \xi - r_0 + 8)}{8x}. \qquad (7.6.2)$$

If q_0 is small then

$$r = \xi_0 - 8 \log(4\sqrt{2}q_0) + \frac{1}{\sqrt{2}q_0} + 8 + O(q_0). \qquad (7.6.3)$$

Define y by $y = 4\sqrt{2}q_0 x$. Then

$$G_{[I,II]} = \frac{-1}{2\sqrt{2}\pi} \cdot \left[\frac{q_0}{q}\right]^3 \int_0^\infty \theta(1 - |\cos\epsilon|) \left\{ \begin{pmatrix} 1 \\ \cos 2\epsilon \end{pmatrix} \right\} \frac{dy}{y^4 \sin\epsilon}$$

$$+ \frac{O(q_0^4)}{q^3} \tag{7.6.4}$$

where now

$$\cos\epsilon = \frac{1 + y^2 - \sqrt{2}q_0(8\log y + \xi - \xi_0)}{2y}. \tag{7.6.5}$$

Now assume that $(\xi - \xi_0)$ is $O(1)$ (so that we are restricting attention to the region immediately surrounding the weak shock). Let $y = 1 + \left[\sqrt{2}q_0\,(\xi - \xi_0)\right]^{\frac{1}{2}} z$. Then

$$\cos\epsilon = 1 + \frac{q_0}{\sqrt{2}}(\xi - \xi_0)(z^2 - 1) + q_0^{3/2} f(z) + O(q_0^2), \tag{7.6.6}$$

where $f(z) = -f(-z)$ (the explicit form of $f(z)$ may be easily found, but it is not important here). Using eqn (7.6.6) it is not difficult to show that

$$G_I = G_{II} = \frac{-1}{2\sqrt{2}} \left[\frac{q_0}{q}\right]^3 + \frac{O(q_0^4)}{q^3} \tag{7.6.7}$$

if $\xi_0 \le \xi$, while if $\xi_0 > \xi$ then they both vanish (there is no $q_0^{7/2}/q^3$ term in eqn (7.6.7) because $f(z)$ is odd).

Let us now examine the behaviour of the source functions $S(q_0, r_0)$ and $T(q_0, r_0)$ in this region (where ξ_0 is $O(1)$ and $q_0 \ll 1$) of the q_0–r_0 plane. From eqn (7.5.4) we have

$$\text{Arg} = \sqrt{2}q_0(8\log y + r_0) - (1 + y^2) + 2y\cos\phi'. \tag{7.6.8}$$

Define $\overline{\text{Arg}} = \text{Arg}/(\sqrt{2}q_0)$ and $x = y/4\sqrt{2}q_0$; then using eqn (7.6.3) we find

$$\overline{\text{Arg}} \simeq 8\log x + \xi_0 + 8 + 8x\cos\phi' - 16\sqrt{2}q_0(x^2 - 1), \tag{7.6.9}$$

and, to lowest order in q_0,

$$e = \frac{-8\sqrt{2}}{\pi q_0} \int_0^\infty \int_0^\pi \cos(2\phi')\theta(\overline{\text{Arg}}) \frac{dx\,d\phi'}{x},$$

$$b = \frac{8\sqrt{2}}{\pi q_0^2} \int_0^\infty \int_0^\pi \cos\phi'\theta(\overline{\text{Arg}}) \frac{dx\,d\phi'}{x^2}, \tag{7.6.10}$$

$$a = \frac{8\sqrt{2}}{\pi q_0^3} \int_0^\infty \int_0^\pi \left[1 - \frac{\overline{\text{Arg}}}{4}\right]\theta(\overline{\text{Arg}}) \frac{dx\,d\phi'}{x^3}$$

(where in each case the terms that have been neglected are of order $q_0 \ln q_0$ times the leading term). Each of the functions in eqn (7.6.10) will possess a series expansion in q_0. That is,

$$e = \frac{1}{q_0}[e_1(\xi_0) + q_0 \log(q_0)e_2(\xi_0) + \cdots], \qquad (7.6.11a)$$

$$b = \frac{1}{q_0^2}[b_1(\xi_0) + q_0 \log(q_0)b_2(\xi_0) + \cdots], \qquad (7.6.11b)$$

$$a = \frac{1}{q_0^3}[a_1(\xi_0) + q_0 \log(q_0)a_2(\xi_0) + \cdots], \qquad (7.6.11c)$$

[Among the functions $a_i(\xi_0)$, only $a_1(\xi_0)$ will be used in this section, and the notation of eqn (7.6.11c) will not be used in any other section. There is thus no risk of confusion with the functions $a_0(\hat{\tau}/\mu)$, $a_2(\hat{\tau}/\mu)$, ... appearing in the series (7.1.2).]

Now

$$\left[\frac{\partial}{\partial q}\right]_r = \left[\frac{\partial}{\partial q}\right]_\xi + \left[\frac{\partial \xi}{\partial q}\right]_r \left[\frac{\partial}{\partial \xi}\right]_q,$$

$$\left[\frac{\partial}{\partial r}\right]_q = \left[\frac{\partial}{\partial \xi}\right]_q, \qquad (7.6.12)$$

and when q is small $(\partial \xi/\partial q)_r = 1/\sqrt{2}q^2 + O(1/q)$. Substituting eqns (7.6.11) and (7.6.12) into eqns (7.5.8) and (7.5.9) we find that in the region of interest

$$S(q_0, r_0) = (q_0)^{-4}\{2b_1(\xi_0)e_1''(\xi_0) - [b_1'(\xi_0)]^2$$
$$+ 2b_1'(\xi_0)e_1'(\xi_0)\} + O\left(\frac{\log(q_0)}{(q_0)^3}\right) \qquad (7.6.13)$$

and

$$T(q_0, r_0) = (q_0)^{-4}\{2a_1(\xi_0)e_1''(\xi_0) + 2b_1(\xi_0)e_1''(\xi_0) - 2b_1'(\xi_0)e_1'(\xi_0)$$
$$+ [b_1'(\xi_0)]^2 - 2e_1(\xi_0)e_1''(\xi_0)\} + O\left(\frac{\log(q_0)}{(q_0)^3}\right), \qquad (7.6.14)$$

and therefore

$$S + T = (q_0)^{-4}\{2[a_1(\xi_0) + 2b_1(\xi_0) - e_1(\xi_0)e_1''](\xi_0)\}$$
$$+ O(\frac{\log(q_0)}{(q_0)^3}). \qquad (7.6.15)$$

This expression may be simplified using the harmonic (de Donder) gauge conditions (6.3.2) relating the first-order metric functions E, B, A defined in eqn (7.5.2). These conditions give

$$A_{,v} + \frac{1}{\rho}B + B_{,\rho} = 0, \qquad B_{,v} - \frac{2}{\rho}E - E_{,\rho} = 0. \qquad (7.6.16)$$

Written in terms of e, b and a, eqn (7.6.16) becomes

$$- \sqrt{2}a_{,r} - 2b - 2qb_{,q} + 8b_{,r} = 0,$$
$$- \sqrt{2}b_{,r} + 2qe_{,q} - 8e_{,r} = 0. \tag{7.6.17}$$

Substituting eqns (7.6.11) and (7.6.12) into the above, we find

$$a_1'(\xi_0) + b_1'(\xi_0) = 0, \qquad b_1'(\xi_0) - e_1'(\xi_0) = 0 \tag{7.6.18}$$

and hence

$$a_1(\xi_0) + 2b_1(\xi_0) - e_1(\xi_0) = \lambda(\text{const.}). \tag{7.6.19}$$

We can find λ most easily by examining the behaviour of e, b and a in eqn (7.6.10) as $\xi_0 \to -\infty$. Now if $\xi_0 \ll -1$ then $\overline{\text{Arg}}$ will be zero unless x is large. This implies that $\lim\limits_{\xi_0 \to -\infty} b_1(\xi_0) = \lim\limits_{\xi_0 \to -\infty} a_1(\xi_0) = 0$, since there are factors of x^{-2} and x^{-3} in the integrands of b and e respectively. From eqn (7.6.10)

$$e = \frac{-4\sqrt{2}}{\pi q_0} \int_{|\cos \epsilon| \leq 1} \sin(2\epsilon) \frac{dx}{x} + o(q_0^{-1}), \tag{7.6.20}$$

where $\cos \epsilon = \frac{1}{8x} \left[16\sqrt{2}q_0 \left(x^2 - 1 \right) - (8 \log x + \xi_0 + 8) \right]$. We recall here that the standard notation $g(z) = o(f(z))$ as $z \to 0$ means that $g(z)/f(z) \to 0$ as $z \to 0$. In eqn (7.6.20), the $o\left(q_0^{-1}\right)$ term refers to the limit $q_0 \to 0$. When $\xi_0 \ll -1$ the lower bound on x is $x_L \simeq -\left(\xi_0 + 8\right)/8$ and the upper bound is $x_u \simeq 1/2\sqrt{2}q_0$. Hence

$$\lim_{\xi_0 \to -\infty} e = \frac{-8\sqrt{2}}{\pi q_0} \left\{ \lim_{\xi_0 \to -\infty} \int_{-(\xi_0+8)/8}^{\infty} \frac{-(\xi_0 + 8)}{8x} \left[1 - \left(\frac{\xi_0 + 8}{8x} \right)^2 \right]^{1/2} \frac{dx}{x} \right.$$
$$\left. + \int_0^{1/2\sqrt{2}q_0} 2\sqrt{2}q_0 x \sqrt{1 - (2\sqrt{2}q_0 x)^2} \frac{dx}{x} \right\} + o(q_0^{-1})$$
$$= \frac{-8\sqrt{2}}{\pi q_0} \left[\int_1^{\infty} y^{-3} \sqrt{(y^2 - 1)} dy + \int_0^1 \sqrt{(1 - z^2)} dz \right] + o(q_0^{-1})$$
$$= \frac{-4\sqrt{2}}{q_0} + o(q_0 - 1). \tag{7.6.21}$$

Thus $\lim\limits_{\xi_0 \to -\infty} e_1(\xi_0) = -4\sqrt{2}$, giving $\lambda = 4\sqrt{2}$ in eqn (7.6.19), and

$$S(q_0, r_0) + T(q_0, r_0) = 8\sqrt{2}e_1''(\xi_0)q_0^{-4}$$
$$+ O[\log(q_0)/(q_0)^3]. \tag{7.6.22}$$

The source term contribution to $d^{(2)} + e^{(2)}$ is

$$\iint_{q_0 < 0, \eta_0 < \eta, \xi_0 < \xi} (G_I S + G_{II} T) \, dq_0 dr_0.$$

In this integral $dq_0 dr_0$ may be replaced by $[\partial (q_0, r_0) / \partial (\xi_0, \eta_0)] d\xi_0 d\eta_0$, where the characteristic coordinates ξ, η are defined in eqns (7.3.10) and (7.3.11). When $q_0 \ll 1$, $\partial (q_0, r_0) / \partial (\xi_0, \eta_0) \simeq \sqrt{2} q_0^2$. Substituting eqns (7.6.7) and (7.6.22) into the source integral, we find the contribution to $d^{(2)} + e^{(2)}$ from the region in which η_0 is large and ξ_0 is $O(1)$ to be

$$\frac{1}{q^3} \int_\Lambda^\eta \int_{-\infty}^\xi [-4\sqrt{2} q_0 e_1''(\xi_0) + o(q_0)] d\xi_0 d\eta_0. \tag{7.6.23}$$

Here ξ and η are the coordinates of the field point and Λ is a lower cut-off which is $O(1)$ (with respect to η) but is sufficiently large that the series expansions derived earlier are valid at $\eta_0 = \Lambda$. Since $\sqrt{2} q_0 = \frac{1}{\eta_0} + o(\eta_0^{-1})$ in the region of interest, (7.6.23) reduces to

$$-\frac{4 \log \eta}{q^3} e_1'(\xi) + \text{terms of } O(q^{-3}) \text{ and less.} \tag{7.6.24}$$

The contribution to $D^{(2)} + E^{(2)}$ from the logarithmic term is therefore

$$-\frac{4 \log \eta}{q^3 \rho^4} e_1'(\xi). \tag{7.6.25}$$

Now

$$\xi = r + 8 \log(4\sqrt{2} q) - \frac{1}{\sqrt{2} q} - 8 + O(q),$$

$$\eta = r + O(\log q),$$

$$r = 8 \log(\nu \rho) - \sqrt{2} v, \qquad q = u \rho^{-2},$$

$$\rho = |\mathbf{r}| \sin \theta, \tag{7.6.26}$$

$$v = \frac{1}{\sqrt{2}} (-2 |\mathbf{r}| \sin^2 \frac{\theta}{2} - \tau),$$

$$u = \frac{1}{\sqrt{2}} (2 |\mathbf{r}| \cos^2 \frac{\theta}{2} + \tau),$$

where $\tau = t - |\mathbf{r}|$ is a retarded time coordinate. Hence (7.6.25) equals

$$-4\sqrt{2} \left[\frac{\log |\mathbf{r}| + \log(2 \sin^2 \theta/2)}{|\mathbf{r}|} \right] \tan^2 \frac{\theta}{2} \sec^2 \frac{\theta}{2} e_1'(T)$$

$$+ o \left[\frac{1}{|\mathbf{r}|} \right], \tag{7.6.27}$$

where

$$T = \tau \sec^2 \frac{\theta}{2} - 8 \log \left(\frac{2 \tan \theta/2}{\nu} \right) + 8 \log 8 - 8.$$

(Note that we are still using dimensionless coordinates here.)

It is easy to see that the contribution to $D^{(2)} + E^{(2)}$ from the source region in which q_0 and r_0 are $O(1)$ is $O\left(1/|\mathbf{r}|\right)$; while that from the region in which η_0 is large and ξ_0 is $O\left(1/q_0\right)$ (i.e. between the weak-shock region and the initial surface) is $o\left(1/|\mathbf{r}|\right)$. The contribution from the surface term (7.5.19) is also $O\left(1/|\mathbf{r}|\right)$. Hence eqn (7.6.27) contains the only $\log r/r$ term in $D^{(2)} + E^{(2}$ (from here on we write r instead of $|\mathbf{r}|$ since there will be no danger of confusion with the two-dimensional coordinate of that name).

We shall now show that this $\log r/r$ term is eliminated from $h^{(2)}_{\phi\phi}$ when we transform to a Bondi coordinate system. In such a coordinate system the metric has the form (Bondi et $al.$ 1962)

$$ds^2 = \nu^2\left\{-\left[1 - \frac{2M}{\hat{r}} + O\left(\frac{1}{\hat{r}^2}\right)\right]d\hat{\tau}^2 - 2\left[1 + O\left(\frac{1}{\hat{r}^2}\right)\right]d\hat{\tau}d\hat{r}\right.$$

$$-2\left[\frac{(\partial c/\partial\hat{\theta}_2 c \cot\hat{\theta})}{\hat{r}} + O\left(\frac{1}{\hat{r}^2}\right)\right]\hat{r}d\hat{\tau}d\hat{\theta}$$

$$+\hat{r}^2\left[1 + \frac{2c}{\hat{r}} + O\left(\frac{1}{\hat{r}^2}\right)\right]d\hat{\theta}^2$$

$$\left.+\hat{r}^2\sin^2\hat{\theta}\left[1 - \frac{2c}{\hat{r}} + O\left(\frac{1}{\hat{r}^2}\right)\right]d\phi^2\right\}. \tag{7.6.28}$$

We shall endeavour to put our metric in this form, to first order in $e^{-2\alpha}$, by searching for an explicit coordinate transformation from our harmonic gauge to a Bondi gauge.

The first-order metric perturbation $h^{(1)}_{ab}$ is given by eqn (7.5.2) in terms of the metric functions $E = \rho^{-2}e$, $B = \rho^{-3}b$ and $A = \rho^{-4}a$ (eqn (7.5.1)). In the asymptotic region of space–time 'near' $\mathcal{I}^+$, $e = e_1\left[(\xi)/q\right] + o\left(q^{-1}\right)$, $b = \left[(e_1(\xi) - e_0)/q^2\right] + o\left(q^{-2}\right)$ and $a = -\left[(e_1(\xi) - e_0)/q^3\right] + o\left(q^{-3}\right)$, where $e_0 = \lim_{\xi\to-\infty} e_1(\xi) = -4\sqrt{2}$. Hence (using eqn (7.6.26))

$$E = \frac{\sec^2\theta/2}{\sqrt{2}r}e_1(T) + o(r^{-1}),$$

$$B = \frac{\tan\theta/2\sec^2\theta/2}{r}[e_1(T) - e_0] + o(r^{-1}), \tag{7.6.29}$$

$$A = -\frac{\sqrt{2}\tan^2\theta/2\sec^2\theta/2}{r}[e_1(T) - e_0] + (r^{-1}),$$

where T was defined after eqn (7.6.27). Now transforming eqn (7.5.2) to coordinates (τ, r, θ, ϕ) in which the background metric is flat space–time in the Bondi form (7.6.28):

$$ds^2 = \nu^2\left[-d\tau^2 - 2d\tau\,dr + r^2\left(d\theta^2 + \sin^2\theta\,d\phi^2\right)\right], \tag{7.6.30}$$

we find

$$h^{(1)}_{\tau\tau} = \frac{-\tan^2\theta/2\sec^2\theta/2}{\sqrt{2}r}[e_1(T) - e_0] + o(r^{-1}),$$

$$h^{(1)}_{\tau r} = o(r^{-1}),$$

$$h^{(1)}_{\tau\theta} = r\left[\frac{\tan\theta/2\sec^2\theta/2}{\sqrt{2}r}[e_1(T) - e_0] + o(r^{-1})\right],$$

$$h^{(1)}_{\tau\phi} = 0,$$

$$h^{(1)}_{rr} = \frac{-2\sqrt{2}\sin^2\theta/2}{r}e_0 + o(r^{-1}),$$

$$h^{(1)}_{r\theta} = r\left[-\frac{\sqrt{2}\tan\theta/2\cos\theta}{r}e_0 + O(r^{-1})\right], \tag{7.6.31}$$

$$h^{(1)}_{r\phi} = 0,$$

$$h^{(1)}_{\theta\theta} = r^2\left[\frac{-\sec^2\theta/2}{\sqrt{2}r}e_1(T) + \frac{2\sqrt{2}\sin^2\theta/2}{r}e_0 + o(r^{-1})\right],$$

$$h^{(1)}_{\theta\phi} = 0,$$

$$h^{(1)}_{\phi\phi} = r^2\sin^2\theta\left[\frac{\sec^2\theta/2}{\sqrt{2}r}e_1(T) + o(r^{-1})\right].$$

If we make a gauge transformation

$$\tau = \hat{\tau} + e^{-2\alpha}\xi_{\hat{\tau}}, \qquad r = \hat{r} + e^{-2\alpha}\xi_{\hat{r}}, \qquad \theta = \hat{\theta} + e^{-2\alpha}\xi_{\hat{\theta}} \tag{7.6.32}$$

of the flat background metric (7.6.30), then (7.6.30) transforms to

$$\begin{aligned}
ds^2 = &-d\hat{\tau}^2 - 2d\hat{\tau}d\hat{r} + \hat{r}^2(d\hat{\theta}^2 + \sin^2\hat{\theta}\,d\phi^2) \\
&+ e^{-2\alpha}[-2(\xi_{\hat{\tau},\hat{\tau}} + \xi_{\hat{r},\hat{\tau}})d\hat{\tau}^2 - 2(\xi_{\hat{\tau},\hat{r}} + \xi_{\hat{r}\hat{r}} + \xi_{\hat{\tau},\hat{\tau}}) \\
&d\hat{r}d\hat{\tau} - 2\xi_{\hat{\tau},\hat{r}}d\hat{r}^2 - 2(\xi_{\hat{\tau},\hat{\theta}} + \xi_{\hat{r},\hat{\theta}} - r^2\xi_{\hat{\theta},\hat{\tau}})d\hat{\tau}d\hat{\theta} \\
&- 2(\xi_{\hat{\tau},\hat{\theta}} - r^2\xi_{\hat{\theta},\hat{r}})d\hat{r}d\hat{\theta} + 2(\hat{r}\xi_{\hat{r}} + \hat{r}^2\xi_{\hat{\theta},\hat{\theta}}) \\
&d\hat{\theta}^2 + 2(\hat{r}\xi_{\hat{r}}\sin^2\hat{\theta} + \hat{r}^2\sin\hat{\theta}\cos\hat{\theta}\,\xi_{\hat{\theta}})d\phi^2] + O(e^{-4\alpha}). \tag{7.6.33}
\end{aligned}$$

If $\xi_{\hat{a}}$ is to transform the metric into Bondi form then $\xi_{\hat{\tau}}$, $\xi_{\hat{r}}$ and $\xi_{\hat{\theta}}$ must all possess series expansions in $\hat{r}$. That is,

$$\begin{aligned}
\xi_{\hat{\tau}} &= f_1(\hat{\tau}, \hat{\theta})\log\hat{r} + f_2(\hat{\tau}, \hat{\theta}) + o(1), \\
\xi_{\hat{r}} &= g_1(\hat{\tau}, \hat{\theta})\log\hat{r} + g_2(\hat{\tau}, \hat{\theta}) + o(1), \\
\xi_{\hat{\theta}} &= h_1(\hat{\tau}, \hat{\theta})\frac{\log\hat{r}}{\hat{r}} + \frac{1}{\hat{r}}h_2(\hat{\tau}, \hat{\theta}) + o(\hat{r}^{-1}).
\end{aligned} \tag{7.6.34}$$

Let $h^{(1)B}_{ab}$ denote the Bondi metric perturbations and $h^{(1)H}_{ab}$ the harmonic ones. Clearly, $h^{(1)\,H}_{rr} - 2\xi_{\hat{\tau},\hat{r}} = h^{(1)B}_{\hat{\tau}\hat{\tau}}$. More explicitly,

$$\frac{f_1(\hat{\tau}, \hat{\theta})}{\hat{r}} = \frac{-\sqrt{2}\sin^2(\hat{\theta}/2)e_0}{r} + o(r^{-1})$$

$$= \frac{-\sqrt{2}\sin^2(\hat{\theta}/2)e_0}{\hat{r}} + O(e^{-2\alpha}), \qquad (7.6.35)$$

and so $f_1\left(\hat{\tau}, \hat{\theta}\right) = -\sqrt{2}e_0 \sin^2\left(\hat{\theta}/2\right)$. (The $O\left(e^{-2\alpha}\right)$ term is irrelevant here since it affects only the second- and higher-order metric perturbations.) Also, $h_{\theta\theta}^{(1)H} + 2\left(\hat{r}\xi_r + \hat{r}^2\xi_{\hat{\theta},\hat{\theta}}\right) = h_{\hat{\theta}\hat{\theta}}^{(1)B}$, which when written out in full is

$$-\frac{\hat{r}}{\sqrt{2}}\sec^2\frac{\hat{\theta}}{2}e_1(T) + 2\sqrt{2}\hat{r}\sin^2\frac{\hat{\theta}}{2}e_0$$

$$+ 2\left[\left(g_1(\hat{\tau}, \hat{\theta}) + \frac{\partial h_1(\hat{\tau}, \hat{\theta})}{\partial\hat{\theta}}\right)\hat{r}\log\hat{r}\right.$$

$$\left. + \left(g_2(\hat{\tau}, \hat{\theta}) + \frac{\partial h_2(\hat{\tau}, \hat{\theta})}{\partial\hat{\theta}}\right)\hat{r}\right] = 2\hat{r}e^{2\alpha}c^{(1)}. \qquad (7.6.36)$$

The corresponding $\phi\phi$ equation is

$$\frac{\hat{r}}{\sqrt{2}}\sin^2\hat{\theta}\sec^2\frac{\hat{\theta}}{2}e_1(T) + 2\sin^2\hat{\theta}\{[g_1(\hat{\tau}, \hat{\theta}) + \cot\hat{\theta}\,h_1(\hat{\tau}, \hat{\theta})]\hat{r}\log\hat{r}$$

$$+ [g_2(\hat{\tau}, \hat{\theta}) + \cot\hat{\theta}\,h_2(\hat{\tau}, \hat{\theta})]\hat{r}\} = -2\hat{r}\sin^2\hat{\theta}\,e^{2\alpha}c^{(1)}. \qquad (7.6.37)$$

The $\hat{r}\log\hat{r}$ terms must vanish in both these equations, and so $\partial h_1(\hat{\tau}, \hat{\theta})/\partial\hat{\theta} = \cot\hat{\theta}\,h_1(\hat{\tau}, \hat{\theta})$, which when integrated yields $h_1(\hat{\tau}, \hat{\theta}) = k(\hat{\tau})\sin\hat{\theta}$ (whence $g_1(\hat{\tau}, \hat{\theta}) = -k(\hat{\tau})\cos\hat{\theta}$). Multiplying eqn (7.6.36) by $\sin^2\hat{\theta}$ and adding it to eqn (7.6.37) leads to

$$\sqrt{2}\sin^2\frac{\hat{\theta}}{2}e_0 + 2g_2(\hat{\tau}, \hat{\theta}) + \frac{\partial h_2(\hat{\tau}, \hat{\theta})}{\partial\hat{\theta}} + \cot\hat{\theta}h_2(\hat{\tau}, \hat{\theta}) = 0, \qquad (7.6.38)$$

while subtracting yields

$$-\sqrt{2}\sec^2\frac{\hat{\theta}}{2}e_1(T) + 2\sqrt{2}\sin^2\frac{\hat{\theta}}{2}e_0$$

$$+ 2\left[\frac{\partial h_2(\hat{\tau}, \hat{\theta})}{\partial\hat{\theta}} - \cot\hat{\theta}h_2(\hat{\tau}, \hat{\theta})\right] = 4e^{2\alpha}c^{(1)}. \qquad (7.6.39)$$

The most general form that $c\left(\hat{\tau}, \hat{\theta}\right)$ can take is

$$c = e^{-2\alpha}\left[\frac{-\sec^2\hat{\theta}/2}{2\sqrt{2}}e_1(T) + \frac{1}{\sqrt{2}}\sin^2\frac{\hat{\theta}}{2}e_0\right.$$

$$+ \frac{1}{2}[\alpha'(\hat{\theta})\cot\hat{\theta} - \alpha''(\hat{\theta})] + O(e^{-4\alpha}). \tag{7.6.40}$$

(The $\hat{\tau}\hat{\theta}$ equation, $\xi_{\hat{\tau},\hat{\theta}} + \xi_{\hat{r},\hat{\theta}} - \hat{r}^2\xi_{\hat{\theta},\hat{\tau}} = O(1)$, implies that the time-varying part of $\xi_{\hat{\theta}}$ is $o(\hat{r}^{-1})$. Then eqn (7.6.38) ensures that the time-dependent part of $\xi_{\hat{r}}$ is $o(1)$. Hence $\lim_{\hat{r}\to\infty} \hat{r}^{-1}(\sin\hat{\theta})^{-2} \partial h_{\phi\phi}^{(1)B}/\partial\hat{\tau} = \lim_{r\to\infty} r^{-1}(\sin\theta)^{-2} \partial h_{\phi\phi}^{(1)H}/\partial\tau$, which proves rigorously that the formula

$$c_0^{(1)} = -\tfrac{1}{2}e^{-2a} \lim_{r\to\infty} r^{-1}(\sin\theta)^{-2} \frac{\partial h_{\phi\phi}^{(1)H}}{\partial\tau}$$

used in Section 6.3 to derive the first-order news function is correct, and leads directly to eqn (7.6.40). In eqn (7.6.40) the derivative of the first term is the news function found previously, the second term is included for convenience, and the third term incorporates the supertranslation freedom (Bondi *et al.* 1962).)

Since the time-dependent parts of $\xi_{\hat{r}}$ and $\xi_{\hat{\theta}}$ are o(1) and $o(\hat{r}^{-1})$ respectively,

$$g_i\left(\hat{\tau},\hat{\theta}\right) = g_i\left(\hat{\theta}\right)$$

and

$$h_i\left(\hat{\tau},\hat{\theta}\right) = h_i\left(\hat{\theta}\right);$$

whence $k(\hat{\tau}) = K$. Now combining eqns (7.6.39) and (7.6.40) we find

$$\frac{\partial h_2(\hat{\theta})}{\partial\hat{\theta}} - \cot\hat{\theta} h_2(\hat{\theta}) = \alpha'(\hat{\theta}) = \alpha'(\hat{\theta})\cot\hat{\theta} - \alpha''(\hat{\theta}). \tag{7.6.41}$$

Thus $h_2(\hat{\theta}) = L\sin\hat{\theta} - \alpha'(\hat{\theta})$ (whence, from eqn (7.6.38), $g_2(\hat{\theta}) = \frac{-1}{\sqrt{2}}\sin^2(\frac{\hat{\theta}}{2})e_0$ $-L\cos\hat{\theta}+\frac{1}{2}\alpha''(\hat{\theta})+\frac{1}{2}\cot\hat{\theta}\alpha'(\hat{\theta}))$. The $\hat{\tau}\hat{\tau}$ equation, $h_{\hat{\tau}\hat{\tau}}^{(1)B} = h_{\tau\tau}^{(1)H} - 2(\xi_{\hat{\tau},\hat{\tau}} + \xi_{\hat{r},\hat{\tau}})$, now implies that $\partial f_i\left(\hat{\tau},\hat{\theta}\right)/\partial\hat{\tau} = 0$, and so $f_2\left(\hat{\tau},\hat{\theta}\right) = f\left(\hat{\theta}\right)$. Making the appropriate substitutions, the $\hat{r}\hat{\theta}$ equation is

$$- \sqrt{2}\tan\frac{\hat{\theta}}{2}\cos\hat{\theta}\, e_0 - 2\left[\frac{-1}{\sqrt{2}}\sin\hat{\theta}\, e_0\log\hat{r} + \frac{\partial f_2(\hat{\theta})}{\partial\hat{\theta}}\right.$$
$$\left. - K\sin\hat{\theta}(1-\log\hat{r}) + L\sin\hat{\theta} - \alpha'(\hat{\theta})\right] = 0. \tag{7.6.42}$$

Therefore $K = e_0/\sqrt{2}$, and

$$\frac{\partial f_2(\hat{\theta})}{\partial\hat{\theta}} = \frac{1}{\sqrt{2}}e_0\tan\frac{\hat{\theta}}{2} - L\sin\hat{\theta} + \alpha'(\hat{\theta}). \tag{7.6.43}$$

Hence $f_2(\hat\theta) = -\sqrt{2}e_0 \log(\cos\frac{\hat\theta}{2}) + L\cos\hat\theta + \alpha(\hat\theta)$, where the constant of integration has been absorbed into $\alpha(\hat\theta)$. If we now examine the forms of $f_2(\hat\theta)$, $g_2(\hat\theta)$, and $h_2(\hat\theta)$, we find that all the terms in L may be eliminated by redefining $\alpha(\hat\theta)$. In sum,

$$\xi_{\hat\tau} = -\sqrt{2}\sin^2\frac{\hat\theta}{2}e_0\log\hat r - \sqrt{2}e_0\log\left[\cos\frac{\hat\theta}{2}\right] + \alpha(\hat\theta) + o(1),$$

$$\xi_{\hat r} = \frac{-1}{\sqrt{2}}e_0\cos\hat\theta\log\hat r - \frac{e_0}{\sqrt{2}}\sin^2\frac{\hat\theta}{2} + \frac{1}{2\sin\hat\theta}\frac{d}{d\hat\theta}[\sin\hat\theta\alpha'(\hat\theta)] + o(1), \qquad (7.6.44)$$

$$\xi_{\hat\theta} = \frac{1}{\sqrt{2}}e_0\sin\hat\theta\frac{\log\hat r}{\hat r} - \frac{\alpha'(\hat\theta)}{\hat r} + o(\hat r^{-1}).$$

(It is obvious that the $\hat\tau\hat\theta$ and $\hat\tau\hat r$ equations are satisfied to the appropriate order in $\hat r$.)

In our harmonic gauge

$$g_{\phi\phi} = \nu^2 r^2 \sin^2\theta[1 + e^{-2\alpha}\frac{\sec^2(\theta/2)e_1(T)}{\sqrt{2}r}$$
$$+ e^{-4\alpha}(D^{(2)} + E^{(2)}) + O(e^{-6\alpha}) + o(r^{-1})]. \qquad (7.6.45)$$

If we now apply the transformation given by eqn (7.6.44), then the new $g_{\phi\phi}$ is

$$g_{\phi\phi} = \nu^2\hat r^2\sin^2\hat\theta\left[1 + e^{-2\alpha}\times\right.$$
$$\times\left(\frac{(1/\sqrt{2})\sec^2(\theta/2)e_1(T) - \sqrt{2}\sin^2(\hat\theta/2)e_0 - [\alpha'(\hat\theta)\cot\hat\theta - \alpha''(\hat\theta)]}{\hat r}\right)$$
$$\left. + e^{-4\alpha}(D^{(2)} + E^{(2)}) + O(e^{-6\alpha}) + o(\hat r^{-1})\right]. \qquad (7.6.46)$$

Let $\hat T = \hat r\sec^2\left(\frac{\hat\theta}{2}\right) - 8\log\left(\frac{2\tan\hat\theta/2}{\nu}\right) + 8\log 8 - 8$. Then

$$T = \hat T + e^{-2\alpha}\sec^2\frac{\hat\theta}{2}\left[-\sqrt{2}\sin^2\frac{\hat\theta}{2}e_0\log\hat r - \sqrt{2}e_0\log\left(\cos\frac{\hat\theta}{2}\right)\right.$$
$$\left. + \alpha(\hat\theta)\right] + o(1) + O(e^{-4\alpha}). \qquad (7.6.47)$$

Hence

$$g_{\phi\phi} = \nu^2 r^2 \sin^2 \hat{\theta} \Bigg[1 + e^{-2\alpha} \times$$

$$\times \left(\frac{(1/\sqrt{2}) \sec^2(\hat{\theta}/2) e_1(\hat{T}) - \sqrt{2} \sin^2(\hat{\theta}/2) e_0 - [\alpha'(\hat{\theta}) \cot \hat{\theta} - a''(\hat{\theta})]}{\hat{r}} \right)$$

$$+ e^{-4\alpha} \Bigg\{ D^{(2)} + E^{(2)} + \frac{1}{\sqrt{2}} \frac{\sec^4(\hat{\theta}/2)}{\hat{r}}$$

$$\times \left[-\sqrt{2} \sin^2 \frac{\hat{\theta}}{2} e_0 \log \hat{r} - \sqrt{2} e_0 \log \left(\cos \frac{\hat{\theta}}{2} \right) + \alpha(\hat{\theta}) \right] e_1'(\hat{T}) \Bigg\}$$

$$+ O(e^{-6\alpha}) + o(\hat{r}^{-1}) \Bigg].$$

$$(7.6.48)$$

The logarithmic term in $D^{(2)} + E^{(2)}$ is (from (7.6.27))

$$e_0 \tan^2 \frac{\hat{\theta}}{2} \sec^2 \frac{\hat{\theta}}{2} e_1'(\hat{T}) \frac{\log \hat{r}}{\hat{r}}, \qquad (7.6.49)$$

which is clearly cancelled by the other $(\log \hat{r})/\hat{r}$ term in eqn (7.6.48). There is also a term of the form "$e_1'(\hat{T})$ times an arbitrary function of $\hat{\theta}$", in the $e^{-4\alpha}$ term in eqn (7.6.48). We postpone to Section 7.8 any discussion of this term, which at first sight seems to make the second-order news function ill-defined.

7.7 Transforming to a Bondi gauge at second order

We shall now investigate the relationship between the harmonic and Bondi gauges at second order in $e^{-2\alpha}$. In other words, we shall consider gauge transformations of the form

$$\tau = \hat{\tau} + e^{-2\alpha} \xi_{\hat{\tau}} + e^{-4\alpha} \xi_{\hat{\tau}}^{(2)},$$
$$r = \hat{r} + e^{-2\alpha} \xi_{\hat{r}} + e^{-4\alpha} \xi_{\hat{r}}^{(2)}, \qquad (7.7.1)$$
$$\theta = \hat{\theta} + e^{-2\alpha} \xi_{\hat{\theta}} + e^{-4\alpha} \xi_{\hat{\theta}}^{(2)},$$

and examine some of the properties that $\xi_{\hat{a}}^{(2)}$ must have if $\left(\hat{\tau}, \hat{r}, \hat{\theta}, \phi \right)$ is to be a Bondi coordinate system. More specifically, we shall show that the time-dependent parts of $\xi_{\hat{\tau}}^{(2)}$ and $\xi_{\hat{\theta}}^{(2)}$ are $o(1)$ and $o\left(\hat{r}^{-1}\right)$ respectively, so that the effect of $\xi_{\hat{a}}^{(2)}$ on the time-varying part of $h_{\phi\phi}^{(2)}$ is $o\left(\hat{r}\right)$.

If we apply eqns (7.7.1) to the flat-space metric (7.6.30), then the $e^{-4\alpha}$ term in the transformed metric in the hatted coordinate system will consist of two terms. The first will be identical to the $e^{-2\alpha}$ term in eqn (7.6.33), except that $\xi_{\hat{a}}$ will be replaced by $\xi_{\hat{a}}^{(2)}$. The second term will have sub-terms that are each quadratic in $\xi_{\hat{a}}$ and its first derivatives. Using eqn (7.6.44) one can show that it has the form

$$e^{-4\alpha}[o(\hat{r}^{-1})d\hat{\tau}^2 + o(\hat{r}^{-1})d\hat{\tau}d\hat{r} + o(1)d\hat{\tau}d\hat{\theta}$$
$$+ o(\hat{r}^{-1})d\hat{r}^2 + o(1)d\hat{r}d\hat{\theta} + o(r)d\hat{\theta}^2 + o(\hat{r})d\phi^2]. \tag{7.7.2}$$

The first-order metric $e^{-2\alpha}h_{ab}^{(1)H}(x^c)$ transforms to

$$e^{-2\alpha}[h_{ab}^{(1)H}(\hat{x}^d) + e^{-2\alpha}h_{ab,e}^{(1)H}(\hat{x}^d)\xi^{\hat{e}}]$$
$$(\delta_c^a + e^{-2\alpha}\xi_{,\hat{c}}^{\hat{\alpha}})(\delta_f^b + e^{-2\alpha}\xi^{\hat{b}}{}_{,f}) + O(e^{-6\alpha}). \tag{7.7.3}$$

If we write out the $e^{-4\alpha}$ term in (7.7.3) in full, we find that it has the form

$$e^{-4\alpha}\{[fn(\hat{\tau},\hat{\theta})\frac{\log\hat{r}}{\hat{r}} + fn(\hat{\tau},\hat{\theta})\frac{1}{\hat{r}} + o(\hat{r}^{-1})]d\hat{\tau}^2$$
$$+ fn(\hat{\tau},\hat{\theta})\log\hat{r} + fn(\hat{\tau},\hat{\theta}) + o(1)]d\hat{\tau}d\hat{\theta} + o(\hat{r}^{-1})d\hat{\tau}d\hat{r}$$
$$+ o(\hat{r}^{-1})d\hat{r}^2 + o(1)d\hat{r}d\hat{\theta} + [A(\hat{\tau},\hat{\theta})\hat{r}\log\hat{r} + B(\hat{\tau},\hat{\theta})\hat{r} + o(\hat{r})]d\hat{\theta}^2$$
$$+ \sin^2\hat{\theta}[-A(\hat{\tau},\hat{\theta})\hat{r}\log\hat{r} - B(\hat{\tau},\hat{\theta})\hat{r} + fn(\hat{\theta})\hat{r}\log\hat{r} + fn(\hat{\theta})\hat{r} + o(\hat{r})]d\phi^2\}, \tag{7.7.4}$$

where the explicit forms of each fn and of A and B can be calculated using eqns (7.6.31) and (7.6.44).

There is, of course, also a contribution to the second-order hatted metric from $h_{ab}^{(2)H}$ itself. Clearly, $e^{-4\alpha}h_{ab}^{(2)H}(x^c)$ transforms to $e^{-4\alpha}h_{ab}^{(2)H}(\hat{x}^c) + O(e^{-6\alpha})$. The second-order metric perturbations $h_{ab}^{(2)H}$ may be written as

$$
\begin{aligned}
h_{tt}^{(2)H} &= A^{(2)}, &\qquad h_{tx}^{(2)H} &= \rho^{-1}xB^{(2)}, \\
h_{ty}^{(2)H} &= \rho^{-1}yB^{(2)}, &\qquad h_{tz}^{(2)H} &= C^{(2)}, \\
h_{zz}^{(2)H} &= G^{(2)}, &\qquad h_{zx}^{(2)H} &= \rho^{-1}xF^{(2)}, \\
h_{zy}^{(2)H} &= \rho^{-1}yF^{(2)}, &\qquad h_{xx}^{(2)H} &= D^{(2)} + (y^2 - x^2)\rho^{-2}E^{(2)}, \\
h_{xy}^{(2)H} &= -2xy\rho^{-2}E^{(2)}, &\qquad h_{yy}^{(2)H} &= D^{(2)} + (x^2 - y^2)\rho^{-2}E^{(2)}.
\end{aligned} \tag{7.7.5}
$$

Each of the functions $A^{(2)},\ldots,G^{(2)}$ in eqn (7.7.5) will possess a series expansion of the form $fn\left(\hat{\tau},\hat{\theta}\right)\left(\frac{\log\hat{r}}{\hat{r}}\right) + \left(fn\left(\hat{\tau},\hat{\theta}\right)\cdot\frac{1}{\hat{r}}\right) + o\left(\hat{r}^{-1}\right)$. When expressed in terms of $A^{(2)},\ldots,G^{(2)}$, the second-order gauge conditions $\overline{h}_{ab}^{(2)}{}^{,b} = 0$ are

$$\frac{\partial}{\partial\tau}(\frac{1}{2}A^{(2)} + D^{(2)} + \frac{1}{2}G^{(2)} + \sin\theta B^{(2)} + \cos\theta C^{(2)}) = o(r^{-1}), \tag{7.7.6a}$$

$$\frac{\partial}{\partial\tau}(-B^{(2)} + \sin\theta E^{(2)} - \cos\theta F^{(2)} + \frac{1}{2}\sin\theta G^{(2)}$$
$$- \frac{1}{2}\sin\theta A^{(2)}) = o(r^{-1}), \tag{7.7.6b}$$

$$\frac{\partial}{\partial \tau}(-C^{(2)} - \sin\theta F^{(2)} - \frac{1}{2}\cos\theta G^{(2)} + \cos\theta D^{(2)}$$
$$- \frac{1}{2}\cos\theta \, A^{(2)}) = o(r^{-1}). \tag{7.7.6c}$$

Now

$$h_{rr}^{(2)} = A^{(2)} + 2\cos\theta C^{(2)} + 2\sin\theta B^{(2)} + \cos^2\theta G^{(2)}$$
$$+ \sin(2\theta)F^{(2)} + \sin^2\theta(D^{(2)} - E^{(2)}). \tag{7.7.7}$$

Multiplying eqn (7.7.6b) by $\sin\theta$ and adding it to $\cos\theta$ times eqn (7.7.6c), we find that

$$h_{rr,\tau}^{(2)H} = o(r^{-1}). \tag{7.7.8}$$

Similarly, one can show that

$$h_{r\theta,\tau}^{(2)H} = o(1), \qquad h_{\phi\phi,\tau}^{(2)H} + \sin^2\theta h_{\theta\theta,\tau}^{(2)H} = o(r). \tag{7.7.9}$$

$\xi_{\hat{a}}^{(2)}$ must be the form

$$\xi_{\hat{\tau}}^{(2)} = fn(\hat{\tau},\hat{\theta})(\log\hat{r})^2 + fn(\hat{\tau},\hat{\theta})\log\hat{r} + fn(\hat{\tau},\hat{\theta}) + o(1),$$
$$\xi_{\hat{r}}^{(2)} = fn(\hat{\tau},\hat{\theta})(\log\hat{r})^2 + fn(\hat{\tau},\hat{\theta})\log\hat{r} + fn(\hat{\tau},\hat{\theta}) + o(1), \tag{7.7.10}$$
$$\xi_{\hat{\theta}}^{(2)} = \frac{1}{\hat{r}}[fn(\hat{\tau},\hat{\theta})(\log\hat{r})^2 + fn(\hat{\tau},\hat{\theta})\log\hat{r} + fn(\hat{\tau},\hat{\theta}) + o(1)].$$

Collecting together all the contributions to the second-order hatted metric, we find that if the hatted coordinate system is to be Bondi then the $e^{-4\alpha}d\hat{\tau}d\hat{\theta}$ term implies that

$$\xi_{\hat{\tau},\hat{\theta}}^{(2)} + \xi_{\hat{r},\hat{\theta}}^{(2)} - \hat{r}^2\xi_{\hat{\theta},\hat{\tau}}^{(2)} = o(\hat{r}), \tag{7.7.11}$$

and so $\xi_{\hat{\theta},\hat{\tau}}^{(2)} = o(\hat{r}^{-1})$. In addition, the $e^{-4\alpha}d\phi^2$ term added to $\sin^2\hat{\theta}$ times the $e^{-4\alpha}d\hat{\theta}^2$ term implies that

$$2\xi_{\hat{\tau}}^{(2)}\sin\hat{\theta} + \hat{r}(\sin\hat{\theta}\xi_{\hat{\theta}}^{(2)})_{,\hat{\theta}} = fn(\hat{\theta})\log\hat{r} + fn(\hat{\theta}) + o(1), \tag{7.7.12}$$

and thus $\xi_{\hat{\tau},\hat{\tau}}^{(2)} = o(1)$. Therefore

$$\lim_{\hat{r}\to\infty} \hat{r}^{-1}(\sin\hat{\theta})^{-2}h_{\phi\phi,\tau}^{(2)B} = \lim_{r\to\infty}[r^{-1}(\sin\theta)^{-2}h_{\phi\phi,\tau}^{(2)H}$$
$$+ \text{ terms which can be calculated using only } h_{\phi\phi}^{(1)H} \text{ and } \xi_{\hat{a}}]. \tag{7.7.13}$$

Thus, when calculating the second-order news function, $\xi_{\hat{a}}^{(2)}$ need not be found explicitly, which is what we set out to prove.

7.8 The ambiguity in the second-order news function caused by the supertranslation freedom

We saw earlier that in addition to the $(\log \hat{r})/\hat{r}$ term (7.6.49), the $(\log \eta)/q^3 \rho^4$ term (7.6.25) contributes

$$-4\sqrt{2}\frac{\log\left(2\sin^2\left(\hat{\theta}/2\right)\right)}{\hat{r}}\tan^2\frac{\hat{\theta}}{2}\sec^2\frac{\hat{\theta}}{2}e_1'\left(\hat{T}\right)$$

to $h^{(2)B}_{\phi\phi}$ (see eqn (7.6.27)). Also, in addition to the $(\log \hat{r})/\hat{r}$ term in eqn (7.6.48), the gauge transformation (7.6.44) introduces a term

$$\frac{1}{\sqrt{2}}\frac{\sec^4(\hat{\theta}/2)}{\hat{r}}\left[8\log\left(\cos\frac{\hat{\theta}}{2}\right)+\alpha\left(\hat{\theta}\right)\right]e_1'\left(\hat{T}\right)$$

into $h^{(2)B}_{\phi\phi}$. There is also a contribution to $h^{(2)B}_{\phi\phi}$ from the surface integral (7.5.19): it is of the form $\tan^2(\hat{\theta}/2)\sec^2(\hat{\theta}/2)fn(\hat{T})/\hat{r}$. The rest of the source integral $\int\int(G_I S + G_{II}T)\,dq_0 dr_0$ of eqn (7.5.20) also contributes a term of the form $\tan^2(\hat{\theta}/2)\sec^2(\hat{\theta}/2)fn(\hat{T})/\hat{r}$. In total, the $1/\hat{r}$ term in $h^{(2)B}_{\phi\phi}$ has the form

$$\hat{r}^2\sin^2\hat{\theta}\left\{\frac{\tan^2(\hat{\theta}/2)\sec^2(\hat{\theta}/2)}{\hat{r}}fn(\hat{T})-\left(\frac{\beta'(\hat{\theta})\cot\hat{\theta}-\beta''(\hat{\theta})}{\hat{r}}\right)\right.$$
$$+\frac{\sec^4(\hat{\theta}/2)}{\hat{r}}\left[-4\sqrt{2}\sin^2\frac{\hat{\theta}}{2}\log\left(2\sin^2\frac{\hat{\theta}}{2}\right)\right.$$
$$\left.\left.+4\sqrt{2}\log\left(\cos\frac{\hat{\theta}}{2}\right)+\alpha(\hat{\theta})\right]e_1'(\hat{T})\right\}. \qquad (7.8.1)$$

The angular dependence of the first term is expected from the analysis in Section 6.4, where we found the form which the $\sin^2\hat{\theta}$ series (1.2) for the news function would take in the boosted frame. The second term, which is time-independent, incorporates the standard supertranslation freedom. The additional terms, however, are somewhat unexpected. The $\log(2\sin^2(\hat{\theta}/2))$ and $\log(\cos(\hat{\theta}/2))$ terms, when transformed to the centre-of-mass frame, do not have an angular dependence of $\sin^2\hat{\theta}$. There is, as well, the term $\frac{\sec^4(\hat{\theta}/2)}{\hat{r}}\alpha(\hat{\theta})e_1'(\hat{T})$ (where $\alpha(\hat{\theta})$ is arbitrary), which seems to make the second-order news function ambiguous.

Moreover, there is an additional problem. We recall from Section 6.3 that $e_1'(\hat{T})$ diverges as $\log|\hat{T}|$ near $\hat{T}=0$. In addition, one can show that the function $fn(\hat{T})$ in the first term in eqn (7.8.1) contains a certain (and calculable—Payne 1983a) amount of $\log|\hat{T}|$ singularity. Hence the second-order news function, which is related directly to the time derivative of eqn (7.8.1), will contain $1/\hat{T}$ singular terms, which are not square-integrable. And yet the news function must be square-integrable, in order that the mass loss be finite.

It is, in fact, not difficult to resolve these puzzles. Consider any Bondi metric in which the various metric functions all have series expansions in powers of some perturbation parameter ϵ. In particular, the function c in eqn (7.6.28) will have the form

$$c(\tau, \theta) = A_0(\tau, \theta) + \epsilon A_1(\tau, \theta) + \epsilon^2 A_2(\tau, \theta) + \cdots. \tag{7.8.2}$$

Now suppose we make a supertranslation

$$\tau = \bar{\tau} + \epsilon f_1(\theta) + \epsilon^2 f_2(\theta) + \cdots \tag{7.8.3}$$

(note that $\bar{\tau} \sim r$, and that $\bar{\theta} = \theta$ on $\mathcal{J}^+$). Then c is unchanged, apart from the addition of a time-independent term (Bondi $et\ al.$ 1962):

$$\bar{c}(\bar{\tau}, \theta) = \sum_{j=0}^{\infty} \epsilon^j A_j \left[\bar{\tau} + \sum_{i=1}^{\infty} \epsilon^i f_i(\theta), \theta \right] + g(\theta), \tag{7.8.4}$$

where

$$g(\theta) = \tfrac{1}{2} \sum_{i=1}^{\infty} \epsilon^i \left[f_i{}'(\theta) \cot \theta - f_i{}''(\theta) \right].$$

But if we now expand out each A_i we find

$$\begin{aligned}
\bar{c}(\bar{\tau}, \theta) = {} & A_0(\bar{\tau}, \theta) + \epsilon[A_1(\bar{\tau}, \theta) + f_1(\theta) A_0'(\bar{\tau}, \theta)], \\
& + \epsilon^2 [A_2(\bar{\tau}, \theta) + \frac{1}{2}(f_1(\theta))^2 A_0''(\bar{\tau}, \theta) \\
& + f_2(\theta) A_0'(\bar{\tau}, \theta) + f_1(\theta) A_1'(\bar{\tau}, \theta)] + \cdots + g(\theta),
\end{aligned} \tag{7.8.5}$$

(where $' = \partial/\partial\tau$).

The barred and unbarred coordinate systems are physically indistinguishable, since it is impossible to tell what supertranslation state one is in. Owing to this complete freedom in the choice of origin of retarded time, only the leading term in the perturbation expansion for c (and the news function c_0) is unambiguously determined—all the higher-order terms being uncertain to the extent shown in eqn (7.8.5). Of course the magnitude of the total news function—given by the sum of the series—remains unchanged; thus the amplitude of the gravitational radiation, which is the physically significant quantity, is well defined. We also note that $\int_{-\infty}^{\infty} (c_0)^2 \, d\tau$ remains invariant at each order in ϵ.

Such behaviour is not limited just to perturbative news functions. A general news function $c_0(\tau, \theta)$ 'supertranslates' to $c_0(\bar{\tau} + f(\theta), \theta)$. We can then (at least formally) expand out in powers of $f(\theta)$, to obtain

$$c_0(\bar{\tau}, \theta) + f(\theta) c_0{}'(\bar{\tau}, \theta) + \tfrac{1}{2} f(\theta)^2 c_0{}''(\bar{\tau}, \theta) + \cdots.$$

(In this way an isotropic distribution of radiation could be made to look non-isotropic!) However, as in the perturbative case, the magnitude of the news function at any given point on $\mathcal{I}^+$ remains unchanged, and none of the additional terms contribute to $\int_{-\infty}^{\infty} (c_0)^2 \, d\tau$.

We see from eqn (7.8.5) that the second-order $g_{\phi\phi}$ may contain an arbitrary multiple of the time derivative of the first-order $g_{\phi\phi}$. This explains the origin of all the $e_1'(\hat{T})$ terms in eqn (7.8.1). The $\log(\hat{T})$ term in $fn(\hat{T})$ must also be due to an $e_1'(\hat{T})$ term that has been introduced by the 'wrong' choice of supertranslation state. All these terms may therefore be eliminated by making an appropriate supertranslation. In fact, since it is the centre-of-mass news function that we would like to be manifestly square-integrable, we shall choose $\alpha(\hat{\theta})$ in eqn (7.8.1) (and eqn (7.6.44)) to ensure that, on matching back to the centre-of-mass frame, the coefficient $a_2(\hat{\tau}/\mu)$ of $\sin^2 \hat{\theta}$ in eqn (7.1.2) contains no $1/\hat{T}$ term near $\hat{T} = 0$.

In a way, it is fortunate that our news function is singular, for otherwise we would have no way of telling how much of a_0' is contained in a_2. The correct amount can be given in analytic form as an integral (Payne 1983a), but here we simply quote a numerical value. Let us describe the radiative part of the second-order gravitational field in terms of the quantity $(\hat{d}^{(2)} + \hat{e}^{(2)}) = q^3 \left(d^{(2)} + e^{(2)} \right)$. We assume that the logarithmic (gauge) part of this quantity near null infinity has been subtracted off, and denote by $\hat{d}^{(2)}(\xi) + \hat{e}^{(2)}(\xi)$ the remaining $O(1)$ part near $\mathcal{J}^+$. This is related to the second-order asymptotic metric function $c^{(2)}$ by

$$c^{(2)} = \frac{-\nu}{\sqrt{2}} e^{-4\alpha} \tan^2 \frac{\hat{\theta}}{2} \sec^2 \frac{\hat{\theta}}{2} \left[\hat{d}^{(2)}(\xi) + \hat{e}^{(2)}(\xi) \right] , \qquad (7.8.6)$$

where

$$\xi = \hat{\tau} \sec^2 \frac{\hat{\theta}}{2} - 8 \log \left[2 \tan \left(\frac{\hat{\theta}}{2} \right) / \nu \right] + 8 \ln 8 - 8.$$

The integral expression in Payne (1983a) shows that the numerical coefficient of $\log|\xi|$ in $\hat{d}^{(2)}(\xi) + \hat{e}^{(2)}(\xi)$ is 4.21867. Further, the coefficient of $\log|\xi|$ in $e_1'(\xi)$ may be shown to be $\sqrt{2}/\pi$. In the numerical calculation we therefore subtract $(\pi/\sqrt{2}) \times 4.21867 \times e_1'(\xi)$ from $\hat{d}^{(2)}(\xi) + \hat{e}^{(2)}(\xi)$ before differentiating to find the news function.

7.9 Comments

In this chapter we have seen how the second-order perturbation problem in the axisymmetric speed-of-light collision can be reduced to a problem in two independent variables, by exploiting the conformal symmetry (7.2.10) at each order of perturbation theory. The second-order metric coefficients can then be expressed in terms of two-dimensional integrals of a Green function multiplying a source function, plus surface contributions. Although the resulting metric is not in a Bondi gauge at null infinity, gauge transformations can be found which put it into Bondi gauge. This allows one to read off the second-order news function, which gives the $\sin^2 \hat{\theta}$ part of the strong-field gravitational radiation pattern (7.1.2), in addition to allowing a further investigation of the new mass-loss formula described in Section 6.5. These results are presented and discussed in the following Chapter 8.

Clearly, a large amount of numerical work is involved in the computation of the integrals giving the $O(1)$ radiative part of $\hat{d}^{(2)} + \hat{e}^{(2)}$ at null infinity, as a function of ξ or τ. We do not discuss this numerical work here; a somewhat detailed treatment is given in Payne (1983a). Nevertheless, we should remark on two of the difficulties which must be overcome numerically. First, the logarithmic terms in $\hat{d}^{(2)} + \hat{e}^{(2)}$ near null infinity must be carefully subtracted off numerically, leaving the $O(1)$ part which carries the information about gravitational waves. Second, the separate contributions from the surface term $\left(d^{(2)} + e^{(2)}\right)_{\text{surf}}$ of eqn (6.5.19) and from the volume contribution to $d^{(2)} + e^{(2)}$ grow exponentially at late times, the exponential terms cancelling each other in the complete $d^{(2)} + e^{(2)}$. This requires very high accuracy in the computation when ξ is only moderately large and positive. This possibility for exponential behaviour, already mentioned in Section 6.2, may actually be realized in some of the higher-order metric perturbations, as will be discussed in Chapter 8.

8

AXISYMMETRIC BLACK-HOLE COLLISIONS AT THE SPEED OF LIGHT: GRAVITATIONAL RADIATION – RESULTS AND CONCLUSIONS

8.1 Introduction

In Chapters 6 and 7, the axisymmetric collision of two black holes was studied in the case in which the black holes approach at the speed of light, each with energy μ in the centre-of-mass frame. A perturbation approach was used, in which a large Lorentz boost is applied to the incoming data, so that the energies λ, ν of the resulting incoming shock waves obey $\lambda \ll \nu$. The subsequent evolution is described by a singular perturbation problem, with small parameter (λ/ν). When one boosts back to the centre-of-mass frame, successive terms of the perturbation series provide information on gravitational radiation which propagates at small angles $\hat{\theta}$ from the symmetry axis $\hat{\theta} = 0$, near the curved shock 2 which has been distorted and deflected in the collision. (By symmetry, the same holds for radiation near the backward direction $\hat{\theta} = \pi$.) The gravitational radiation is described by the news function (Bondi *et al.* 1962) which is expected to have the convergent expansion (eqn (6.1.3))

$$c_0\left(\hat{\tau}, \hat{\theta}\right) = \sum_{n=0}^{\infty} a_{2n}\left(\hat{\tau}/\mu\right) \sin^{2n} \hat{\theta} , \qquad (8.1.1)$$

where $\hat{\tau}$ is a retarded time coordinate. First-order perturbation theory gave $a_0\left(\hat{\tau}/\mu\right)$ in Chapter 6, in agreement with the results of the calculation (Chapter 5) of radiation emitted near the forward direction when two black holes approach, each with large Lorentz factor γ in the centre-of-mass frame, and collide. In that case, the news function has an asymptotic expansion

$$c_0\left(\overline{\tau}, \hat{\theta} = \gamma^{-1}\psi\right) \sim \sum_{n=0}^{\infty} \gamma^{-2n} Q_{2n}\left(\overline{\tau}, \psi\right), \qquad (8.1.2)$$

valid as $\gamma \to \infty$ with $\overline{\tau}, \psi$ fixed. The function $a_0\left(\hat{\tau}/\mu\right)$ is found from the limiting form of $Q_0\left(\overline{\tau}, \psi\right)$ as $\psi \to \infty$.

For the speed-of-light collision, the higher-order perturbation theory in (λ/ν) can be considerably simplified following the analysis in Chapter 7. There is a conformal symmetry at each order in perturbation theory, and hence all metric perturbations can be expressed in terms of functions of two variables, rather than the three variables which would be expected given only axisymmetry. This leads to an integral expression for $a_2 (\hat{\tau}/\mu)$, using second-order perturbation theory, which is numerically tractable.

In Section 8.2, we evaluate the time integral $\int_{-\infty}^{\infty} a_2 (\hat{\tau}/\mu) \, d\hat{\tau}$, which appears in the new mass-loss formula of Section 6.5. This formula gives an expression for the mass of the final black hole resulting from the collision, assuming that at late times there is only one Schwarzschild black hole at rest, and that the gravitational radiation obeys a certain uniformity condition (6.5.12). The new mass-loss formula turns out to give a 'final mass' exceeding 3.5μ. The most probable explanation is that the uniformity condition fails. This is expected to happen if, as seems likely, there is a 'second burst' of gravitational radiation, generated near the centre of the space–time, which will be delayed relative to the 'first burst', described by eqn (8.1.1), by an amount $\Delta\hat{\tau} \simeq |8\mu \log \hat{\theta}|$ for small angles $\hat{\theta}$. The requirement of matching of the late-time radiation pattern of the 'first burst' with the early-time pattern of the 'second burst' shows that the radiation at such times is given by a sum of terms proportional to $e^{j\hat{\tau}/4\mu} \sin^{2n} \hat{\theta}$, where j and n are positive integers. Thus the presence of a 'second burst' will be signalled by exponential growth of this type at late times in the news function (8.1.1), where the 'first burst' occurs near $\hat{\tau} = 0$. Such exponential behaviour is indeed liable to occur because of the singular nature of the perturbation problem, in which the initial data become large at late times on the characteristic initial surface just to the future of the strong shock 1. It only fails to occur in $a_2 (\hat{\tau}/\mu)$ because of the cancellation between exponentially growing volume and surface contributions to a_2.

In Section 8.3, the result of the numerical calculation of $a_2 (\hat{\tau}/\mu)$ is presented (Fig 8.4), following the analytical simplifications of Chapter 7. A crude estimate of the mass loss in gravitational waves can be found by keeping only the first two terms in the series (8.1.1), i.e. approximating $c_0(\hat{\tau}, \hat{\theta})$ by $a_0 (\hat{\tau}/\mu) + a_2 (\hat{\tau}/\mu) \sin^2 \hat{\theta}$. With this truncation, an energy loss of $0.328 \, \mu$ was found from the conventional Bondi formula (Bondi *et al.* 1962) corresponding to an efficiency of 16.4% for gravitational wave generation. Further discussion is given in Section 8.4.

8.2   Prediction of new mass-loss formula

In this section we insert the calculated values of the relevant quantities into the mass-loss formula in the limit $\gamma \to \infty$ (eqn (6.5.20)):

$$
\begin{aligned}
m_{\text{final}} = & -\int_{-\infty}^{\infty} [a_0(\hat{\tau}/\mu)]^2 \, d\hat{\tau} \\
& + \int_{-\infty}^{\infty} [4a_2(\hat{\tau}/\mu) - a_0(\hat{\tau}/\mu)] d\hat{\tau} + o(1) \\
= & -\frac{\mu}{2} + \int_{-\infty}^{\infty} [4a_2(\hat{\tau}/\mu) - a_0(\hat{\tau}/\mu)] d\hat{\tau} + o(1), \qquad (8.2.1)
\end{aligned}
$$

which was derived in Section 6.5 for the high-speed axisymmetric black-hole collision. Here $a_0 (\hat{\tau}/\mu)$ and $a_2 (\hat{\tau}/\mu)$ are the first two coefficients in the $\sin^2 \hat{\theta}$

expansion for the news function on angular scales of $O(1)$ in the centre-of-mass frame:

$$c_0(\hat{\tau}, \hat{\theta}) = \sum_{n=0}^{\infty} a_{2n}(\hat{\tau}/\mu) \sin^{2n} \hat{\theta} + O(\gamma^{-1}). \qquad (8.2.2)$$

The metric function $c(\hat{\tau}, \hat{\theta})$, the retarded time derivative of which is the news function, has a corresponding expansion

$$c(\hat{\tau}, \hat{\theta}) = \sum_{n=0}^{\infty} b_{2n}(\hat{\tau}/\mu) \sin^{2n} \hat{\theta}$$
$$+ \text{time-independent terms} + O(\gamma^{-1}), \qquad (8.2.3)$$

where clearly

$$b_{2n}(x) = \mu \int_{-\infty}^{x} a_{2n}(y) dy + \text{const.} \qquad (8.2.4)$$

Hence

$$m_{\text{final}} = \frac{-\mu}{2} + [4b_2(\hat{\tau}/\mu) - b_0(\hat{\tau}/\mu)] \mid_{-\infty}^{\infty} + o(1), \qquad (8.2.5)$$

where, again, the $o(1)$ correction denotes a term tending to zero as $\gamma \to \infty$. For convenience, as will be seen shortly, we have left b_2 and b_0 in the combination $(4b_2 - b_0)$, although it was shown in Section 6.5 that $b_0 \mid_{-\infty}^{\infty} = -2\mu$.

In the boosted frame we have computed $\hat{d}^{(2)} + \hat{e}^{(2)}$, as described in Chapter 7, from which the notation is taken, where

$$h_{\phi\phi}^{(2)} = r^2 \sin^2 \theta (D^{(2)} + E^{(2)})$$
$$= r^2 \sin^2 \theta \left[\frac{\hat{d}^{(2)} + \hat{e}^{(2)}}{q^3 \rho^4} \right] \qquad (8.2.6)$$

(here r is the radial coordinate). As described in Section 7.8, it is assumed that the $(\log r)/r$ part of the transverse second-order metric perturbations has been subtracted off by a gauge transformation, before $(\hat{d}^{(2)} + \hat{e}^{(2)})$ is evaluated. By $(\hat{d}^{(2)} + \hat{e}^{(2)})$ we mean simply the remaining $0(1)$ part of this quantity as $r \to \infty$. Since $c = -(1/2) \lim_{r \to \infty} (h_{\phi\phi}/r \sin^2 \theta)$ (in a Bondi gauge), the $e^{-4\alpha}$ term in c in the boosted frame is (using eqn (7.6.26))

$$c^{(2)} = -\frac{\nu e^{-4\alpha}}{\sqrt{2}} \tan^2 \frac{\theta}{2} \sec^2 \frac{\theta}{2} [\hat{d}^{(2)}(\xi) + \hat{e}^{(2)}(\xi)], \qquad (8.2.7)$$

where $\xi = (\tau/\nu) \sec^2(\theta/2) - 8 \log \left(\frac{2 \tan \theta/2}{\nu} \right) + 8 \log 8 - 8$. Further, as explained in Sections 7.6 and 7.7, we choose the supertranslation state so that $\hat{d}^{(2)}(\xi) + \hat{e}^{(2)}(\xi)$ contains no $\log|\xi|$ term in an expansion about $\xi = 0$. Using eqns (6.3.24) and

(6.4.3) and $c = \hat{c}/K^2$, where $K(\theta) = \cosh\alpha + \sinh\alpha\cos\theta$ (Bondi *et al.* 1962), it is easy to show that $b_0\,(\hat{\tau}/\mu) + b_2\,(\hat{\tau}/\mu)\sin^2\hat{\theta}$ transforms to

$$e^{-\alpha}\sec^2\frac{\theta}{2}b_0\left(\frac{\tau}{\nu}\sec^2\frac{\theta}{2}\right) + e^{-3\alpha}\tan^2\frac{\theta}{2}\sec^2\frac{\theta}{2}\times$$

$$\times\left[4b_2\left(\frac{\tau}{\nu}\sec^2\frac{\theta}{2}\right) - b_0\left(\frac{\tau}{\nu}\sec^2\frac{\theta}{2}\right) - \frac{\tau}{\nu}\sec^2\frac{\theta}{2}b_0'\left(\frac{\tau}{\nu}\sec^2\frac{\theta}{2}\right)\right] \quad (8.2.8)$$

in the boosted frame. Hence

$$4b_2\left(\frac{\tau}{\nu}\sec^2\frac{\theta}{2}\right) - b_0\left(\frac{\tau}{\nu}\sec^2\frac{\theta}{2}\right) - \frac{\tau}{\nu}\sec^2\frac{\theta}{2}b_0'\left(\frac{\tau}{\nu}\sec^2\frac{\theta}{2}\right)$$

$$= -\frac{\nu e^{-\alpha}}{\sqrt{2}}\left[\hat{d}^{(2)}\left(\frac{\tau}{\nu}\sec^2\frac{\theta}{2}\right) + \hat{e}^{(2)}\left(\frac{\tau}{\nu}\sec^2\frac{\theta}{2}\right)\right]. \quad (8.2.9)$$

Note also, following the remarks above, that this implies that $b_2\,(\hat{\tau}/\mu)$ contains no $\log|\hat{\tau}/\mu|$ term in an expansion about $\hat{\tau} = 0$. Therefore

$$\left[4b_2\left(\frac{\hat{\tau}}{\mu}\right) - b_0\left(\frac{\hat{\tau}}{\mu}\right)\right]\Big|_{-\infty}^{+\infty} = \frac{-\mu}{\sqrt{2}}[\hat{d}^{(2)}(\xi) + \hat{e}^{(2)}(\xi)]\,|_{-\infty}^{\infty}. \quad (8.2.10)$$

As $\xi \to -\infty$, $\hat{d}^{(2)} + \hat{e}^{(2)} \to 0$. This may be verified by noting that the integration region in the source integral (7.5.20) tends to zero as $\xi \to -\infty$. Inspection of eqn (7.5.19) shows that the contribution from the surface term is also zero in this limit. The asymptotic behaviour of $\hat{d}^{(2)} + \hat{e}^{(2)}$ for large positive ξ is shown in Figs 8.1(a) and (b). It is expected to be of the form

$$(\hat{d}^{(2)} + \hat{e}^{(2)}) \sim \kappa + \sum_{n=1}^{\infty}\frac{R_n(\log|\xi|)}{\xi^n} \quad (8.2.11)$$

as $\xi \to \infty$, where the R_n are polynomials. Terms of this form appear in the first-order news function, as one can check by a detailed asymptotic analysis of the integral expression (6.3.22) for $a_0\,(\hat{\tau}/\mu)$, there denoted by $H_0\,(\hat{\tau}/\mu)$. One can reasonably expect them here too, in which case $\lim_{\xi\to\infty}\left(\hat{d}^{(2)} + \hat{e}^{(2)}\right) = \kappa$. We have only computed $\hat{d}^{(2)} + \hat{e}^{(2)}$ out to $\xi = 49.5$, since, as explained in Section 7.9, it is not possible to go to larger values of ξ because there are large cancellations between the source and surface term contributions to $\hat{d}^{(2)} + \hat{e}^{(2)}$. In fact, the source and surface contributions to $\hat{d}^{(2)} + \hat{e}^{(2)}$ separately have parts which grow exponentially at late retarded times, at rates $\exp(\ell\hat{\tau}/8\mu)$ for suitable integer ℓ. In the case of the surface contribution, these arise from the exponential growth of the initial data for the second-order metric perturbations at large negative coordinate $\hat{v}$ on the characteristic initial surface $\hat{u} = 0$ (the strong shock 1). This late-time growth of the characteristic initial data is responsible for the singular nature of the perturbation theory for this space–time, as discussed at the end of Section 6.2; in particular, see eqn (6.3.26). The very accurate numerical cancellation between the source and surface contributions (as seen in Figs 8.1(a) and (b)) leads one to expect the late-time behaviour of eqn (8.2.11) for $\hat{d}^{(2)} + \hat{e}^{(2)}$. To determine κ accurately, we must do a least-squares fit of the form (8.2.11) to the computed $\hat{d}^{(2)} + \hat{e}^{(2)}$. However, because there is no way of telling which terms are actually present in eqn (8.2.11), this method proves incapable of determining κ very accurately. We find $\kappa \simeq -6.3$, with an estimated error of 5%.

Fortunately, a rough estimate is quite sufficient for our needs. Since the change in $\hat{d}^{(2)} + \hat{e}^{(2)}$ in going from $\xi = 25$ to $\xi = 50$ will be of roughly the same magnitude as in going from $\xi = 50$ to $\xi = \infty$, inspection of Fig. 8.1(b) shows that κ will certainly be less than $-4\sqrt{2}$. Hence

$$\left[4b_2 \left(\frac{\hat{\tau}}{\mu} \right) - b_0 \left(\frac{\hat{\tau}}{\mu} \right) \right] \Big|_{-\infty}^{+\infty} > 4\mu, \tag{8.2.12}$$

and substituting this into eqn (8.2.5) we find

$$m_{\text{final}} > 3.5\mu. \tag{8.2.13}$$

Thus eqn (8.2.1) predicts that the final mass will be considerable greater than the initial energy 2μ. This result cannot, of course, be correct—energy must be conserved—and therefore one (or more) of the assumptions that went into the derivation of eqn (8.2.1) must be wrong.

The first of those assumptions was that the final mass aspect of the system is isotropic. This enabled us to replace $M(\infty, \hat{\theta})$ by m_{final} on the left-hand side of eqn (6.5.9) and in all the equations derived from it, such as eqn (8.2.1). This assumption seems very reasonable. If cosmic censorship holds, the event horizon that is formed in the collision cannot bifurcate (Hawking and Ellis 1973) and hence there must be a connected 'object', symmetrical about the equator $\hat{\theta} = \pi/2$ in the centre-of-mass frame, at the centre of the space–time, which is bounded by a horizon that always has an area greater than $32\pi\mu^2$, the area of the apparent horizon found by Penrose (1974) (as mentioned in Section 6.1). This object should shed its non-spherical perturbations and decay to a Schwarzschild geometry. Assuming all this to be true, the only way the final mass aspect can be non-isotropic is for there to be other bodies present, and they must be moving relative to the centre-of-mass frame. For example, in addition to the central black hole, one could (just) envisage two small black holes being formed by the focusing of the incoming waves, which might then fly off in opposite directions along the axis of symmetry, thereby concentrating the mass aspect around this axis, so that $M(\hat{\tau} = \infty, \hat{\theta} = 0, \pi) > m_{\text{final}}$. But this seems very unlikely, since any focusing that is strong enough to produce horizons should take place within the central trapped region.

Second, we assumed that each $Q_{2n}(\overline{\tau}, \psi)$, in the asymptotic expansion (8.1.2) for the news function at angles $\hat{\theta} = \gamma^{-1}\psi$ close to the axis in the large finite-γ

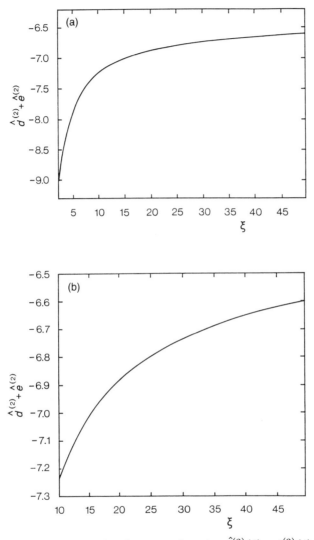

FIG. 8.1. The asymptotic second-order metric function $\hat{d}^{(2)}(\xi) + \hat{e}^{(2)}(\xi)$ is shown for moderately large values of ξ, where $\mu\xi$ is a measure of retarded time. Numerical problems prevented its accurate evaluation for larger values of ξ. The limiting value of $\hat{d}^{(2)}(\xi) + \hat{e}^{(2)}(\xi)$ as $\xi \to \infty$ yields $-(\sqrt{2}/\mu)$ times the change in the quantity $[4b_2(\hat{\tau}/\mu) - b_0(\hat{\tau}/\mu)]$ from early to late retarded times, and hence leads to an expression for the final mass in eqn (8.2.5), following the assumptions of the new mass-loss formula.

collision, tends to a limiting form given by $a_{2n}(\hat{\tau}/\mu)$ as $\psi \to \infty$. In other words, the radiation pattern in the high-speed collision tends to that produced by the speed-of-light collision as $\gamma \to \infty$. This seems intuitively obvious, has been explicitly shown for $n = 0$, by combining the arguments of Chapter 5 and of the end of Section 6.3, and we see no reason to doubt it.

Instead, the reason for the non-physical answer in eqn (8.2.13) is probably that at late times, close to the axis of symmetry, there is another burst of radiation, making an additional contribution to the integral $\int_{-\infty}^{\infty} c_0(\overline{\tau}, \psi) d\overline{\tau}$, which when substituted into eqn (6.5.9) will modify eqn (8.2.1) and bring m_{final} down to some sensible value below 2μ. In other words, we are suggesting that $c(\overline{\tau}, \psi)$ will not in fact satisfy the 'uniformity' condition (6.5.12) which, we recall, had to be assumed in order to derive eqn (8.2.1).

One can see how this might happen. At both finite and infinite γ one uses perturbation theory to calculate the radiation in the neighbourhood of the axis of symmetry (in the centre-of-mass frame) produced by the focusing of the far fields (where $\rho \gg \mu$) of the incident shock waves as they pass through each other during the collision, and one then matches out to larger angular scales. But if one is close to the axis $\hat{\theta} = 0, \pi$, as is necessary when doing perturbation theory, then one must wait a long time after the peak of the first burst of radiation has come by before one can see non-zero initial data at $\rho \sim \mu$ (see Figs 5.3 and 6.1). The effectiveness of perturbation theory is due to this delay by an amount proportional to $\log(\rho_0)$, where $(\rho_0)^2 = (x_0)^2 + (y_0)^2$ gives the distance from the axis at the moment when the shocks collide, as described in Sections 6.1 and 6.2. This gives the far field of each shock a large head start over its near-field counterpart, enabling one to study its essentially wavelike self-interaction without having to concern oneself with the details of the highly non-linear region at the centre of the space–time. And yet it is in this non-linear region where the shocks come through each other at $\rho \sim \mu$, and in which we expect a black hole to be formed, that additional radiation might be produced. This central black hole would certainly not be formed 'cleanly', and could shed its non-spherical perturbations only by emitting gravitational waves.

Using the parametric representation (6.2.29) for the weak shock, it is easy to show that in the speed-of-light collision in the centre-of-mass frame the two bursts would be separated by a time interval $\Delta \hat{\tau} \simeq 8\mu \log[\mu/2 \tan(\hat{\theta}/2)]$ when $\hat{\theta} \ll 1$. In the finite-γ collision the time delay $\Delta \hat{\tau}$ has the same form as long as $\gamma^{-1} \ll \hat{\theta} \ll 1$, but when $\hat{\theta}$ becomes of $0(\gamma^{-1})$ it flattens off to a value $\Delta \hat{\tau} \simeq 8\mu \log(\gamma/\mu)$ (i.e. $\Delta \hat{\tau} \not\to \infty$ as $\hat{\theta} \to 0$), reflecting the fact that the radiation has detailed angular structure in this region. Since $\Delta \hat{\tau}$ is large, one would expect the 'influence' of the second burst on the first to be small. However, the series (8.1.1) and (8.1.2) provide exact or asymptotic descriptions of their respective space–time metrics in the vicinity of the first burst of radiation. One would expect to be able to find traces of any second burst somewhere in these series. We shall demonstrate this for the speed-of-light collision; there is a completely analogous argument at finite γ.

Suppose that the supertranslation state in the speed-of-light collision is such that the first burst is centred around $\hat{\tau} = 8\mu \log\left(\frac{2\tan\hat{\theta}/2}{\mu}\right)$ when $\hat{\theta}$ is small. (We are therefore keeping the logarithmic term which appears quite naturally in the argument of the news function in Section 6.3; see, in particular, eqn (6.3.26) and lines following.) From the discussion in the previous paragraph it is clear that the second burst would be centred near $\hat{\tau} = 0$. Let us restrict attention to angles $\hat{\theta} \ll 1$. One might, at first sight, guess that the fall-off of the news function as one moves back in time, away from the centre of the second burst, is given by an asymptotic expansion of a type similar to eqn (8.2.11), going as

$$\sin^2\hat{\theta}\left[\sum_{i=1 \text{ or } 2}^{\infty} \frac{R_i[\log(-\hat{\tau}/\mu)]}{(\hat{\tau}/\mu)^i}\right]$$

$$+ \text{ analogous terms proportional}$$
$$\text{to } \sin^4\hat{\theta}, \sin^6\hat{\theta}, \dots , \tag{8.2.14}$$

where each R_i is a polynomial. (There will be no $\sin^0\hat{\theta}$ term in the second burst: it is only the very special focusing of the far fields of the incoming shocks that causes the $a_0 \sin^0\hat{\theta}$ term in eqn (8.1.1).) Define $\hat{u}$ by

$$\hat{u} = \hat{\tau} - 8\mu \log\left(\frac{2\tan\hat{\theta}/2}{\mu}\right),$$

and note that the right hand side of eqn (8.1.1) has the form $\sum_{n=0}^{\infty} a_{2n}(\hat{u}/\mu)\sin^{2n}\hat{\theta}$ in the supertranslation state we have chosen. Then expressing (8.2.14) in terms of $\hat{u}$ we obtain

$$\sin^2\hat{\theta}\left[\sum_{i=1 \text{ or } 2}^{\infty} \frac{R_i\left\{\log\left[-\frac{\hat{u}}{\mu} - 8\log\left((2\tan\hat{\theta}/2)/\mu\right)\right]\right\}}{\left[\frac{\hat{u}}{\mu} + 8\log\left((2\tan\hat{\theta}/2)/\mu\right)\right]^i}\right]$$

$$+ \text{ analogous terms proportional to } \sin^4\hat{\theta}, \sin^6\hat{\theta}, \dots \tag{8.2.15}$$

If we now fix $\hat{u}$ and let $\hat{\theta} \to 0$ then all the resulting terms in eqn (8.2.15) should appear in eqn (8.1.1) (with $\hat{\tau}$ replaced by $\hat{u}$), since perturbation theory is certainly valid in this limit. The first term in eqn (8.2.15) becomes

$$\sin^2\hat{\theta}\sum_i \frac{R_i\left[\log(-8\tan\hat{\theta}/2)Q_i\left((\hat{u} + \text{const})/8\mu\log(\tan\hat{\theta}/2)\right)\right]}{[8\log(\tan\hat{\theta}/2)]^i}$$

$$\times V_i\left(\frac{\hat{u} + \text{const.}}{8\mu\log(\tan\hat{\theta}/2)}\right), \tag{8.2.16}$$

where Q_i and V_i are appropriate convergent power series. This certainly does not match with the $\sin^2\hat{\theta}$ expansion (8.1.1) (again with $\hat{\tau}$ replaced by $\hat{u}$) for the news function, and therefore the second burst cannot have a tail with an inverse power-law decay, as described by eqn (8.2.14).

We must instead find a suitable form for the behaviour of the news function at times early compared to the second burst, but late compared to the first burst. The form $8\mu \log[\mu/2 \tan(\hat{\theta}/2)]$ of the time-delay between the first and second bursts, together with the property (1.1) that the radiation admits a convergent series expansion in powers of $\sin^2 \hat{\theta}$, shows that the news function, at times early compared to the second burst, should have the $\hat{\tau} \to -\infty$ fall-off

$$\sin^2 \hat{\theta} \sum_{j=1}^{\infty} A_j e^{j\hat{\tau}/4\mu}$$

+ analogous terms proportional to $\sin^4 \hat{\theta}$, $\sin^6 \hat{\theta}, \dots$,   (8.2.17)

where the A_j are constants. The particular form of the exponentials in eqn (8.2.17) can be seen as follows to be necessary. First, note that $\tan(\hat{\theta}/2)$ can be expressed as a power series in $\sin\hat{\theta}$, as $\tan\left(\hat{\theta}/2\right) = \sin\hat{\theta} \sum_{k=0}^{\infty} B_k \sin^{2k} \hat{\theta}$. Hence the first term in eqn (8.2.17), which can be written as

$$\sin^2 \hat{\theta} \sum_{j=1}^{\infty} A_j e^{j\hat{u}/4\mu}(\mu^{-1} \tan(\hat{\theta}/2))^{2j},$$

can be rewritten as

$$\sin^2 \hat{\theta} \left[\sum_{j=1}^{\infty} A_j e^{j\hat{u}/4\mu} \sin^{2j} \hat{\theta} \left(\mu^{-1} \sum_{k=0}^{\infty} B_k \sin^{2k} \hat{\theta} \right)^{2j} \right] .   (8.2.18)$$

This matches with the form (8.1.1) of the radiation, as viewed relative to the first burst using retarded coordinate $\hat{u}$, as $\hat{u} \to \infty$. Indeed, only the particular exponential fall-off of eqn (8.2.17) as $\hat{\tau} \to -\infty$ will 'unscramble' the logarithmic time-delay $8\mu \log[\mu/2 \tan(\hat{\theta}/2)]$ between the two bursts, so as to give an expression such as eqn (8.2.18) in a power series in $\sin^2 \hat{\theta}$. Exactly the same applies to the terms in eqn (8.2.17) with higher powers of $\sin^2 \hat{\theta}$.

In summary, we are led to conclude that the presence of a second burst of radiation will be indicated by exponentially growing terms at late times in the $a_{2n} (\hat{u}/\mu)$ in eqn (8.1.1) (with $\hat{\tau}$ replaced by $\hat{u}$) beginning at third, or perhaps some higher, order. (We recall that there were such terms in the source and surface contributions to $a_2 (\hat{u}/\mu)$ which fortunately cancelled.) The important point is that the second burst cannot be detected directly at either first or second order in perturbation theory; only indirectly through eqn (8.2.1) and an examination of the assumptions under which it holds.

As mentioned above, a similar analysis may be carried out for the finite-γ collision, leading to an identical result for the perturbation expansion (8.1.2) in powers of γ^{-2}: that the imprints of the second burst will be exponentially growing terms beginning at third $(Q_4(\bar{\tau}, \psi))$, or some higher, order. This does, we feel, provide a satisfactory explanation of the unphysical result in eqn (8.2.13), since if the $Q_{2n}(\bar{\tau}, \psi)$ contain exponentially growing terms then eqn (6.5.12) will be violated, and so the chain of reasoning leading to eqn (8.2.13) will be broken.

To find the true loss of mass (assuming final isotropy) it would only be necessary to compute the time integral of the coefficient of $\sin^2 \hat{\theta}$ in the news function describing the second burst. This would, presumably, make a negative contribution to eqn (6.5.9), thus modifying eqn (8.2.1), and lead to a sensible value for m_{final}. However, since no perturbation theory will be able to describe the strong-field region from which the second burst emanates, the task of calculating the complete $\sin^2 \hat{\theta}$ term is rather formidable.

As already outlined in this section, the time delay in the speed-of-light collision between the centres of the two bursts will be $\Delta \hat{\tau} \simeq 8\mu |\log\left((2\tan\hat{\theta}/2)/\mu\right)|$ when $\hat{\theta}$ is small. At finite γ the form of the time delay will be $\Delta \hat{\tau} \simeq 8\mu \log(\gamma/\mu)$ if $\hat{\theta}$ is $0(\gamma^{-1})$, and $\Delta \hat{\tau} \simeq 8\mu (\log \gamma - \log(\psi/\mu))$ when ψ^{-1} and $\gamma^{-1}\psi$ are $o(1)$ (i.e. in the matching region in which $\gamma^{-1} \ll \hat{\theta} \ll 1$). This means that the 'two' bursts of radiation will be truly separate only in the limit $\hat{\theta} \to 0$ in the speed-of-light collision, and for $\gamma \to \infty$ with $\hat{\theta} = o(1)$ in the 'finite'-γ collision. Away from the axis of symmetry they will merge, and cannot be thought of as physically distinct. As long as eqn (8.1.1) is convergent it should provide the exact speed-of-light news function—even if the perturbation theory has broken down near the initial surface to the past of the point in question.

Because of the possible exponentially growing terms in $a_4(\hat{u}/\mu)$, $a_6(\hat{u}/\mu)$, ..., one might doubt whether the series (8.1.1) for the speed-of-light news function (with $\hat{\tau}$ replaced by $\hat{u}$) could converge at late retarded times $\hat{u}$. However, at least if one considers the time-integrated quantity $c(\hat{u}, \hat{\theta})$, which is expected to be a continuous function for $0 \leq \hat{\theta} \leq \pi$, symmetrical about $\hat{\theta} = \pi/2$, then by the Stone–Weierstrass theorem (Dieudonné 1960) it will admit the convergent expansion

$$c\left(\hat{u}, \hat{\theta}\right) = \sum_{n=0}^{\infty} b_{2n}\left(\hat{u}/\mu\right) \sin^{2n} \hat{\theta} \qquad (8.2.19)$$

as a power series in $\sin^2 \hat{\theta}$. The analogous expansion should also hold for the function c when regarded as a function of $\hat{\tau}$ and $\hat{\theta}$, where the retarded-time coordinates $\hat{\tau}$ and $\hat{u}$ are related by the supertranslation $\hat{u} = \hat{\tau} - 8\mu \log[2\tan(\hat{\theta}/2)/\mu]$. Thus

$$c\left(\hat{u} = \hat{\tau} - 8\mu \log\left[2\tan\left(\hat{\theta}/2\right)/\mu\right], \hat{\theta}\right) = \sum_{n=0}^{\infty} D_{2n}\left(\hat{\tau}/\mu\right) \sin^{2n} \hat{\theta}, \qquad (8.2.20)$$

where the b_{2n} and D_{2n} are of course different functions. It is only the simultaneous validity of eqns (8.2.19) and (8.2.20), for this particular supertranslation, which is powerful enough to lead us to the form (8.2.17) of the early-$\hat{\tau}$ expansion of $c_0\left(\hat{\tau},\hat{\theta}\right)$.

8.3 The second-order news function

In this section, we return to the convention that the supertranslation state is such that the radiation is centred on the retarded time $\hat{\tau}=0$.

From eqn (8.2.9) we have

$$4b_2\left(\frac{\hat{\tau}}{\mu}\right) = \frac{-\mu}{\sqrt{2}}\left[\hat{d}^{(2)}\left(\frac{\hat{\tau}}{\mu}\right) + \hat{e}^{(2)}\left(\frac{\hat{\tau}}{\mu}\right) - \frac{\sqrt{2}}{\mu}b_0\left(\frac{\hat{\tau}}{\mu}\right) \right.$$
$$\left. - \frac{\sqrt{2}}{\mu}\left(\frac{\hat{\tau}}{\mu}\right)b_0'\left(\frac{\hat{\tau}}{\mu}\right)\right], \tag{8.3.1}$$

where the b_{2n} are the functions appearing in eqn (8.2.3). Using eqns (7.6.29) and (8.2.8), we see that

$$-2\sec^2\frac{\theta}{2}b_0\left(\frac{\tau}{\nu}\sec^2\frac{\theta}{2}\right) = \frac{\mu}{\sqrt{2}}\sec^2\frac{\theta}{2}e_1\left(\frac{\tau}{\nu}\sec^2\frac{\theta}{2}\right). \tag{8.3.2}$$

The function $e_1(\xi)$ was defined in eqn (7.6.11), and from eqn (7.6.10) has the explicit form

$$e_1(\xi) = \frac{8\sqrt{2}}{\pi}\int_0^\infty \int_0^\pi dx\,d\phi'x^{-2}\cos\phi'\times$$
$$\times\,\theta(8\log x + \xi + 8 + 8x\cos\phi') - 4\sqrt{2}, \tag{8.3.3}$$

where we have used the gauge conditions (7.6.18). Hence

$$b_0\left(\hat{\tau}/\mu\right) = -\left(\mu/2\sqrt{2}\right)e_1\left(\hat{\tau}/\mu\right)$$

and

$$b_2\left(\frac{\hat{\tau}}{\mu}\right) = \frac{-\mu}{4\sqrt{2}}[\hat{d}^{(2)}(\xi) + \hat{e}^{(2)}(\xi) + \frac{1}{2}e_1(\xi)$$
$$+ \frac{1}{2}\xi e_1'(\xi)]\,|_{\xi=\hat{\tau}/\mu}. \tag{8.3.4}$$

Therefore $a_2\left(\hat{\tau}/\mu\right)$, the coefficient of $\sin^2\hat{\theta}$ in eqn (8.2.2) or eqn (8.1.1), is

$$a_2\left(\frac{\hat{\tau}}{\mu}\right) = \frac{-1}{4\sqrt{2}}\left(\frac{d}{d\xi}[\hat{d}^{(2)}(\xi) + \hat{e}^{(2)}(\xi) + \frac{1}{2}e_1(\xi)\right.$$
$$\left. + \frac{1}{2}\xi e_1'(\xi)]\right)\Bigg|_{\xi=\hat{\tau}/\mu} \tag{8.3.5}$$

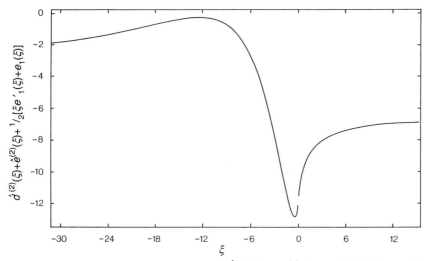

FIG. 8.2. The asymptotic metric function $\hat{d}^{(2)}(\xi) + \hat{e}^{(2)}(\xi) + \frac{1}{2}[\xi e_1'(\xi) + e_1(\xi)]$ is shown. Its ξ derivative, evaluated at $\xi = \hat{\tau}/\mu$, gives $-4\sqrt{2}a_2(\hat{\tau}/\mu)$, where the news function is given by $c_0(\hat{\tau}, \hat{\theta}) = \sum_{n=0}^{\infty} a_{2n}(\hat{\tau}/\mu)\sin^{2n}\hat{\theta} + O(\gamma^{-1})$. There is a gap in the computational results near the singular point $\xi = 0$.

The function we have computed—at a number of discrete values of ξ—is $\hat{d}^{(2)} + \hat{e}^{(2)}$, and so to calculate $a_2(\hat{\tau}/\mu)$ we shall have to do some numerical differentiation. A graph of the computed $\hat{d}^{(2)}(\xi) + \hat{e}^{(2)}(\xi) + \frac{1}{2}[e_1(\xi) + \xi e_1'(\xi)]$ is shown in Fig 8.2. It has two stationary points, in contrast to the first-order metric function $e_1(\xi)$, shown in Fig. 8.3, which has only one. We cannot compute $\hat{d}^{(2)} + \hat{e}^{(2)}$ or e_1 too close to the singular point $\xi = 0$, and so there is a gap there.

In each region $\xi < 0$ and $\xi > 0$, we interpolate

$$\hat{d}^{(2)}(\xi) + \hat{e}^{(2)}(\xi) + \frac{1}{2}[e_1(\xi) + \xi e_1'(\xi)]$$

using the cubic spline that passes exactly through all the data points (a cubic spline is a $C^{(2)}$ function made up of cubic polynomial segments). It is the continuity of its derivatives that is the great advantage of the spline here; in addition it does not develop nasty fluctuations between data points near the end-points of the region of interpolation, as ordinary polynomial interpolants are apt to do (for the unpleasant things that can happen with polynomials, see Wendroff (1969)). To check the accuracy of interpolation, we compare each of our splines with another that has only 2/3 as many cubic segments (this second spline is a best least-squares fit, since it cannot pass exactly through all the data points). The maximum difference between the pairs of splines is about 5×10^{-5}, giving a conservative estimate for the true accuracy of interpolation.

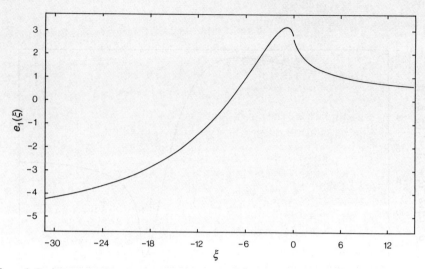

FIG. 8.3. The asymptotic metric function $e_1(\xi)$, the ξ derivative of which, evaluated at $\xi = \hat{\tau}/\mu$, gives $-2\sqrt{2}a_0(\hat{\tau}/\mu)$. There is again a gap in the computational results near the singular point $\xi = 0$.

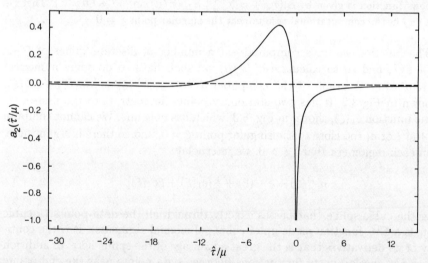

FIG. 8.4. The contribution $a_2(\hat{\tau}/\mu)$ to the series $c_0(\hat{\tau}, \hat{\theta}) = \sum_{n=0}^{\infty} a_{2n}(\hat{\tau}/\mu) \sin^{2n} \hat{\theta}$ for the speed-of-light news function.

To find $a_2(\hat{\tau}/\mu)$ we differentiate the spline interpolant and divide by $-4\sqrt{2}$. The result of this is shown in Fig. 8.4. The singularity at $\hat{\tau} = 0$ is of the form

$$\sum_{n=0}^{\infty} \left(\frac{\hat{\tau}}{\mu}\right)^n \left[E_n \left(\log |\frac{\hat{\tau}}{\mu}| \right)^2 + F_n \log |\frac{\hat{\tau}}{\mu}| + G_n \right]. \tag{8.3.6}$$

Using the conventional formula for the energy emitted in gravitational radiation (Bondi *et al.* 1962):

$$\Delta m = \frac{1}{2} \int_{-\infty}^{\infty} \int_0^{\pi} (c_0)^2 \sin\hat\theta \, d\hat\tau d\hat\theta, \qquad (8.3.7)$$

we can derive an estimate for the mass loss, assuming that the total news function is given only by $a_0\,(\hat\tau/\mu) + a_2\,(\hat\tau/\mu)\sin^2\hat\theta$. From the computed $a_0\,(\hat\tau/\mu)$ and $a_2\,(\hat\tau/\mu)$ we find

$$\int_{-\infty}^{\infty} (a_0(x))^2 \, dx = 0.500 \ ,$$

$$\int_{-\infty}^{\infty} a_0(x)a_2(x)dx = -0.586 \ ,$$

$$\int_{-\infty}^{\infty} (a_2(x))^2 \, dx = 1.14 \ . \qquad (8.3.8)$$

Therefore, if the news function were

$$c_0 = a_0\left(\frac{\hat\tau}{\mu}\right) + a_2\left(\frac{\hat\tau}{\mu}\right)\sin^2\hat\theta \ , \qquad (8.3.9)$$

then the mass loss would be

$$\Delta m = \frac{1}{2}\int_{-\infty}^{\infty}\int_0^{\pi}\left[a_0\left(\frac{\hat\tau}{\mu}\right) + a_2\left(\frac{\hat\tau}{\mu}\right)\sin^2\hat\theta\right]^2 \sin\hat\theta \, d\hat\tau d\hat\theta = 0.328\mu \qquad (8.3.10)$$

or about 16.4% of the initial energy (this may be compared with the 25% that one obtains when using just the isotropic term). The angular dependence of $dE/d\Omega$, the energy radiated per unit solid angle, for the news function (8.3.9) is shown in Fig. 8.5. It is interesting to note that the magnitude of $dE/d\Omega$ is never greater than its isotropic part $(1/4\pi)\int_{-\infty}^{\infty}(a_0\,(\hat\tau/\mu))^2\,d\hat\tau$.

Of course, this estimate must be taken with several grains of salt, since we have truncated the series (8.1.1). It is nevertheless useful as an order-of-magnitude estimate.

8.4   Discussion

In previous work (e.g. Curtis 1978a, b; D'Eath 1978) on high-speed black-hole collisions, both at finite and infinite γ, it has been tacitly assumed that the product of the collision is a single black hole plus out-going gravitational radiation, where the burst of radiation is that produced by the focusing of the far fields of the colliding holes—and its continuation to larger angular scales. If this were the case, then by solving the perturbation theory to all orders one could determine the entire news function.

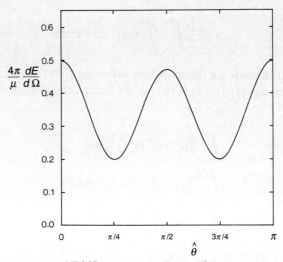

FIG. 8.5. The energy $dE/d\Omega$ (in units of $\mu/4\pi$) which would be radiated per unit solid angle, if the news function had the truncated form
$c_0(\hat{\tau}, \hat{\theta}) = a_0(\hat{\tau}/\mu) + a_2(\hat{\tau}/\mu)\sin^2\hat{\theta}$.

Using the mass-loss formula (8.2.1), or eqn (6.5.9) from which it was derived,
we have shown that the true picture is not so simple. We have argued that there
will be an additional burst of radiation produced during the decay to equilibrium
(i.e. Schwarzschild) of the central black hole formed by the collision, and that
this will be manifested by the appearance at high orders in perturbation theory
of terms growing exponentially with time. The two bursts are truly separate
only in the limit $\hat{\theta} \to 0$ or π, merging together away from this axis. Because
of this, and because the 'second' burst cannot be treated using perturbation
theory, originating as it does in a highly non-linear part of the space–time, we
cannot calculate the news function away from the axis of symmetry, except at
early times. The expressions that have been derived for the news function—by
previous authors and in this book—are only valid either in the vicinity of the
first burst close to the axis of symmetry, or at early times away from this axis.

It is likely that further analytical progress on this problem will be limited.
Computational limitations will prevent higher-order calculations in perturbation theory—and there is little point in attempting them, at least within the
framework presented here. The speed-of-light and large-γ space–times are algebraically general (Curtis 1978a, b) (i.e. their Riemann tensor does not become
simple with respect to any null tetrad) and, given the complexity of just the news
function, one certainly cannot expect to find an exact solution. The combined
analytical–numerical approach used here must be regarded as complementary to
the numerical construction of these speed-of-light space–times.

9

CONCLUSION

9.1 Summary

The examples of black-hole interactions treated in this book show the power gained by the use of matched asymptotic expansions in general relativity. This technique, in fact, seems very well suited to a variety of problems involving the interaction of strongly self-gravitating systems, and in one case—as in Chapters 5–8—to a strong-field interaction between strong-field systems, namely the collision or close encounter of two black holes at or near the speed of light.

In Chapter 3, matched asymptotic expansions were needed to generalize from a black hole in isolation, with various weak-field perturbations which die away at infinity, to the case of a black hole in a background universe, where the near-field perturbations grow as one moves outward from the black hole. Here the black hole generates perturbations of the background, as one sees in the external perturbation scheme; symmetrically, the background generates perturbations of the black hole, as in the internal scheme. By pursuing these methods, one can build up a good understanding of the space–time geometry around the black hole. From this, one can deduce that, in the lowest approximation, the black hole moves on a geodesic and that its spin is parallel transported. This may be of astrophysical relevance for black holes at the centre of galactic nuclei (Begelman *et al.* 1980). Such a black hole may emit, owing to accretion and other astrophysical causes, a jet along the axis of rotation. Some sources emit an S-shaped jet, and one may conjecture that this is due to the interaction with another black hole.

The space–time structure for two black holes in Newtonian orbit around one another was discussed in Chapter 4. This is again found by matching, and leads to relativistic corrections to the equations of motion and spin propagation. These equations agree, in the limit that one black hole is much less massive than another, with the equations of motion and spin propagation for a test particle in the far field of a rotating gravitational source (Wald 1972). Further, the equations of motion at post-Newtonian order agree with the Einstein–Infeld–Hoffmann equations of motion (Einstein *et al.* 1938).

At higher speeds, at which the impact parameter is such that one has scattering of black holes, one can derive the gravitational radiation generated by using a post-linear formalism (Kovács and Thorne 1978). The form of the radiation becomes qualitatively different at speeds close to that of light. This is discussed in Chapters 5–8. Chapter 5 was concerned with the interaction or collision at speeds close to, but not equal to, the speed of light. It was found that, for impact parameters of the order of the incoming centre-of-mass energy, there is no beam in the radiation, but that the radiation is spread over the celestial sphere. If the radiation were isotropic with the form found here near the axis, then 25% of the incoming energy would have been converted into gravitational waves. For impact parameters of the order of γ times the incoming energy, there is a beam, which contains strong-field radiation over an angular width of order γ^{-1}.

The speed-of-light collision was considered in Chapters 6–8. For simplicity, the axisymmetric collision only was studied. A perturbation scheme is set up in which one makes a large Lorentz boost, so that one has a weak shock propagating on a strong shock. Because of the properties of the geometry under boosts, there is a conformal symmetry in the perturbation theory, which allows the reduction of the perturbation theory to two independent variables. This then allows a tractable numerical calculation of the $\sin^2 \hat\theta$ part of the radiation. A new mass-loss formula is derived, showing that knowledge of the first two terms above in the angular decomposition of the radiation, together with a certain uniformity assumption, lead to an expression for the final mass, and hence the amount of energy radiated. This is also of interest because Penrose (1974) obtained an upper bound of 29% on the fraction radiated, assuming cosmic censorship. Thus one has a potentially interesting test of cosmic censorship. In fact, the new mass-loss formula leads to a final mass greater than the incoming energy 2μ, so giving a contradiction. The uniformity condition above is probably the reason for this; the contradiction may well occur because we have tried to assume that all the gravitational radiation occurs in one burst. This suggests, then, that there should be a second burst of radiation. This seems very reasonable, since the first burst is generated in the far-field curved shocks, while the second burst is generated in the centre of the space–time, and there is a logarithmic time-separation between them. The second burst would leave imprints on the first burst at late retarded times, but is not obviously accessible to perturbation theory. Nevertheless, one can obtain a rough approximation to the gravitational wave output, by applying the Bondi mass loss formula to the known first two terms in the $\sin^2 \hat\theta$ expansion. This gives an estimate of 16.8% for the efficiency of gravitational wave generation in the head-on speed-of-light collision. The limits of the capacity of analytic work on this problem may have been reached, and further progress may well only be feasible by a numerical treatment.

9.2 Quantum directions

Here, following 't Hooft (1987), we consider the quantum scattering process for two pointlike particles at centre-of-mass energies of the order of the Planck energy or higher. This is explicitly calculable, since in the language of particle physics graviton exchange dominates over all other interaction processes. When the energy is taken to be much higher than the Planck energy, one has the quantum version of the work described in Chapters 5–8, with the production of a black hole and the coherent emission of gravitons.

Assume here that all particles considered are electrically neutral (this can be relaxed). Consider the scattering of two particles with rest-masses $m^{(1)}, m^{(2)} \ll M_{\text{Planck}}$. First take a coordinate system in which the ingoing particle (1) is at rest or moves slowly. Assume that the second particle arrives along the trajectory

$$(x^{(2)}, y^{(2)}) =_{\text{def}} \tilde{x}^{(2)} = 0, \qquad z^{(2)} = -t^{(2)}, \tag{9.2.1}$$

with energy

$$\frac{1}{2} p_-^{(2)} = p_0^{(2)} = -p_3^{(2)} = O(1/Gm^{(1)}). \tag{9.2.2}$$

Here G is Newton's constant, taking units in which $\hbar = c = 1$. Because of the assumption $m^{(i)} \ll M_{\text{Planck}}$, the particle (2) may be assumed to move at the speed of light. The gravitational field is the shock wave found in Chapter 5. The join across the null shock of particle (2) may be summarized by

$$x_{\mu(+)} = x_{\mu(-)} - 2G p_\mu^{(2)} \log(\tilde{x}^2/C), \tag{9.2.3}$$

where C is an irrelevant constant. Here $\tilde{x}$ denotes the transverse coordinates x and y.

For simplicity of exposition, assume that the particle (1) is spinless. Before it meets shock 2, its wave function is

$$\begin{aligned}
\psi_{(-)}^{(1)} &= \exp(i\tilde{p}^{(1)}\tilde{x} + ip_3^{(1)} z - ip_0^{(1)} t) \\
&= \exp(i\tilde{p}^{(1)}\tilde{x} - ip_{(+)}^{(1)} u - ip_{(-)}^{(1)} v,)
\end{aligned} \tag{9.2.4}$$

where $u = (t-z)/2$ and $v = (t+z)/2$ are double null coordinates. As in Chapters 5–8, the wave function is shifted along the null shock generators, following eqn (9.2.3). This leads to the wave function just above shock 2 (i.e. at $v = 0$):

$$\psi_{(+)}^{(1)} = \exp\{i\tilde{p}^{(1)}\tilde{x} - ip_+^{(1)}[u + 2G p_0^{(2)} \log(\tilde{x}^2/C)]\}. \tag{9.2.5}$$

This can be expanded in plane waves as

$$\psi_{(+)}^{(1)} = \int A(k_+, \tilde{k}) dk_+ d^2\tilde{k} \exp(i k \tilde{x} - ik_+ u - ik_- v), \tag{9.2.6}$$

where

$$k_- = (\tilde{k}^2 + m^{(1)2})/k_+. \tag{9.2.7}$$

Here

$$A(k_+, \tilde{k}) = \delta(k_+ - p_+^{(1)}) \times$$
$$\times \frac{1}{(2\pi)^2} \int d^2\tilde{x} \exp[(\tilde{p}^{(1)} - \tilde{k}^{(1)})\tilde{x} - 2iG p_+^{(1)} p_0^{(2)} \log(\tilde{x}^2/C)]. \tag{9.2.8}$$

The integral can be done, and the scattering amplitude takes the form, apart from the factor $(k_+/k_0)\delta(\Sigma k - \Sigma p)$, where $p^{(1)}$ denotes the incoming momentum:

$$U(s, t) = \frac{\Gamma(1 - iGs)}{4\pi\Gamma(iGs)} \left(\frac{4}{-t}\right)^{1-iGs}, \tag{9.2.9}$$

where $Gs = -2G(p^{(1)} \cdot p^{(2)}) = 2Gp_+^{(1)}p_0^{(2)}$, and s is the standard Mandelstam variable (Eden *et al.* 1966). Also there is an exchange of momentum

$$q = k^{(1)} - p^{(1)},\tag{9.2.10}$$

where the Mandelstam variable t is defined by $t = -q^2$. The corresponding cross-section is given by

$$\sigma(\tilde{p}^{(1)} \to \tilde{k}^{(1)})d^2\tilde{k} = 4G^2(s^2/t^2)d^2\tilde{k}.\tag{9.2.11}$$

This cross-section is elastic, since with our assumptions there is no particle production, and corresponds to the exchange of a single graviton. Thus particle scattering in this régime is explicitly calculable by putting together well-known properties of gravitation and quantum field theory, rather as with the emission of radiation by black holes (Hawking 1975).

If Gs becomes much larger than one, then one has instead a nearly classical problem, as in chapters 5–8. The quantum amplitude is expected to have the approximate form

$$\langle | \rangle \sim \exp(iS),\tag{9.2.12}$$

where S is the action of a classical solution of the Einstein-matter equations, if such exists, obeying the boundary conditions. Here we expect the classical solution to be approximately a speed-of-light collision space–time, as in Chapters 5–8, but in general non-axisymmetric, corresponding to a non-zero impact parameter. As suggested in Chapters 5–8, for a range of impact parameters a black hole should be formed, together with copious gravitational radiation, but for larger impact parameters only the radiation will be produced. One might also investigate the behaviour of quantum loops, for very high loop momentum, to see if the above ideas can be helpful.

't Hooft's work has generated a large amount of interest; some further references are Schücker (1990), Deibel and Schücker (1991), Verlinde and Verlinde (1992), Amati *et al.* (1993) and Fabbrichesi *et al.* (1994). This work shows the profound connection between quantum theory in the high-energy (Planckian) limit and classical structure appearing in a high-energy black-hole collision.

REFERENCES

1. Abramowitz, M. and Stegun I. A. (1972). *Handbook of mathematical functions.* Dover Publications, New York.
2. Aichelburg, P. C. and Sexl, R. U. (1971). On the gravitational field of a massless particle. *Gen. Relativ. Grav.*, **2**, 303–12.
3. Amati, D., Ciafaloni, M., and Veneziano, G. (1993). Effective action and all-order gravitational eikonal at Planckian energies. *Nucl. Phys.*, **403**, 707–24.
4. Anderson, J. L. (1985). Gravitational radiation, source behaviour and the method of matched asymptotic expansions. *Foundat. Phys.*, **15**, 411–8.
5. Anderson, J. L. (1987). Gravitational radiation damping in systems with compact components. *Phys. Rev. D*, **36**, 2301–13.
6. Anderson, J. L. and Kegeles, L. S. (1980). Causes and cures for the infinities in slow-motion expansions in general relativity. *Gen. Relativ. Grav.*, **12**, 633–47.
7. Anderson, J. L., Kates, R. E., Kegeles, L. S., and Madonna, R. G. (1982). On the divergent integrals of post-Newtonian gravity: nonanalytic terms in the near-zone expansion found by matching. *Phys. Rev. D*, **25**, 2038–48.
8. Anninos, P., Hobill, A., Seidel, E., Smarr, L., and Suen, W.-M. (1993). Collision of two black holes. *Phys. Rev. Lett.*. **71**, 2851–4.
9. Anninos, P., Camarda, K., Masso, J., Seidel, E., Suen, W.-M., and Towns, J. (1995*a*). Three-dimensional numerical relativity: the evolution of black holes. Unpublished.
10. Anninos, P., Price, R. H., Pullin, J., Seidel, E., and Suen, W.-M. (1995*b*). Head-on collision of two black holes: comparison of different approaches. Unpublished.
11. Atiyah, M. F. and Hitchin, N. J. (1985). Low-energy scattering of non-abelian monopoles. *Phys. Lett. A*, **107**, 21–5.
12. Barker, B. M. and O'Connell, R. F. (1970). Derivation of the equations of motion of a gyroscope from the quantum theory of gravitation. *Phys. Rev. D*, **2**, 1428–35.
13. Barker, B. M. and O'Connell, R. F. (1975). Gravitational two-body problem with arbitrary masses, spins and quadrupole moments. *Phys. Rev. D*, **12**, 329–35.
14. Barker, B. M. and O'Connell, R. F. (1976). Lagrangian–Hamiltonian formalism for the gravitational two-body problem with spin and parametrized post-Newtonian parameters γ and β. *Phys. Rev. D*, **14**, 861–9.
15. Barker, B. M. and O'Connell, R. F. (1987). On the completion of the post-Newtonian gravitational two-body problem with spin. *J. Math. Phys.*, **28**, 661–7.
16. Barrow, J. D. and Carr, B. J. (1978). Primordial black hole formation in an anisotropic universe. *Mon. Not. R. Astron. Soc.*, **182**, 537–58.

17. Begelman, M., Blandford, R. D., and Rees, M. (1980). Massive black hole binaries in active galactic nuclei. *Nature (London)*, **287**, 307–9.
18. Begelman, M., Blandford, R. D., and Rees, M. (1984). Theory of extragalactic radio sources. *Rev. Mod. Phys.*, **56**, 255–351.
19. Bishop, N. T. (1988). The event horizons of two Schwarzschild black holes. *Gen. Relativ. Grav.*, **20**, 573–81.
20. Blandford, R. D. and Thorne, K. S. (1979). Black hole astrophysics, in *General relativity. An Einstein centenary survey* (ed. S. W. Hawking and W. Israel). Cambridge University Press, Cambridge.
21. Bondi, H., van der Burg, M. G. J., and Metzner, A. W. K. (1962). Gravitational waves in general relativity VII. Waves from axi-symmetric isolated systems. *Proc. R. Soc. London A*, **269**, 21–52.
22. Börner, G., Ehlers, J., and Rudolph, E. (1975). Relativistic spin precession in two-body systems. *Astron. Astrophys.*, **44**, 417–20.
23. Bowen, J. M. and York, J. W. (1980). Time-asymmetric initial data for black holes and black-hole collisions. *Phys. Rev. D.*, **21**, 2047–56.
24. Boyer, R. H. and Lindquist, R. W. (1967). Maximal analytic extension of the Kerr metric. *J. Math. Phys*, **8**, 265–81.
25. Brans, C. and Dicke, R. H. (1961). Mach's principle and a relativistic theory of gravitation. *Phys. Rev.* **124**, 925–35.
26. Breuer, R. A. and Rudolph, E. (1982). The force law for the dynamic two-body problem in the second post-Newtonian approximation of general relativity. *Gen. Relativ. Grav.*, **14**, 181–211.
27. Brill, D. R., Chrzanowski, P. L., Pereira, C. M., Fackerell, E. D., and Ipser, J. R. (1972). Solution of the scalar wave equation in a Kerr background by separation of variables. *Phys. Rev. D*, **5**, 1913-5.
28. Brill, D. R. and Lindquist, R. W. (1963). Interaction energy in geometrostatics. *Phys. Rev.*, **131**, 471–6.
29. Brumberg, V. A. and Kopejkin, S. M. (1989). Relativistic reference systems and motion of test bodies in the vicinity of the earth. *Nuovo Cim. B*, **103**, 63–98.
30. Burke, W. L. (1971). Gravitational radiation damping of slowly moving systems calculated using matched asymptotic expansions. *J. Math. Phys.*, **12**, 402–18.
31. Carter, B. (1968). Global structure of the Kerr family of gravitational fields. *Phys. Rev.*, **174**, 1559–71.
32. Carter, B. (1972). Unpublished notes.
33. Carter, B. (1973). Black hole equilibrium states. In *Black holes* (ed. C. M. DeWitt and B. S. DeWitt). Gordon and Breach, New York.
34. Carter, B. (1979). The general theory of the mechanical, electromagnetic and thermodynamic properties of black holes. In *General relativity. An Einstein centenary survey* (ed. S. W. Hawking and W. Israel). Cambridge University Press, Cambridge.
35. Chan, L.-H. and O'Connell, R. F. (1977). Two-body problem – a unified, classical and simple treatment of spin-orbit effects. *Phys. Rev. D*, **15**, 3058–9.

36. Chandrasekhar, S. (1939). *An introduction to the theory of stellar structure.* University of Chicago Press, Chicago.

37. Chandrasekhar, S. (1965). The post-Newtonian equations of hydrodynamics in general relativity. *Astrophys. J.*, **142**, 1488–512.

38. Chandrasekhar, S. (1979). An introduction to the theory of the Kerr metric and its perturbations. In *General relativity. An Einstein centenary survey* (ed. S. W. Hawking and W. Israel). Cambridge University Press, Cambridge.

39. Chandrasekhar, S. (1983). *The mathematical theory of black holes.* Clarendon Press, Oxford.

40. Chandrasekhar, S. and Esposito, F. P. (1970). The $2\frac{1}{2}$-post-Newtonian equations of hydrodynamics and radiation reaction in general relativity. *Astrophys. J.*, **160**, 153–80.

41. Chapman, G. I. (1979). Unpublished.

42. Cho, C. F. and Hari Dass, N. D. (1976). Gravitational two-body problem—a source theory viewpoint. *Ann. Phys. (N.Y.)*, **96**, 406–28.

43. Christodoulou, D. and Klainerman, S. (1993). *The global nonlinear stability of Minkowski space.* Princeton University Press, Princeton, New Jersey.

44. Chrzanowski, P. L. (1975). Vector potential and metric perturbations of a rotating black hole. *Phys. Rev. D*, **11**, 2042–62.

45. Corkill, R. W. (1983). Ph.D. thesis (unpublished), Cambridge University.

46. Corkill, R. W. and Stewart, J. M. (1983). Numerical relativity II. Numerical methods for the characteristic initial value problem problem and the evolution of the vacuum field equations for space–times with two Killing vectors. *Proc. R. Soc. London A*, **386**, 373–91.

47. Crowley, R. J. and Thorne, K. S. (1977). The generation of gravitational waves II. The postlinear formalism revisited. *Astrophys. J.*, **215**, 624–35.

48. Curtis, G. E. (1975). D.Phil. thesis, Oxford University.

49. Curtis, G. E. (1978a). Twistors and linearized Einstein theory on plane-fronted impulsive wave backgrounds. *Gen. Relativ. Grav.*, **9**, 987-97.

50. Curtis, G. E. (1978b). Ultrarelativistic black-hole encounters. *Gen. Relativ. Grav.*, **9**, 999–1008.

51. Damour, T. (1983). Gravitational radiation and the motion of compact bodies. In *Gravitational radiation* (ed. N. Deruelle and T. Piran). North-Holland, Amsterdam.

52. Damour, T. (1987). The problem of motion in Newtonian and Einsteinian gravity. In *300 years of gravitation* (ed. S. W. Hawking and W. Israel). Cambridge University Press, Cambridge.

53. Damour, T. and Taylor, J. H. (1992). Strong-field tests of relativistic gravity and binary pulsars. *Phys. Rev. D*, **45**, 1840–68.

54. Damour, T., Soffel, M., and Xu, C. (1991). General-relativistic celestial mechanics I. Method and definition of reference frames. *Phys. Rev. D*, **43**, 3273–307.

55. Damour, T., Soffel, M., and Xu, C. (1992). General-relativistic celestial mechanics II. Translational equations of motion. *Phys. Rev. D*, **45**, 1017–44.

56. Damour, T., Soffel, M., and Xu, C. (1993). General-relativistic celestial mechanics III. Rotational equations of motion. *Phys. Rev. D*, **47**, 3124–35.

57. Davis, M., Ruffini, R., Press, W. H., and Price, R. H. (1971). Gravitational radiation from a particle falling radially into a Schwarzschild black hole. *Phys. Rev. Lett.*, **27**, 1466–9.

58. D'Eath, P. D. (1975a). Dynamics of a small black hole in a background universe. *Phys. Rev. D*, **11**, 1387–403.

59. D'Eath, P. D. (1975b). The interaction of two black holes in the slow-motion limit. *Phys. Rev. D*, **12**, 2183–99.

60. D'Eath, P. D. (1978). High-speed encounters and gravitational radiation. *Phys. Rev. D*, **18**, 990–1019.

61. D'Eath, P. D. (1979a). Perturbation methods for interactions between strongly self-gravitating systems. In *Proceedings of the International School of Physics Enrico Fermi, course LXVII* (ed. J. Ehlers) North-Holland, Amsterdam.

62. D'Eath, P. D. (1979b). Gravitational radiation from hyperbolic encounters. In *Sources of gravitational radiation* (ed. L. Smarr). Cambridge University Press, Cambridge.

63. D'Eath, P. D. and Payne, P. N. (1992a). Gravitational radiation in black-hole collisions at the speed of light: I. Perturbation treatment of the axisymmetric collision. *Phys. Rev. D*, **46**, 658–74.

64. D'Eath, P. D. and Payne, P. N. (1992b). Gravitational radiation in black-hole collisions at the speed of light: II. Reduction to two independent variables and calculation of the second-order news function. *Phys. Rev. D*, **46**, 675–93.

65. D'Eath, P. D. and Payne, P. N. (1992c). Gravitational radiation in black-hole collisions at the speed of light: III. Results and conclusions. *Phys. Rev. D*, **46**, 694–701.

66. Deibel, S. and Schücker, T. (1991). Classical versus quantum scattering: gravitational cross sections. *Class. Quantum Grav.*, **8**, 1949–53.

67. Demiański, M. and Grishchuk, L. P. (1974). Note on the motion of black holes. *Gen. Relativ. Grav.*, **5**, 673–9.

68. Detweiler, S. L. (1979). Black holes and gravitational waves: perturbation analysis. In *Sources of gravitational radiation* (ed. L. Smarr). Cambridge University Press, Cambridge.

69. DeWitt, B. S. and Brehme, R. W. (1960). Radiation damping in a gravitational field. *Ann. Phys. (N.Y.)*, **9**, 220–59.

70. Dieudonné, J. (1960). *Foundations of modern analysis*. Academic Press, New York.

71. Dixon, W. G. (1979). The problem of causality in slow-motion approximations. In *Proceedings of the third Gregynog relativity workshop: gravitational radiation theory* (ed. M. Walker). Preprint MPI-PAE/Astro 204.

72. Dray, T. and 't Hooft, G. (1985a). The gravitational shock wave of a massless particle. *Nucl. Phys. B*, **253**, 173–88.

73. Dray, T. and 't Hooft, G. (1985b). The effect of spherical shells of matter on the Schwarzschild black hole. *Commun. Math. Phys.*, **99**, 613–25.

74. Dray, T. and 't Hooft, G. (1986). The gravitational effect of colliding planar sheets of matter. *Class. Quantum Grav.*, **3**, 825–40.

75. Edelstein, L. A. and Vishveshwara, C. V. (1970). Differential equations for perturbations on the Schwarzschild metric. *Phys. Rev. D*, **1**, 3514–7.

76. Ehlers, J. and Kundt, W. (1962). Exact solutions of the gravitational field equations. In *Gravitation: an introduction to current research* (ed. L. Witten). John Wiley, New York.

77. Ehlers, J. and Rudolph, E. (1977). Dynamics of extended bodies in general relativity. Centre-of-mass description and quasirigidity. *Gen. Relativ. Grav.*, **8**, 197–217.

78. Ehlers, J., Rosenblum, A., Goldberg, J. N., and Havas, P. (1976). Comments on gravitational radiation damping and energy loss in binary systems. *Astrophys. J.*, **L208**, L77–81.

79. Einstein, A. and Infeld, L. (1940). The gravitational equations and the problem of motion. II. *Ann. Math.*, **41**, 455–64.

80. Einstein, A. and Infeld, L. (1948). On the motion of particles in general relativity theory. *Can. J. Math.*, **1**, 209–41.

81. Einstein, A. and Rosen, N. (1937). On gravitational waves. *J. Franklin Inst.*, **223**, 43–54.

82. Einstein, A. Infeld, L., and Hoffmann, B. (1938). The gravitational equations and the problem of motion. *Ann. Math.*, **39**, 65–100.

83. Ellis, G. F. R. (1971). Relativistic cosmology. In *General relativity and cosmology, proceedings of course XLVII of the International School of Physics Enrico Fermi* (ed. R. K. Sachs). Academic Press, New York.

84. Epstein, R. and Wagoner, R. V. (1975). Post-Newtonian generation of gravitational waves. *Astrophys. J.*, **197**, 717–23.

85. Erdélyi, A. (ed.) (1953). *Higher transcendental functions*, Vol. I. McGraw-Hill, New York.

86. Fabbrichesi, M., Pettorino, R., Veneziano, G., and Vilkovisky, G. A. (1994). Planckian energy scattering and surface terms in the gravitational action. *Nucl. Phys. B*, **419**, 147–88.

87. Fitchett, M. J. (1981). The influence of gravitational wave momentum losses on the centre of mass motion of a Newtonian binary system. *Mon. Not. R. Astron. Soc.*, **203**, 1049–62.

88. Fitchett, M. J. (1984). Enhanced gravitational wave emission from perturbed wide binaries. Preprint.

89. Fitchett, M. J. and Detweiler, S. L. (1984). Linear momentum and gravitational waves: circular orbits around a Schwarzschild black hole. *Mon. Not. R. Astron. Soc.*, **211**, 933–42.

90. Fock, V. (1964). *The theory of space, time and gravitation*. Pergamon, Oxford.

91. Friedlander, F. G. (1975). *The wave equation on a curved space–time*. Cambridge University Press, Cambridge.

92. Friedrich, H. and Stewart, J. M. (1983). Characteristic initial data and wavefront singularities in general relativity. *Proc. R. Soc. London A* **385**, 345–71.

93. Futamase, T. (1983). Gravitational radiation reaction in the Newtonian limit. *Phys. Rev. D*, **28**, 2373–81.

94. Futamase, T. (1985). Point-particle limit and the far-zone quadrupole limit in general relativity. *Phys. Rev. D*, **32**, 2566–74.

95. Futamase, T. (1987). Strong-field point-particle limit and the equations of motion of the binary pulsar. *Phys. Rev. D*, **36**, 321–9.

96. Futterman, J. A. H., Handler, F. A., and Matzner, R. A. (1987). *Scattering from black holes*. Cambridge University Press, Cambridge.

97. Geroch, R. (1969). Limits of spacetimes. *Commun. Math Phys.*, **13**, 180–93.

98. Gibbons, G. W. and Ruback, P. J. (1986). The motion of extreme Reissner–Nordström black holes in the low velocity limit. *Phys. Rev. Lett.*, **57**, 1492–5.

99. Goldberg, J. N., Macfarlane, A. T., Newman, E. T., Rohrlich, F., and Sudarshan, E. C. G. (1967). Spin-s spherical harmonics and ð. *J. Math. Phys.*, **8**, 2155–61.

100. Goldstein, H. (1980). *Classical mechanics* (2nd edn). Addison-Wesley, Reading, Mass.

101. Griffiths, J. B. (1991). *Colliding plane waves in general relativity*. Clarendon Press, Oxford.

102. Hansen, R. O. (1972). Post-Newtonian gravitational radiation from point masses in a hyperbolic Kepler orbit. *Phys. Rev. D*, **5**, 1021–3.

103. Hartle, J. B. (1974). Tidal shapes and shifts on rotating black holes. *Phys. Rev. D*, **9**, 2749–59.

104. Hartle, J. B. (1978). Bounds on the mass and moment of inertia of non-rotating neutron stars. *Phys. Rep. C*, **46**, 201-47.

105. Hartle, J. B. and Hawking, S. W. (1976). Path-integral derivation of black hole radiance. *Phys. Rev. D*, **13**, 2188–203.

106. Hawking, S. W. (1971). Gravitational radiation from colliding black holes. *Phys. Rev. Lett.*, **26**, 1344-6.

107. Hawking, S. W. (1972a). Black holes in general relativity. *Commun. Math. Phys.*, **25**, 152–66.

108. Hawking, S. W. (1972b). Black holes in the Brans–Dicke theory of gravitation. *Commun. Math. Phys.*, **25**, 167–71.

109. Hawking, S. W. (1973). The event horizon. In *Black holes* (ed. C. M. DeWitt and B. S. DeWitt). Gordon and Breach, New York.

110. Hawking, S. W. (1975). Particle creation by black holes. *Commun. Math. Phys.*, **43**, 199–220.

111. Hawking, S. W. and Ellis, G. F. R. (1973). *The large scale structure of space–time*. Cambridge University Press, Cambridge.

112. Hawking, S. W. and Hartle, J. B. (1972). Energy and angular momentum flow into a black hole. *Commun. Math. Phys.*, **27**, 283–90.

113. Hobill, D. W. (1984). Axially symmetric gravitational radiation from isolated sources. *J. Math. Phys.*, **25**, 3527–37.

114. Hulse, R. A. and Taylor, J. H. (1975). Discovery of a pulsar in a binary system. *Astrophys. J.*, **L 195**, L51–3.

115. Infeld, L. and Schild, A. (1949). On the motion of test particles in general relativity. *Rev. Mod. Phys.*, **21**, 408–13.

116. Isaacson, R. A. and Winicour, J. (1968). Harmonic and null descriptions of gravitational radiation. *Phys. Rev.*, **168**, 1451–6.

117. Kates, R. E. (1980a). Motion of a small body through an external field in general relativity calculated by matched asymptotic expansions. *Phys. Rev. D*, **22**, 1853–70.

118. Kates, R. E. (1980b). Gravitational radiation damping of a binary system containing compact objects calculated using matched asymptotic expansions. *Phys. Rev. D*, **22**, 1871–8.

119. Kates, R. E. (1980c). Motion of an electrically or magnetically charged body with possible strong internal gravity through external electromagnetic and gravitational fields. *Phys. Rev. D*, **22**, 1879–81.

120. Kates, R. E. (1981). Underlying structure of singular perturbations on manifolds. *Ann. Phys. (N.Y.)*, **132**, 1–17.

121. Kates, R. E. and Kegeles, L. S. (1982). Nonanalytic terms in the slow-motion expansion of a radiating scalar field on a Schwarzschild background. *Phys. Rev. D*, **25**, 2030–48.

122. Kates, R. E. and Madonna, R. G. (1982). Gravitational radiation reaction from small-angle scattering of slowly moving compact bodies, calculated by matched asymptotic expansions. *Phys. Rev. D*, **25**, 2499–508.

123. Kerr, R. P. (1963). Gravitational field of a spinning mass as an example of algebraically special metrics. *Phys Rev. Lett.*, **11**, 237–8.

124. Kinnersley, W. (1969). Type D vacuum metrics. *J. Math. Phys.*, **10**, 1195–203.

125. Klioner, S. A. and Voinov, A. V. (1993). Relativistic theory of astronomical reference systems in closed form. *Phys. Rev. D*, **48**, 1451–61.

126. Kojima, Y. and Nakamura, T. (1983). Gravitational radiation from a particle with zero orbital angular momentum plunging into a Kerr black hole. *Phys. Lett. A*, **96**, 335–8.

127. Kopejkin, S. M. (1988). Celestial coordinate reference systems in curved space–time. *Celest. Mech.*, **44**, 87–115.

128. Kovács, S. J. and Thorne, K. S. (1977). The generation of gravitational waves III. Derivation of bremsstrahlung formulae. *Astrophys. J.*, **217**, 252–80.

129. Kovács, S. J. and Thorne, K. S. (1978). The generation of gravitational waves IV. Bremsstrahlung. *Astrophys. J.*, **224**, 62–85.

130. Kruskal, M. D. (1960). Maximal extension of Schwarzschild metric. *Phys. Rev.*, **119**, 1743–5.

131. Landau, L. D. and Lifschitz, E. M. (1962). *The classical theory of fields*. Addison-Wesley, Reading, Mass.

132. Mackie, A. G. (1965). *Boundary value problems*. Oliver and Boyd, Edinburgh.

133. Madore, J. (1970a). Gravitational radiation from a bounded source I. *Ann. Inst. H. Poincaré A*, **XII**, 285–305.

134. Madore, J. (1970b). Gravitational radiation from a bounded source II. *Ann. Inst. H. Poincaré A*, **XII**, 365–92.

135. Majumdar, S. D. (1947). A class of exact solutions of Einstein's field equations. *Phys. Rev.*, **72**, 390–8.

136. Manasse, F. K. (1963). Distortion in the metric of a small centre of gravitational attraction due to its proximity to a very large mass. *J. Math. Phys.*, **4**, 746–61.

137. Manton, N. S. (1977). The force between 't Hooft–Polyakov monopoles. *Nucl. Phys. B*, **126**, 525–41.

138. Matzner, R. A. and Nutku, Y. (1974). On the method of virtual quanta and gravitational radiation. *Proc. R. Soc. London A*, **336**, 285–305.

139. Misner, C. W. (1960). Wormhole initial conditions. *Phys. Rev.*, **118**, 1110–11.

140. Misner, C. W., Thorne, K. S., and Wheeler, J. A. (1973). *Gravitation*. Freeman, San Francisco.

141. Moncrief, V., Cunningham, C. T., and Price, R. H. (1979). Radiation from slightly nonspherical models of gravitational collapse. In *Sources of gravitational radiation* (ed. L. Smarr). Cambridge University Press, Cambridge.

142. Nakamura, T. and Haugan, M. P. (1983). Gravitational radiation from particles falling along the symmetry axis into a Kerr black hole: the momentum radiated. *Astrophys. J.*, **269**, 292–6.

143. Nayfeh, A. H. (1973). *Perturbation methods*. John Wiley, New York.

144. Newman, E. T. and Penrose, R. (1962). An approach to gravitational radiation by a method of spin coefficients. *J. Math. Phys.*, **3**, 566–78.

145. Newman, E. T., Couch, E., Chinnapared, K., Exton, A., Prakash, A., and Torrence, R. (1965). Metric of a rotating charged mass. *J. Math. Phys.*, **6**, 918–9.

146. Novikov, I. D. and Frolov, V.P. (1989). *Physics of black holes*. Kluwer, Dordrecht.

147. Oohara, K.-I. (1984). Excitation of the free oscillation of a Schwarzschild black hole by the gravitational waves from a scattered test particle. Preprint KUNS 702.

148. Oohara, K.-I. and Nakamura, T. (1983). Energy, momentum and angular momentum of gravitational radiation from a particle plunging into a nonrotating black hole. *Phys. Lett. A*, **94**, 349–52.

149. Oppenheimer, J. R. and Snyder, H. (1939). On continued gravitational contraction. *Phys. Rev.*, **56**, 455–9.

150. Papapetrou, A. (1947). A static solution of the equations of the gravitational field for an arbitrary charge-distribution. *Proc. R. Irish Acad. A*, **51**, 191–204.

151. Papapetrou, A. (1951). Spinning test particles in general relativity I. *Proc. R. Soc. London A*, **209**, 248–58.

152. Payne, P. N. (1983a). Ph.D. thesis, University of Cambridge.

153. Payne, P. N. (1983b). Smarr's zero-frequency-limit calculation. *Phys. Rev. D*, **28**, 1894–7.

154. Penrose, R. (1968). Structure of spacetime. In *Battelle rencontres* (ed. C. M. DeWitt and J. A. Wheeler). Benjamin, New York.

155. Penrose, R. (1971). Unpublished notes.

156. Penrose, R. (1972a). *Techniques of differential topology in relativity*. SIAM, Philadelphia.

157. Penrose, R. (1972b). In *General relativity* (ed. L. O'Raifeartaigh). Clarendon Press, Oxford.

158. Penrose, R. (1974). Seminar at Cambridge University (unpublished).

159. Penrose, R. (1980). Null hypersurface initial data for classical fields of arbitrary spin and for general relativity. *Gen. Relativ. Grav.*, **12**, 225–64.

160. Penrose, R. and Rindler, W. (1984). *Spinors and space–time*, Vol. 1. Cambridge University Press, Cambridge.

161. Penrose, R. and Rindler, W. (1986). *Spinors and space–time*, Vol. 2. Cambridge University Press, Cambridge.

162. Peters, P. C. (1964). Gravitational radiation and the motion of two point masses. *Phys. Rev.*, **136**, 1224–32.

163. Peters, P. C. (1970). Relativistic gravitational bremsstrahlung. *Phys. Rev. D.*, **1**, 1559–71.

164. Peters, P. C. and Mathews, J. (1963). Gravitational radiation from point masses in a Keplerian orbit. *Phys. Rev.*, **131**, 435–40.

165. Pirani, F. A. E. (1959). Gravitational waves in general relativity IV. The gravitational field of a fast-moving particle. *Proc. R. Soc. London A*, **252**, 96–101.

166. Polyakov, A. M. (1974). *JETP Letters*, **20**, 194–8.

167. Price, R. H. (1972). Nonspherical perturbations of relativistic gravitational collapse I: Scalar and gravitational perturbations. *Phys. Rev. D*, **5**, 2419–38.

168. Regge, T. and Wheeler, J. A. (1957). Stability of a Schwarzschild singularity. *Phys. Rev.*, **108**, 1063–9.

169. Robertson, H. P. (1937). Test corpuscles in general relativity. *Proc. Edinb. Math. Soc.*, **5**, 63–81.

170. Robertson, H. P. (1938). Note on the preceding paper: the two body problem in general relativity. *Ann. Math.*, **39**, 101–4.

171. Robinson, D. C. (1975). Uniqueness of the Kerr black hole. *Phys. Rev. Lett.*, **34**, 905–6.

172. Rüdinger, R. (1983). Conserved quantities of spinning test particles in general relativity. II. *Proc. R. Soc. A*, **385**, 229–39.

173. Ruffini, R. (1973). On the energetics of black holes. In *Black holes* (ed. C. M. DeWitt and B. S. DeWitt). Gordon and Breach, New York.

174. Ruffini, R. and Wheeler, J. A. (1971). Relativistic cosmology and space platforms. In *The significance of space research for fundamental physics*. European Space Research Organization book Sp-52, Paris.

175. Sachs, R. K. (1962). Gravitational waves in general relativity, VIII. Waves in asymptotically flat space–time. *Proc. R. Soc. London A*, **270**, 103–26.

176. Schücker, T. (1990). The gravitational cross section of massless particles. *Class. Quantum Grav.*, **7**, 503–9.

177. Sciama, D. W., Waylen, P. C., and Gilman, R. C. (1969). Generally covariant integral form of Einstein's field equations. *Phys. Rev.*, **187**, 1762–6.

178. Smarr, L. (1973). Surface geometry of a charged rotating black hole. *Phys. Rev. D*, **7**, 289–95.

179. Smarr, L. (1977a). Gravitational radiation from distant encounters and from head-on collisions of black holes: the zero-frequency limit. *Phys. Rev. D*, **15**, 2069–77.

180. Smarr, L. (1977b). Spacetimes generated by computers: black holes with gravitational radiation. *Ann. N. Y. Acad. Sci.*, **302**, 569–604.

181. Smarr, L. (1979). Gauge conditions, radiation formulae and the two black hole collision. In *Sources of gravitational radiation* (ed. L. Smarr). Cambridge University Press, Cambridge.

182. Smarr, L., Cadez, A., DeWitt, B. S., and Eppley, K. (1976). Collision of two black holes: theoretical framework. *Phys. Rev. D*, **14**, 2443–52.

183. Stewart, J. M. and Friedrich, H. (1982). Numerical relativity I. The characteristic initial value problem. *Proc. R. Soc. London A*, **384**, 427–54.

184. Szekeres, P. (1972). Colliding gravitational waves. *J. Math. Phys.* **13**, 286–94.

185. Teukolsky, S. A. (1973). Perturbations of a rotating black hole. I. Fundamental equations for gravitational, electromagnetic and neutrino-field perturbations. *Astrophys. J.*, **185**, 635–47.

186. 't Hooft, G. (1974). Magnetic monopoles in unified gauge theories. *Nucl. Phys. B*, **79**, 276–84.

187. 't Hooft, G. (1987). Graviton dominance in ultra-high-energy scattering. *Phys. Lett. B*, **198**, 61–3.

188. Thorne, K. S. (1977). The generation of gravitational waves: a review of computational techniques. In *Proceedings of the international school of cosmology and gravitation*, at Erice, Sicily. Cal Tech preprint OAP-495.

189. Thorne, K. S. (1980a). Gravitational-wave research: current status and future prospects. *Rev. Mod. Phys.*, **52**, 285–97.

190. Thorne, K. S. (1980b). Multipole expansions of gravitational radiation. *Rev. Mod. Phys.*, **52**, 299–339.

191. Thorne, K. S. (1987). Gravitational radiation. In *300 years of gravitation* (ed. S. W. Hawking and W. Israel). Cambridge University Press, Cambridge.

192. Thorne, K. S. and Hartle, J. B. (1985). Laws of motion and precession for black holes and other bodies. *Phys. Rev. D*, **31**, 1815–37.

193. Thorne, K. S. and Kovács, S. J. (1975). The generation of gravitational waves I. Weak-field sources. *Astrophys. J.*, **200**, 245–62.

194. Thorne, K. S., Price, R. H., and Macdonald, D. A. (1986). *Black holes. The membrane paradigm*. Yale University Press, New Haven.

195. Tomimatsu, A. (1989). Gravitational field in a collision of two Schwarzschild black holes. *Gen. Relativ. Grav.*, **21**, 1233–47.

196. Turner, M. and Will, C. M. (1978). Post-Newtonian gravitational bremsstrahlung. *Astrophys. J.*, **220**, 1107–24.

197. Verlinde, E. and Verlinde, H. (1992). Scattering at Planckian energies. *Nucl. Phys. B*, **371**, 246–68.

198. Vishveshwara, C. V. (1970). Stability of the Schwarzschild metric. *Phys. Rev. D*, **1**, 2870–9.

199. Wagoner, R. V. and Will, C. M. (1976). Post-Newtonian gravitational radiation from orbiting point masses. *Astrophys. J.*, **210**, 764–75.

200. Wald, R. M. (1972). Gravitational spin interaction. *Phys. Rev. D*, **6**, 406–13.

201. Wald, R. M. (1973). On perturbations of a Kerr black hole. *J. Math. Phys.*, **14**, 1453–61.

202. Wald, R. M. (1978). Construction of solutions of gravitational, electromagnetic or other perturbation equations from solutions of decoupled equations. *Phys. Rev. Lett.*, **41**, 203–6.

203. Wald, R. M. (1984). *General relativity*. University of Chicago Press, Chicago.

204. Wendroff, B. (1969). *First principles of numerical analysis*. Addison-Wesley, Reading, Mass.

205. Westpfahl, K. (1985). High-speed scattering of charged and uncharged particles in general relativity. *Fortschr. Phys.*, **33**, 417–93.

206. Whiting, B. F. (1989). Mode stability of the Kerr black hole. *J. Math. Phys.*, **30**, 1301–5.

207. Will, C. M. (1993). *Theory and experiment in gravitational physics* (2nd edn). Cambridge University Press, Cambridge.

208. Zerilli, F. J. (1970). Gravitational field of a particle falling in a Schwarzschild geometry analyzed in harmonics. *Phys. Rev. D*, **2**, 2141–60.

GENERAL INDEX